PHYSICS DEMYSTIFIED

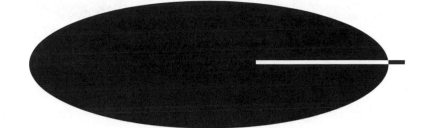

PHYSICS
DEMYSTIFIED

STAN GIBILISCO

McGraw-Hill
New York   Chicago   San Francisco   Lisbon   London
Madrid   Mexico City   Milan   New Delhi   San Juan
Seoul   Singapore   Sydney   Toronto

The McGraw·Hill Companies

Cataloging-in-Publication Data is on file with the Library of Congress.

Copyright © 2002 by The McGraw-Hill Companies, Inc. All rights reserved. Printed in the United States of America. Except as permitted under the United States Copyright Act of 1976, no part of this publication may be reproduced or distributed in any form or by any means, or stored in a data base or retrieval system, without the prior written permission of the publisher.

4 5 6 7 8 9 0   AGM/AGM   0 8 7 6 5 4 3 2

ISBN 0-07-138201-1

The sponsoring editor for this book was Scott Grillo, the editing supervisor was Caroline Levine, and the production supervisor was Sherri Souffrance. It was set in Times Roman by Wayne A. Palmer and Deirdre Sheean of McGraw-Hill Professional's Hightstown, N.J., composition unit.

Quebecor/Martinsburg was printer and binder.

McGraw-Hill books are available at special quantity discounts to use as premiums and sales promotions, or for use in corporate training programs. For more information, please write to the Director of Special Sales, Professional Publishing, McGraw-Hill, Two Penn Plaza, New York, NY 10121-2298. Or contact your local bookstore.

DEDICATION

To Samuel, Tony, and Tim
from Uncle Stan

CONTENTS

PART ONE   Classical Physics

CONTENTS

PREFACE

This book is for people who want to learn basic physics without taking a formal course. It can also serve as a supplemental text in a classroom, tutored, or home-schooling environment. I recommend that you start at the beginning of this book and go straight through, with the possible exception of Part Zero.

If you are confident about your math ability, you can skip Part Zero. But take the Part Zero test anyway, to see if you are actually ready to jump into Part One. If you get 90 percent of the answers correct, you're ready. If you get 75 to 90 percent correct, skim through the text of Part Zero and take the chapter-ending quizzes. If you get less than three-quarters of the answers correct in the quizzes and the section-ending test, find a good desk and study Part Zero. It will be a drill, but it will get you "in shape" and make the rest of the book easy.

In order to learn physics, you *must* have some mathematical skill. Math is the language of physics. If I were to tell you otherwise, I'd be cheating you. Don't get intimidated. None of the math in this book goes beyond the high school level.

This book contains an abundance of practice quiz, test, and exam questions. They are all multiple choice, and are similar to the sorts of questions used in standardized tests. There is a short quiz at the end of every chapter. The quizzes are "open-book." You may (and should) refer to the chapter texts when taking them. When you think you're ready, take the quiz, write down your answers, and then give your list of answers to a friend. Have the friend tell you your score, but not which questions you got wrong. The answers are listed in the back of the book. Stick with a chapter until you get most of the answers right.

This book is divided into three major sections after Part Zero. At the end of each section is a multiple choice test. Take these tests when you're done with the respective sections and have taken all the chapter quizzes. The section tests are "closed-book." Don't look back at the text when taking them. The questions are not as difficult as those in the quizzes, and they don't require that you memorize trivial things. A satisfactory score is three-quarters of the answers correct. Again, answers are in the back of the book.

There is a final exam at the end of this course. The questions are practical, and are less mathematical than those in the quizzes. The final exam contains questions drawn from Parts One, Two, and Three. Take this exam when you have finished all the sections, all the section tests, and all of the chapter quizzes. A satisfactory score is at least 75 percent correct answers.

With the section tests and the final exam, as with the quizzes, have a friend tell you your score without letting you know which questions you missed. That way, you will not subconsciously memorize the answers. You might want to take each test, and the final exam, two or three times. When you have gotten a score that makes you happy, you can check to see where your knowledge is strong and where it is not so keen.

I recommend that you complete one chapter a week. An hour or two daily ought to be enough time for this. Don't rush yourself; give your mind time to absorb the material. But don't go too slowly either. Take it at a steady pace and keep it up. That way, you'll complete the course in a few months. (As much as we all wish otherwise, there is no substitute for "good study habits.") When you're done with the course, you can use this book, with its comprehensive index, as a permanent reference.

Suggestions for future editions are welcome.

Stan Gibilisco

ACKNOWLEDGMENTS

Illustrations in this book were generated with CorelDRAW. Some clip art is courtesy of Corel Corporation, 1600 Carling Avenue, Ottawa, Ontario, Canada K1Z 8R7.

I extend thanks to Mary Kaser, who helped with the technical editing of the manuscript for this book.

A Review of Mathematics

CHAPTER 1

Equations, Formulas, and Vectors

An *equation* is a mathematical expression containing two parts, one on the left-hand side of an equals sign ($=$) and the other on the right-hand side. A *formula* is an equation used for the purpose of deriving a certain value or solving some practical problem. A *vector* is a special type of quantity in which there are two components: *magnitude* and *direction*. Physics makes use of equations, formulas, and vectors. Let's jump in and immerse ourselves in them. Why hesitate? You won't drown in this stuff. All you need is a little old-fashioned perseverance.

Notation

Equations and formulas can contain *coefficients* (specific numbers), *constants* (specific quantities represented by letters of the alphabet), and/or *variables* (expressions that stand for numbers but are not specific). Any of the common arithmetic operations can be used in an equation or formula. These include addition, subtraction, multiplication, division, and raising to a power. Sometimes functions are also used, such as logarithmic functions, exponential functions, trigonometric functions, or more sophisticated functions.

Addition is represented by the plus sign ($+$). Subtraction is represented by the minus sign ($-$). Multiplication of specific numbers is represented either by a plus sign rotated 45 degrees ($\times$) or by enclosing the numerals in parentheses and writing them one after another. Multiplication involving a coefficient and one or more variables or constants is expressed by writing the coefficient followed by the variables or constants with no symbols in between. Division is represented by a forward slash (/) with the numerator on the left and the denominator on the right. In complicated expressions, a horizontal line is used to denote division, with the numerator on the top and the denominator on the bottom. Exponentiation (raising to a power) is expressed by writing the base value, followed by a superscript indicating the power to which the base is to be raised. Here are some examples:

Two plus three	$2 + 3$
Four minus seven	$4 - 7$
Two times five	2×5 or $(2)(5)$
Two times x	$2x$
Two times $(x + 4)$	$2(x + 4)$
Two divided by x	$2/x$
Two divided by $(x + 4)$	$2/(x + 4)$
Three to the fourth power	3^4
x to the fourth power	x^4
$(x + 3)$ to the fourth power	$(x + 3)^4$

SOME SIMPLE EQUATIONS

Here are some simple equations containing only numbers. Note that these are true no matter what.

$$3 = 3$$

$$3 + 5 = 4 + 4$$

$$1{,}000{,}000 = 10^6$$

$$- (-20) = 20$$

Once in a while you'll see equations containing more than one equals sign and three or more parts. Examples are

$$3 + 5 = 4 + 4 = 10 - 2$$

$$1{,}000{,}000 = 1{,}000 \times 1{,}000 = 10^3 \times 10^3 = 10^6$$

$$-(-20) = -1 \times (-20) = 20$$

All the foregoing equations are obviously true; you can check them easily enough. Some equations, however, contain variables as well as numbers. These equations are true only when the variables have certain values; sometimes such equations can never be true no matter what values the variables attain. Here are some equations that contain variables:

$$x + 5 = 8$$

$$x = 2y + 3$$

$$x + y + z = 0$$

$$x^4 = y^5$$

$$y = 3x - 5$$

$$x^2 + 2x + 1 = 0$$

Variables usually are represented by italicized lowercase letters from near the end of the alphabet.

Constants can be mistaken for variables unless there is supporting text indicating what the symbol stands for and specifying the units involved. Letters from the first half of the alphabet often represent constants. A common example is c, which stands for the speed of light in free space (approximately 299,792 if expressed in kilometers per second and 299,792,000 if expressed in meters per second). Another example is e, the exponential constant, whose value is approximately 2.71828.

SOME SIMPLE FORMULAS

In formulas, we almost always place the quantity to be determined all by itself, as a variable, on the left-hand side of an equals sign and some mathematical expression on the right-hand side. When denoting a formula, it is important that every constant and variable be defined so that the reader knows what the formula is used for and what all the quantities represent.

One of the simplest and most well-known formulas is the formula for finding the area of a rectangle (Fig. 1-1). Let b represent the length (in meters) of the base of a rectangle, and let h represent the height (in meters) measured perpendicular to the base. Then the area A (in square meters) of the rectangle is

$$A = bh$$

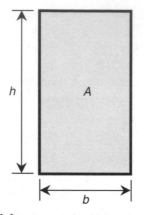

Fig. 1-1.   A rectangle with base length *b*,
height *h*, and area *A*.

A similar formula lets us calculate the area of a triangle (Fig. 1-2). Let *b* represent the length (in meters) of the base of a triangle, and let *h* represent the height (in meters) measured perpendicular to the base. Then the area *A* (in square meters) of the triangle is

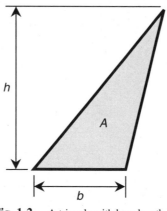

Fig. 1-2.   A triangle with base length *b*,
height *h*, and area *A*.

$$A = bh/2$$

Consider another formula involving distance traveled as a function of time and speed. Suppose that a car travels at a constant speed *s* (in meters per second) down a straight highway (Fig. 1-3). Let *t* be a specified length of time (in seconds). Then the distance *d* (in meters) that the car travels in that length of time is given by

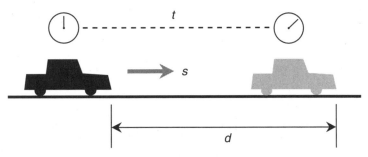

Fig. 1-3.   A car traveling down a straight highway over distance d at constant speed s for a length of time t.

$$d = st$$

If you're astute, you will notice something that all three of the preceding formulas have in common: All the units "agree" with each other. Distances are always given in meters, time is given in seconds, and speed is given in meters per second. The preceding formulas for area will not work as shown if A is expressed in square inches and d is expressed in feet. However, the formulas can be converted so that they are valid for those units. This involves the insertion of constants known as *conversion factors*.

CONVERSION FACTORS

Refer again to Fig. 1-1. Suppose that you want to know the area A in square inches rather than in square meters. To derive this answer, you must know how many square inches comprise one square meter. There are about 1,550 square inches in one square meter. Thus we can restate the formula for Fig. 1-1 as follows: Let b represent the length (in meters) of the base of a rectangle, and let h represent the height (in meters) measured perpendicular to the base. Then the area A (in square inches) of the rectangle is

$$A = 1,550bh$$

Look again at Fig. 1-2. Suppose that you want to know the area in square inches when the base length and the height are expressed in feet. There are exactly 144 square inches in one square foot, so we can restate the formula for Fig. 1-2 this way: Let b represent the length (in feet) of the base of a triangle, and let h represent the height (in feet) measured perpendicular to the base. Then the area A (in square inches) of the triangle is

$$A = 144bh/2$$
$$= (144/2)\ bh$$
$$= 72bh$$

Give Fig. 1-3 another look. Suppose that you want to know how far the car has traveled in miles when its speed is given in feet per second and the time is given in hours. To figure this out, you must know the relationship between miles per hour and feet per second. To convert feet per second approximately to miles per hour, it is necessary to multiply by 0.6818. Then the units will be consistent with each other: The distance will be in miles, the speed will be in miles per hour, and the time will be in hours. The formula for Fig. 1-3 can be rewritten: Suppose that a car travels at a constant speed s (in feet per second) down a straight highway (see Fig. 1-3). Let t be a certain length of time (in hours). Then the distance d (in miles) that the car travels in that length of time is given by

$$d = 0.6818st$$

You can derive these conversion factors easily. All you need to know is the number of inches in a meter, the number of inches in a foot, the number of feet in a mile, and the number of seconds in an hour. As an exercise, you might want to go through the arithmetic for yourself. Maybe you'll want to derive the factors to greater precision than is given here.

Conversion factors are not always straightforward. Fortunately, databases abound in which conversion factors of all kinds are listed in tabular form. You don't have to memorize a lot of data. You can simply look up the conversion factors you need. The Internet is a great source of this kind of information. At the time of this writing, a comprehensive conversion database for physical units was available at the following location on the Web:

http://www.physics.nist.gov/Pubs/SP811/appenB8.html

If you've used the Web very much, you know that uniform resource locators (URLs) are always changing. If the preceding URL does not guide you to conversion factors, point your browser to the National Institute of Standards and Technology (NIST) home page and search the site for tables of conversion factors:

http://www.nist.gov

If the manner in which units are expressed on academic Web sites seems unfathomable, don't worry. As you work your way through this book, you will get used to scientific notation, and such expressions will evolve from arcane to mundane.

One-Variable First-Order Equations

In algebra, it is customary to classify equations according to the highest exponent, that is, the highest power to which the variables are raised. A *one-variable first-order equation,* also called a *first-order equation in one variable,* can be written in the following standard form:

$$ax + b = 0$$

where *a* and *b* are constants, and *x* is the variable. Equations of this type always have one *real-number* solution.

WHAT'S A "REAL" NUMBER?

A *real number* can be defined informally as any number that appears on a *number line* (Fig. 1-4). Pure mathematicians would call that an oversimplification, but it will do here. Examples of real numbers include 0, 5, −7, 22.55, the square root of 2, and π.

If you wonder what a "nonreal" number is like, consider the square root of −1. What real number can you multiply by itself and get −1? There is no such number. All the negative numbers, when squared, yield positive numbers; all the positive numbers also yield positive numbers; zero squared equals zero. The square root of −1 exists, but it lies somewhere other than on the number line shown in Fig. 1-4.

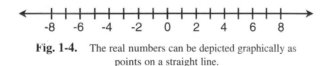

Fig. 1-4.   The real numbers can be depicted graphically as points on a straight line.

Later in this chapter you will be introduced to *imaginary numbers* and *complex numbers,* which are, in a certain theoretical sense, "nonreal." For now, however, let's get back to the task at hand: first-order equations in one variable.

SOME EXAMPLES

Any equation that can be converted into the preceding standard form is a one-variable first-order equation. Alternative forms are

$$cx = d$$

$$x = m/n$$

where c, d, m, and n are constants and $n \neq 0$. Here are some examples of single-variable first-order equations:

$$4x - 8 = 0$$

$$-\pi x = 22$$

$$3ex = c$$

$$x = \pi/c$$

In these equations, π, e, and c are known as *physical constants,* representing the circumference-to-diameter ratio of a circle, the natural exponential base, and the speed of light in free space, respectively. The constants π and e are not specified in units of any sort. They are plain numbers, and as such, they are called *dimensionless constants:*

$$\pi \approx 3.14159$$

$$e \approx 2.7\,1828$$

The squiggly equals sign means "is approximately equal to." The constant c does not make sense unless units are specified. It must be expressed in speed units of some kind, such as miles per second (mi/s) or kilometers per second (km/s):

$$c \approx 186,282 \text{ mi/s}$$

$$c \approx 299,792 \text{ km/s}$$

HOW TO SOLVE

To solve a single-variable equation, it must in effect be converted into a formula. The variable should appear all by itself on the left-hand side of the equals sign, and the expression on the right-hand side should be reducible to a specific number. There are several techniques for getting such an equation into the form of a statement that expressly tells you the value of the variable:

• Add the same quantity to each side of the equation.

• Subtract the same quantity from each side of the equation.

- Multiply each side of the equation by the same quantity.
- Divide each side of the equation by the same quantity.

The quantity involved in any of these processes can contain numbers, constants, variables—anything. There's one restriction: You can't divide by zero or by anything that can equal zero under any circumstances. The reason for this is simple: Division by zero is not defined.

Consider the four equations mentioned a few paragraphs ago. Let's solve them. Listed them again, they are

$$4x - 8 = 0$$

$$-\pi x = 22$$

$$3ex = c$$

$$x = \pi/c$$

The first equation is solved by adding 8 to each side and then dividing each side by 4:

$$4x - 8 = 0$$

$$4x = 8$$

$$x = 8/4 = 2$$

The second equation is solved by dividing each side by π and then multiplying each side by -1:

$$-\pi x = 22$$

$$-x = 22/\pi$$

$$x - -22/\pi$$

$$x \approx -22/3.14159$$

$$x \approx -7.00282$$

The third equation is solved by first expressing c (the speed of light in free space) in the desired units, then dividing each side by e (where $e \approx 2.71828$), and finally dividing each side by 3. Let's consider c in kilometers per second; $c \approx 299{,}792$ km/s. Then

$$3ex = c$$

$$(3 \times 2.71828)\, x \sim 299{,}792 \text{ km/s}$$

$$3x \approx (299{,}792/2.71828) \text{ km/s} \approx 110{,}287 \text{ km/s}$$

$$x \approx (110{,}287/3) \text{ km/s} \approx 36{,}762.3 \text{ km/s}$$

Note that we must constantly keep the units in mind here. Unlike the first two equations, this one involves a variable having a dimension (speed).

The fourth equation doesn't need solving for the variable, except to divide out the right-hand side. However, the units are tricky! Consider the speed of light in miles per second for this example; $c \approx 186{,}282$ mi/s. Then

$$x = \pi/c$$

$$x \approx 3.14159/ (186{,}282 \text{ mi/s})$$

When units appear in the denominator of a fractional expression, as they do here, they must be inverted. That is, we must take the reciprocal of the unit involved. In this case, this means changing miles per second into seconds per mile (s/mi). This gives us

$$x \approx (3.14159/186{,}282) \text{ s/mi}$$

$$x \approx 0.0000168647 \text{ s/mi}$$

This is not the usual way to express speed, but if you think about it, it makes sense. Whatever "object x" might be, it takes about 0.0000168647 s to travel 1 mile.

One-Variable Second-Order Equations

A *one-variable second-order equation,* also called a *second-order equation in one variable* or, more often, a *quadratic equation,* can be written in the following standard form:

$$ax^2 + bx + c = 0$$

where a, b, and c are constants, and x is the variable. (The constant c here does not stand for the speed of light.) Equations of this type can have two real-number solutions, one real-number solution, or no real-number solutions.

SOME EXAMPLES

Any equation that can be converted into the preceding form is a quadratic equation. Alternative forms are

$$mx^2 + nx = p$$

$$qx^2 = rx + s$$

$$(x + t)(x + u) = 0$$

where m, n, p, q, r, s, t, and u are constants. Here are some examples of quadratic equations:

$$x^2 + 2x + 1 = 0$$

$$-3x^2 - 4x = 2$$

$$4x^2 = -3x + 5$$

$$(x + 4)(x - 5) = 0$$

GET IT INTO FORM

Some quadratic equations are easy to solve; others are difficult. The first step, no matter what scheme for solution is contemplated, is to get the equation either into standard form or into factored form.

The first equation above is already in standard form. It is ready for an attempt at solution, which, we will shortly see, is rather easy.

The second equation can be reduced to standard form by subtracting 2 from each side:

$$-3x^2 - 4x = 2$$

$$-3x^2 - 4x - 2 = 0$$

The third equation can be reduced to standard form by adding $3x$ to each side and then subtracting 5 from each side:

$$4x^2 = -3x + 5$$

$$4x^2 + 3x = 5$$

$$4x^2 + 3x - 5 = 0$$

The fourth equation is in factored form. Scientists and engineers like this sort of equation because it can be solved without having to do any work. Look at it closely:

$$(x + 4)(x - 5) = 0$$

The expression on the left-hand side of the equals sign is zero if either of the two factors is zero. If $x = -4$, then the equation becomes

$$(-4 + 4)\,(-4 - 5) = 0$$

$$0 \times -9 = 0 \qquad \text{(It works)}$$

If $x = 5$, then the equation becomes

$$(5 + 4)\,(5 - 5) = 0$$

$$9 \times 0 = 0 \qquad \text{(It works again)}$$

It is the height of simplicity to "guess" which values for the variable in a factored quadratic will work as solutions. Just take the additive inverses (negatives) of the constants in each factor.

There is one possible point of confusion that should be cleared up. Suppose that you run across a quadratic like this:

$$x\,(x + 3) = 0$$

In this case, you might want to imagine it this way:

$$(x + 0)\,(x + 3) = 0$$

and you will immediately see that the solutions are $x = 0$ or $x = -3$.

In case you forgot, at the beginning of this section it was mentioned that a quadratic equation may have only one real-number solution. Here is an example of the factored form of such an equation:

$$(x - 7)\,(x - 7) = 0$$

Mathematicians might say something to the effect that, theoretically, this equation has two real-number solutions, and they are both 7. However, the physicist is content to say that the only real-number solution is 7.

THE QUADRATIC FORMULA

Look again at the second and third equations mentioned a while ago:

$$-3x^2 - 4x = 2$$

$$4x^2 = -3x + 5$$

These were reduced to standard form, yielding these equivalents:

$$-3x^2 - 4x - 2 = 0$$

$$4x^2 + 3x - 5 = 0$$

You might stare at these equations for a long time before you get any ideas about how to factor them. You might never get a clue. Eventually, you might wonder why you are wasting your time. Fortunately, there is a formula you can use to solve quadratic equations in general. This formula uses "brute force" rather than the intuition that factoring often requires.

Consider the standard form of a one-variable second-order equation once again:

$$ax^2 + bx + c = 0$$

The solution(s) to this equation can be found using this formula:

$$x = [-b \pm (b^2 - 4ac)^{1/2}]/2a$$

A couple of things need clarification here. First, the symbol $\pm$. This is read "plus or minus" and is a way of compacting two mathematical expressions into one. It's sort of a scientist's equivalent of computer data compression. When the preceding "compressed equation" is "expanded out," it becomes two distinct equations

$$x = [-b + (b^2 - 4ac)^{1/2}]/2a$$

$$x = [-b - (b^2 - 4ac)^{1/2}]/2a$$

The second item to be clarified involves the fractional exponent. This is not a typo. It literally means the $\frac{1}{2}$ power, another way of expressing the square root. It's convenient because it's easier for some people to write than a radical sign. In general, the zth root of a number can be written as the $1/z$ power. This is true not only for whole-number values of z but also for all possible values of z except zero.

PLUGGING IN

Examine this equation once again:

$$-3x^2 - 4x - 2 = 0$$

Here, the coefficients are

$$a = -3$$
$$b = -4$$
$$c = -2$$

Plugging these numbers into the quadratic formula yields

$$x = \{4 \pm [(-4)^2 - (4 \times -3 \times -2)]^{1/2}\}/(2 \times -3)$$

$$= 4 \pm (16 - 24)^{1/2}/-6$$

$$= 4 \pm (-8)^{1/2}/-6$$

We are confronted with the square root of -8 in the solution. This a "non-real" number of the sort you were warned about a while ago.

THOSE "NONREAL" NUMBERS

Mathematicians symbolize the square root of -1, called the *unit imaginary number,* by using the lowercase italic letter i. Scientists and engineers more often symbolize it using the letter j, and henceforth, that is what we will do.

Any imaginary number can be obtained by multiplying j by some real number q. The real number q is customarily written after j if q is positive or zero. If q happens to be a negative real number, then the absolute value of q is written after $-j$. Examples of imaginary numbers are $j3$, $-j5$, $j2.787$, and $-j\pi$.

The set of imaginary numbers can be depicted along a number line, just as can the real numbers. In a sense, the real-number line and the imaginary-number line are identical twins. As is the case with human twins, these two number lines, although they look similar, are independent. The sets of imaginary and real numbers have one value, zero, in common. Thus

$$j0 = 0$$

A *complex number* consists of the sum of some real number and some imaginary number. The general form for a complex number k is

$$k = p + jq$$

where p and q are real numbers.

Mathematicians, scientists, and engineers all denote the set of complex numbers by placing the real-number and imaginary-number lines at right angles to each other, intersecting at zero. The result is a rectangular coordinate plane (Fig. 1-5). Every point on this plane corresponds to a unique complex number; every complex number corresponds to a unique point on the plane.

Now that you know a little about complex numbers, you might want to examine the preceding solution and simplify it. Remember that it contains

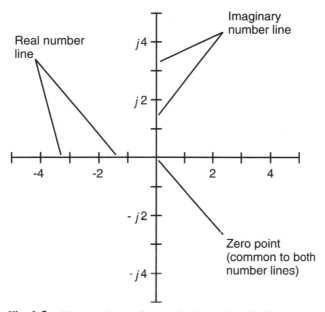

Fig. 1-5. The complex numbers can be depicted graphically as points on a plane, defined by two number lines at right angles.

$(-8)^{1/2}$. An engineer or physicist would write this as $j8^{1/2}$, so the solution to the quadratic is

$$x = 4 \pm j8^{1/2}/-6$$

BACK TO "REALITY"

Now look again at this equation:

$$4x^2 + 3x - 5 = 0$$

Here, the coefficients are

$$a = 4$$
$$b = 3$$
$$c = -5$$

Plugging these numbers into the quadratic formula yields

$$x = \{-3 \pm [3^2 - (4 \times 4 \times -5)]^{1/2}\}/(2 \times 4)$$
$$= -3 \pm (9 + 80)^{1/2}/8$$
$$= -3 \pm (89)^{1/2}/8$$

The square root of 89 is a real number but a messy one. When expressed in decimal form, it is nonrepeating and nonterminating. It can be approximated but never written out precisely. To four significant digits, its value is 9.434. Thus

$$x \approx -3 \pm 9.434/8$$

If you want to work this solution out to obtain two plain numbers without any addition, subtraction, or division operations in it, go ahead. However, it's more important that you understand the process by which this solution is obtained. If you are confused on this issue, you're better off reviewing the last several sections again and not bothering with arithmetic that any calculator can do for you mindlessly.

One-Variable Higher-Order Equations

As the exponents in single-variable equations become larger and larger, finding the solutions becomes an ever more complicated and difficult business. In the olden days, a lot of insight, guesswork, and tedium were involved in solving such equations. Today, scientists have the help of computers, and when problems are encountered containing equations with variables raised to large powers, brute force is the method of choice. We'll define *cubic equations, quartic equations, quintic equations,* and *nth-order equations* here but leave the solution processes to the more advanced pure-mathematics textbooks.

THE CUBIC

A *cubic equation,* also called a *one-variable third-order equation* or a *third-order equation in one variable,* can be written in the following standard form:

$$ax^3 + bx^2 + cx + d = 0$$

where a, b, c, and d are constants, and x is the variable. (Here, c does not stand for the speed of light in free space but represents a general constant.) If you're lucky, you'll be able to reduce such an equation to factored form to find real-number solutions r, s, and t:

$$(x - r)(x - s)(x - t) = 0$$

Don't count on being able to factor a cubic equation into this form. Sometimes it's easy, but usually it is exceedingly difficult and time-consuming.

THE QUARTIC

A *quartic equation,* also called a *one-variable fourth-order equation* or a *fourth-order equation in one variable,* can be written in the following standard form:

$$ax^4 + bx^3 + cx^2 + dx + e = 0$$

where a, b, c, d, and e are constants, and x is the variable. (Here, c does not stand for the speed of light in free space, and e does not stand for the exponential base; instead, these letters represent general constants in this context.) There is an outside chance that you'll be able to reduce such an equation to factored form to find real-number solutions r, s, t, and u:

$$(x - r)(x - s)(x - t)(x - u) = 0$$

As is the case with the cubic, you will be lucky if you can factor a quartic equation into this form and thus find four real-number solutions with ease.

THE QUINTIC

A *quintic equation,* also called a *one-variable fifth-order equation* or a *fifth-order equation in one variable,* can be written in the following standard form:

$$ax^5 + bx^4 + cx^3 + dx^2 + ex + f = 0$$

where a, b, c, d, e, and f are constants, and x is the variable. (Here, c does not stand for the speed of light in free space, and e does not stand for the exponential base; instead, these letters represent general constants in this context.) There is a remote possibility that if you come across a quintic, you'll be able to reduce it to factored form to find real-number solutions r, s, t, u, and v:

$$(x - r)\,(x - s)\,(x - t)\,(x - u)\,(x - v) = 0$$

As is the case with the cubic and the quartic, you will be lucky if you can factor a quintic equation into this form. The "luck coefficient" goes down considerably with each single-number exponent increase.

THE *n*TH-ORDER *EQUATION*

A one-variable *n*th-order equation can be written in the following standard form:

$$a_1 x^n + a_2 x^{n-1} + a_3 x^{n-2} + \ldots + a_{n-2} x^2 + a_{n-1} x + a_n = 0$$

where $a_1, a_2, \ldots, a_n$ are constants, and x is the variable. We won't even think about trying to factor an equation like this in general, although specific cases may lend themselves to factorization. Solving equations like this requires the use of a computer or else a masochistic attitude.

Vector Arithmetic

As mentioned at the beginning of this chapter, a vector has two independently variable properties: magnitude and direction. Vectors are used commonly in physics to represent phenomena such as force, velocity, and acceleration. In contrast, real numbers, also called *scalars,* are one-dimensional (they can be depicted on a line); they have only magnitude. Scalars are satisfactory for representing phenomena or quantities such as temperature, time, and mass.

VECTORS IN TWO DIMENSIONS

Do you remember *rectangular coordinates,* the familiar *xy* plane from your high-school algebra courses? Sometimes this is called the *cartesian plane* (named after the mathematician Rene Descartes.) Imagine two vectors in that plane. Call them **a** and **b.** (Vectors are customarily written in boldface, as opposed to variables, constants, and coefficients, which are usually written in italics). These two vectors can be denoted as rays from the origin (0, 0) to points in the plane. A simplified rendition of this is shown in Fig. 1-6.

Suppose that the end point of **a** has values (x_a, y_a) and the end point of **b** has values (x_b, y_b). The magnitude of **a,** written |a|, is given by

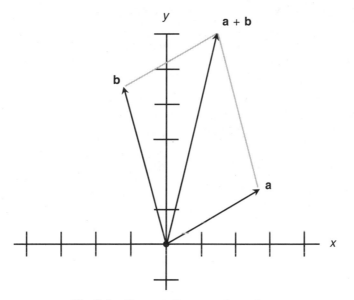

Fig. 1-6.   Vectors in the rectangular xy plane.

$$|\mathbf{a}| = (x_a^2 + y_a^2)^{1/2}$$

The sum of vectors **a** and **b** is

$$\mathbf{a} + \mathbf{b} = [(x_a + x_b), (y_a + y_b)]$$

This sum can be found geometrically by constructing a parallelogram with **a** and **b** as adjacent sides; then **a** + **b** is the diagonal of this parallelogram.

The *dot product*, also known as the *scalar product* and written **a** · **b**, of vectors **a** and **b** is a real number given by the formula

$$\mathbf{a} \cdot \mathbf{b} = x_a x_b + y_a y_b$$

The *cross product*, also known as the *vector product* and written **a** × **b**, of vectors **a** and **b** is a vector perpendicular to the plane containing **a** and **b**. Suppose that the angle between vectors **a** and **b**, as measured counterclockwise (from your point of view) in the plane containing them both, is called q. Then **a** × **b** points toward you, and its magnitude is given by the formula

$$|\mathbf{a} \times \mathbf{b}| = |\mathbf{a}|\,|\mathbf{b}|\,\sin q$$

VECTORS IN THREE DIMENSIONS

Now expand your mind into three dimensions. In rectangular xyz space, also called *cartesian three-space,* two vectors **a** and **b** can be denoted as rays from the origin $(0, 0, 0)$ to points in space. A simplified illustration of this is shown in Fig. 1-7.

Suppose that the end point of **a** has values (x_a, y_a, z_a) and the end point of **b** has values (x_b, y_b, z_b). The magnitude of **a,** written $|a|$, is

$$|a| = (x_a^2 + y_a^2 + z_a^2)^{1/2}$$

The sum of vectors **a** and **b** is

$$a + b = [(x_a + x_b), (y_a + y_b), (z_a + z_b)]$$

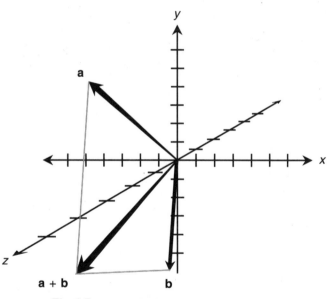

Fig. 1-7. Vectors in three-dimensional xyz space.

This sum can, as in the two-dimensional case, be found geometrically by constructing a parallelogram with **a** and **b** as adjacent sides. The sum **a** + **b** is the diagonal.

The dot product **a** · **b** of two vectors **a** and **b** in xyz space is a real number given by the formula

$$a \cdot b = x_a x_b + y_a y_b + z_a z_b$$

The cross product $\mathbf{a} \times \mathbf{b}$ of vectors $\mathbf{a}$ and $\mathbf{b}$ in xyz space is a little more complicated to envision. It is a vector perpendicular to the plane P containing both $\mathbf{a}$ and $\mathbf{b}$ and whose magnitude is given by the formula

$$|\mathbf{a} \times \mathbf{b}| = |\mathbf{a}|\,|\mathbf{b}|\sin q$$

where $\sin q$ is the sine of the angle q between $\mathbf{a}$ and $\mathbf{b}$ as measured in P. The direction of the vector $\mathbf{a} \times \mathbf{b}$ is perpendicular to plane P. If you look at $\mathbf{a}$ and $\mathbf{b}$ from some point on a line perpendicular to P and intersecting P at the origin, and q is measured counterclockwise from $\mathbf{a}$ to $\mathbf{b}$, then the vector $\mathbf{a} \times \mathbf{b}$ points toward you.

Some Laws for Vectors

When it comes to rules, vectors are no more exalted than ordinary numbers. Here are a few so-called laws that all vectors obey.

MULTIPLICATION BY SCALAR

When any vector is multiplied by a real number, also known as a *scalar,* the vector magnitude (length) is multiplied by that scalar. The direction remains unchanged if the scalar is positive but is reversed if the scalar is negative.

COMMUTATIVITY OF ADDITION

When you add two vectors, it does not matter in which order the sum is performed. If $\mathbf{a}$ and $\mathbf{b}$ are vectors, then

$$\mathbf{a} + \mathbf{b} = \mathbf{b} + \mathbf{a}$$

COMMUTATIVITY OF VECTOR-SCALAR MULTIPLICATION

When a vector is multiplied by a scalar, it does not matter in which order the product is performed. If $\mathbf{a}$ is a vector and k is a real number, then

$$k\mathbf{a} = \mathbf{a}k$$

COMMUTATIVITY OF DOT PRODUCT

When the dot product of two vectors is found, it does not matter in which order the vectors are placed. If **a** and **b** are vectors, then

$$\mathbf{a} \cdot \mathbf{b} = \mathbf{b} \cdot \mathbf{a}$$

NEGATIVE COMMUTATIVITY OF CROSS PRODUCT

The cross product of two vectors reverses direction when the order in which the vectors are "multiplied" is reversed. That is,

$$\mathbf{b} \times \mathbf{a} = -(\mathbf{a} \times \mathbf{b})$$

ASSOCIATIVITY OF ADDITION

When you add three vectors, it makes no difference how the sum is grouped. If **a, b,** and **c** are vectors, then

$$(\mathbf{a} + \mathbf{b}) + \mathbf{c} = \mathbf{a} + (\mathbf{b} + \mathbf{c})$$

ASSOCIATIVITY OF VECTOR-SCALAR MULTIPLICATION

Let **a** be a vector, and let k_1 and k_2 be real-number scalars. Then the following equation holds:

$$k_1 (k_2 \mathbf{a}) = (k_1 k_2) \mathbf{a}$$

DISTRIBUTIVITY OF SCALAR MULTIPLICATION OVER SCALAR ADDITION

Let **a** be a vector, and let k_1 and k_2 be real-number scalars. Then the following equations hold:

$$(k_1 + k_2) \mathbf{a} = k_1 \mathbf{a} + k_2 \mathbf{a}$$

$$\mathbf{a} (k_1 + k_2) = \mathbf{a}k_1 + \mathbf{a}k_2 = k_1 \mathbf{a} + k_2 \mathbf{a}$$

DISTRIBUTIVITY OF SCALAR MULTIPLICATION OVER VECTOR ADDITION

Let **a** and **b** be vectors, and let k be a real-number scalar. Then the following equations hold:

$$k\,(\mathbf{a} + \mathbf{b}) = k\mathbf{a} + k\mathbf{b}$$

$$(\mathbf{a} + \mathbf{b})\,k = \mathbf{a}k + \mathbf{b}k = k\mathbf{a} + k\mathbf{b}$$

DISTRIBUTIVITY OF DOT PRODUCT OVER VECTOR ADDITION

Let **a, b,** and **c** be vectors. Then the following equations hold:

$$\mathbf{a} \cdot (\mathbf{b} + \mathbf{c}) = \mathbf{a} \cdot \mathbf{b} + \mathbf{a} \cdot \mathbf{c}$$

$$(\mathbf{b} + \mathbf{c}) \cdot \mathbf{a} = \mathbf{b} \cdot \mathbf{a} + \mathbf{c} \cdot \mathbf{a} = \mathbf{a} \cdot \mathbf{b} + \mathbf{a} \cdot \mathbf{c}$$

DISTRIBUTIVITY OF CROSS PRODUCT OVER VECTOR ADDITION

Let **a, b,** and **c** be vectors. Then the following equations hold:

$$\mathbf{a} \times (\mathbf{b} + \mathbf{c}) = \mathbf{a} \times \mathbf{b} + \mathbf{a} \times \mathbf{c}$$

$$(\mathbf{b} + \mathbf{c}) \times \mathbf{a} = \mathbf{b} \times \mathbf{a} + \mathbf{c} \times \mathbf{a}$$

$$= -(\mathbf{a} \times \mathbf{b}) - (\mathbf{a} \times \mathbf{c})$$

$$= (\mathbf{a} \times \mathbf{b} + \mathbf{a} \times \mathbf{c})$$

DOT PRODUCT OF CROSS PRODUCTS

Let **a, b, c,** and **d** be vectors. Then the following equation holds:

$$(\mathbf{a} \times \mathbf{b}) \cdot (\mathbf{c} \times \mathbf{d}) = (\mathbf{a} \cdot \mathbf{c})(\mathbf{b} \cdot \mathbf{d}) - (\mathbf{a} \cdot \mathbf{d})(\mathbf{b} \cdot \mathbf{c})$$

These are only a few examples of the rules vectors universally obey. If you have trouble directly envisioning how these rules work, you are not alone. Some vector concepts are impossible for mortal humans to see with the "mind's eye." This is why we have mathematics. Equations and formulas like the ones in this chapter allow us to work with "beasts" that would otherwise forever elude our grasp.

Quiz

Refer to the text in this chapter if necessary. A good score is eight correct. Answers are in the back of the book.

1. The equation $(x - 4)(x + 5)(x - 1) = 0$ is an example of
   (a) a first-order equation.
   (b) a second-order equation.
   (c) a third-order equation.
   (d) a fourth-order equation.

2. The real-number solutions to the equation in problem 1 are
   (a) -4, 5, and -1.
   (b) 4, -5, and 1.
   (c) There are no real-number solutions to this equation.
   (d) There is not enough information to tell.

3. Suppose that there are two vectors in the xy plane as follows:

$$\mathbf{a} = (x_a, y_a) = (3, 0)$$

$$\mathbf{b} = (x_b, y_b) = (0, 4)$$

What is the length of the sum of these vectors?
   (a) 5 units
   (b) 7 units
   (c) 12 units
   (d) There is not enough information to tell.

4. Consider two vectors $\mathbf{a}$ and $\mathbf{b}$, where $\mathbf{a}$ points east and $\mathbf{b}$ points north. In what direction does $\mathbf{a} \cdot \mathbf{b}$ point?
   (a) Northeast
   (b) Straight up
   (c) Straight down
   (d) Irrelevant question! The dot product is not a vector.

5. Consider the two vectors $\mathbf{a}$ and $\mathbf{b}$ of problem 4. In what direction does $\mathbf{a} \times \mathbf{b}$ point?
   (a) Northeast

    (b) Straight up

    (c) Straight down

    (d) Irrelevant question! The cross product is not a vector.

6. When dividing each side of an equation by a quantity, what must you be careful to avoid?

    (a) Dividing by a constant

    (b) Dividing by a variable

    (c) Dividing by anything that can attain a value of zero

    (d) Dividing each side by the same quantity

7. Consider a second-order equation of the form $ax^2 + bx + c = 0$ in which the coefficients have these values:

$$a = 2$$
$$b = 0$$
$$c = 8$$

    What can be said about the solutions to this equation?

    (a) They are real numbers.

    (b) They are pure imaginary numbers.

    (c) They are complex numbers.

    (d) There are no solutions to this equation.

8. Consider the equation $4x + 5 = 0$. What would be a logical first step in the process of solving this equation?

    (a) Subtract 5 from each side.

    (b) Divide each side by x.

    (c) Multiply each side by x.

    (d) Multiply each side by 0.

9. When two vectors **a** and **b** are added together, which of the following statements holds true in all situations?

    (a) The composite is always longer than either **a** or **b**.

    (b) The composite points in a direction midway between **a** and **b**.

    (c) The composite is perpendicular to the plane containing **a** and **b**.

    (d) None of the above.

10. An equation with a variable all by itself on the left-hand side of the equals sign and having an expression not containing that variable on the right-hand side of the equals sign and that is used to determine a physical quantity is

    (a) a formula.

    (b) a first-order equation.

    (c) a coefficient.

    (d) a constant.

CHAPTER 2

Scientific Notation

Now that you've refreshed your memory on how to manipulate unspecified numbers (variables), you should know about *scientific notation,* the way in which physicists and engineers express the extreme range of values they encounter. How many atoms are in the earth? What is the ratio of the volume of a marble to the volume of the sun? These numbers can be approximated pretty well, but in common decimal form they are difficult to work with.

Subscripts and Superscripts

Subscripts are used to modify the meanings of units, constants, and variables. A subscript is placed to the right of the main character (without spacing), is set in smaller type than the main character, and is set below the baseline.

Superscripts almost always represent *exponents* (the raising of a base quantity to a power). Italicized lowercase English letters from the second half of the alphabet (n through z) denote variable exponents. A superscript is placed to the right of the main character (without spacing), is set in smaller type than the main character, and is set above the baseline.

EXAMPLES OF SUBSCRIPTS

Numeric subscripts are never italicized, but alphabetic subscripts sometimes are. Here are three examples of subscripted quantities:

Z_0 read "Z sub nought"; stands for characteristic impedance of a transmission line

R_{out} read "R sub out"; stands for output resistance in an electronic circuit

y_n read "y sub n"; represents a variable

Ordinary numbers are rarely, if ever, modified with subscripts. You are not likely to see expressions like this:

$$3_5$$

$$-9.7755_\pi$$

$$-16_x$$

Constants and variables, however, can come in many "flavors." Some physical constants are assigned subscripts by convention. An example is m_e, representing the mass of an electron at rest. You might want to represent points in three-dimensional space by using ordered triples like (x_1, x_2, x_3) rather than (x, y, z). This subscripting scheme becomes especially convenient if you're talking about points in a higher-dimensional space, for example, $(x_1, x_2, x_3, ..., x_{11})$ in cartesian 11-space. Some cosmologists believe that there are as many as 11 dimensions in our universe, and perhaps more, so such applications of subscripts have real-world uses.

EXAMPLES OF SUPERSCRIPTS

Numeric superscripts are never italicized, but alphabetic superscripts usually are. Examples of superscripted quantities are

2^3 read "two cubed"; represents $2 \times 2 \times 2$

e^x read "e to the xth"; represents the exponential function of x

$y^{1/2}$ read "y to the one-half"; represents the square root of y

There is a significant difference between 2^3 and 2! There is also a difference, both quantitative and qualitative, between the expression e that symbolizes the natural-logarithm base (approximately 2.71828) and e^x, which can represent e raised to a variable power and which is sometimes used in place of the words *exponential function*.

Power-of-10 Notation

Scientists and engineers like to express extreme numerical values using an exponential technique known as *power-of-10 notation*. This is usually what is meant when they talk about "scientific notation."

STANDARD FORM

A numeral in *standard power-of-10 notation* is written as follows:

$$m.n \times 10^z$$

where the dot (.) is a period written on the baseline (not a raised dot indicating multiplication) and is called the *radix point* or *decimal point*. The value m (to the left of the radix point) is a positive integer from the set {1, 2, 3, 4, 5, 6, 7, 8, 9}. The value n (to the right of the radix point) is a nonnegative integer from the set {0, 1, 2, 3, 4, 5, 6, 7, 8, 9}. The value z, which is the power of 10, can be any integer: positive, negative, or zero. Here are some examples of numbers written in standard scientific notation:

$$2.56 \times 10^6$$
$$8.0773 \times 10^{-18}$$
$$1.000 \times 10^0$$

ALTERNATIVE FORM

In certain countries and in many books and papers written before the middle of the twentieth century, a slight variation on the preceding theme is used. The *alternative power-of-10 notation* requires that $m = 0$. When the preceding quantities are expressed this way, they appear as decimal fractions larger than 0 but less than 1, and the value of the exponent is increased by 1 compared with the standard form:

$$0.256 \times 10^7$$
$$0.80773 \times 10^{-17}$$
$$0.1000 \times 10^1$$

These are the same three numerical values as the previous three; the only difference is the way in which they're expressed. It's like saying that you're driving down a road at 50,000 meters per hour rather than at 50 kilometers per hour.

THE "TIMES SIGN"

The multiplication sign in a power-of-10 expression can be denoted in various ways. Most scientists in America use the cross symbol ($\times$), as in the preceding examples. However, a small dot raised above the baseline ($\cdot$) is sometimes used to represent multiplication in power-of-10 notation. When written this way, the preceding numbers look like this in the standard form:

$$2.56 \cdot 10^6$$

$$8.0773 \cdot 10^{-18}$$

$$1.000 \cdot 10^0$$

This small dot should not be confused with a radix point, as in the expression

$$m.n \cdot 10^z$$

in which the dot between m and n is a radix point and lies along the baseline, whereas the dot between n and 10^z is a multiplication symbol and lies above the baseline. The dot symbol is preferred when multiplication is required to express the dimensions of a physical unit. An example is kilogram-meter per second squared, which is symbolized $kg \cdot m/s^2$ or $kg \cdot m \cdot s^{-2}$.

When using an old-fashioned typewriter or a word processor that lacks a good repertoire of symbols, the lowercase nonitalicized letter x can be used to indicate multiplication. But this can cause confusion because it's easy to mistake this letter x for a variable. Thus, in general, it's a bad idea to use the letter x as a "times sign." An alternative in this situation is to use an asterisk (*). This is why occasionally you will see numbers written like this:

$$2.56*10^6$$

$$8.0773*10^{-18}$$

$$1.000*10^0$$

PLAIN-TEXT EXPONENTS

Once in awhile you will have to express numbers in power-of-10 notation using plain, unformatted text. This is the case, for example, when transmitting information within the body of an e-mail message (rather than as an attachment). Some calculators and computers use this system. The uppercase letter E indicates 10 raised to the power of the number that follows. In this format, the preceding quantities are written

$$2.56E6$$

$$8.0773E - 18$$

$$1.000E0$$

Sometimes the exponent is always written with two numerals and always includes a plus sign or a minus sign, so the preceding expressions appear as

$$2.56E + 06$$

$$8.0773E - 18$$

$$1.000E + 00$$

Another alternative is to use an asterisk to indicate multiplication, and the symbol ^ to indicate a superscript, so the expressions look like this:

$$2.56*10^6$$

$$8.0773*10^ - 18$$

$$1.000*10^0$$

In all these examples, the numerical values represented are identical. Respectively, if written out in full, they are

$$2,560,000$$

$$0.000000000000000080773$$

$$1.000$$

ORDERS OF MAGNITUDE

As you can see, power-of-10 notation makes it possible to easily write down numbers that denote unimaginably gigantic or tiny quantities. Consider the following:

$$2.55 \times 10^{45,589}$$

$$-9.8988 \times 10^{-7,654,321}$$

Imagine the task of writing either of these numbers out in ordinary decimal form! In the first case, you'd have to write the numerals 255 and then follow them with a string of 45,587 zeros. In the second case, you'd have to write a minus sign, then a numeral zero, then a radix point, then a string of 7,654,320 zeros, and then the numerals 9, 8, 9, 8, and 8.

Now consider these two numbers:

$$2.55 \times 10^{45,592}$$

$$-9.8988 \times 10^{-7,654,318}$$

These look a lot like the first two, don't they? However, both these new numbers are 1,000 times larger than the original two. You can tell by looking at the exponents. Both exponents are larger by 3. The number 45,592 is 3 more than 45,589, and the number $-7,754,318$ is 3 larger than $-7,754,321$. (Numbers grow larger in the mathematical sense as they become more positive or less negative.) The second pair of numbers is three *orders of magnitude* larger than the first pair of numbers. They look almost the same here, and they would look essentially identical if they were written out in full decimal form. However, they are as different as a meter is from a kilometer.

The order-of-magnitude concept makes it possible to construct number lines, charts, and graphs with scales that cover huge spans of values. Three examples are shown in Fig. 2-1. Part *a* shows a number line spanning three orders of magnitude, from 1 to 1,000. Part *b* shows a number line spanning 10 orders of magnitude, from 10^{-3} to 10^7. Part *c* shows a graph whose horizontal scale spans 10 orders of magnitude, from 10^{-3} to 10^7, and whose vertical scale extends from 0 to 10.

If you're astute, you'll notice that while the 0-to-10 scale is the easiest to envision directly, it covers more orders of magnitude than any of the others: infinitely many. This is so because no matter how many times you cut a nonzero number to $^1\!/_{10}$ its original size, you can never reach zero.

PREFIX MULTIPLIERS

Special verbal prefixes, known as *prefix multipliers,* are used commonly by physicists and engineers to express orders of magnitude. Flip ahead to Chapter 6 for a moment. Table 6.1 shows the prefix multipliers used for factors ranging from 10^{-24} to 10^{24}.

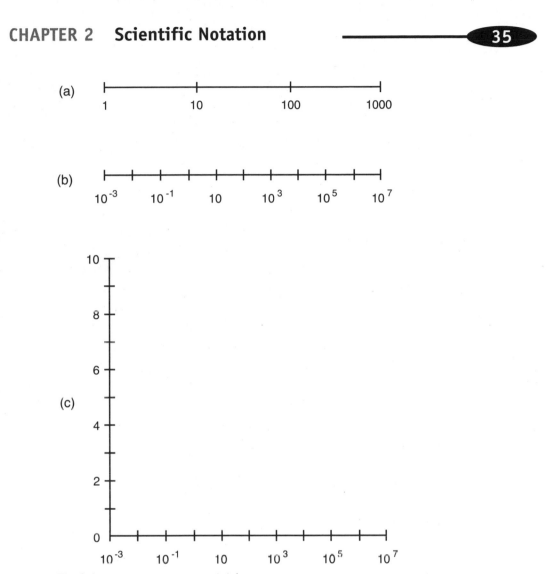

Fig. 2-1. (*a*) A number line spanning three orders of magnitude. (*b*) A number line spanning 10 orders of magnitude. (*c*) A coordinate system whose horizontal scale spans 10 orders of magnitude and whose vertical scale extends from 0 to 10.

Rules for Use

In printed literature, power-of-10 notation generally is used only when the power of 10 is large or small. If the exponent is between -2 and 2 inclusive, numbers are written out in plain decimal form as a rule. If the exponent is -3 or 3, numbers are sometimes written out and are sometimes written in power-of-10 notation. If the exponent is -4 or smaller, or if it is 4 or larger, values are expressed in power-of-10 notation as a rule.

Some calculators, when set for power-of-10 notation, display all numbers that way. This can be confusing, especially when the power of 10 is zero and the calculator is set to display a lot of digits. Most people understand the expression 8.407 more easily than 8.407000000E + 00, for example, even though they represent the same number.

With this in mind, let's see how power-of-10 notation works when we want to do simple arithmetic using extreme numbers.

ADDITION

Addition of numbers is best done by writing numbers out in ordinary decimal form if at all possible. Thus, for example,

$$(3.045 \times 10^5) + (6.853 \times 10^6) = 304{,}500 + 6{,}853{,}000$$
$$= 7{,}157{,}500$$
$$= 7.1575 \times 10^6$$

$$(3.045 \times 10^{-4}) + (6.853 \times 10^{-7)} = 0.0003045 + 0.0000006853$$
$$= 0.0003051853$$
$$= 3.051853 \times 10^{-4}$$

$$(3.045 \times 10^5) + (6.853 \times 10^{-7}) = 304{,}500 + 0.0000006853$$
$$= 304{,}500.0000006853$$
$$= 3.045000000006853 \times 10^5$$

SUBTRACTION

Subtraction follows the same basic rules as addition:

$$(3.045 \times 10^5) - (6.853 \times 10^6) = 304{,}500 - 6{,}853{,}000$$
$$= -6{,}548{,}500$$
$$= -6.548500 \times 10^6$$

$$(3.045 \times 10^{-4}) - (6.853 \times 10^{-7}) = 0.0003045 - 0.0000006853$$

$$= 0.0003038147$$
$$= 3.038147 \times 10^{-4}$$

$$(3.045 \times 10^5) - (6.853 \times 10^{-7}) = 304{,}500 - 0.0000006853$$
$$= 304{,}499.9999993147$$
$$= 3.044999999993147 \times 10^5$$

Power-of-10 notation may, at first, seem to do more harm than good when it comes to addition and subtraction. However, there is another consideration: the matter of *significant figures*. These make addition and subtraction, in the inexact world of experimental physics, quite easy and sometimes trivial. If the absolute values of two numbers differ by very many orders of magnitude, the one with the smaller absolute value (that is, the one closer to zero) can vanish into insignificance and, for practical purposes, can be ignored. We'll look at this phenomenon later in this chapter.

MULTIPLICATION

When numbers are multiplied in power-of-10 notation, the decimal numbers (to the left of the multiplication symbol) are multiplied by each other. Then the powers of 10 are added. Finally, the product is reduced to standard form. Here are three examples, using the same three number pairs as before:

$$(3.045 \times 10^5) \times (6.853 \times 10^6) = (3.045 \times 6.853) \times (10^5 \times 10^6)$$
$$= 20.867385 \times 10^{(5+6)}$$
$$= 20.867385 \times 10^{11}$$
$$= 2.0867385 \times 10^{12}$$

$$(3.045 \times 10^{-4}) \times (6.853 \times 10^{-7}) = (3.045 \times 6.853) \times (10^{-4} \times 10^{-7})$$
$$= 20.867385 \times 10^{[-4+(-7)]}$$
$$= 20.867385 \times 10^{-11}$$
$$= 2.0867385 \times 10^{-10}$$

$$(3.045 \times 10^5) \times (6.853 \times 10^{-7}) = (3.045 \times 6.853) \times (10^5 \times 10^{-7})$$
$$= 20.867385 \times 10^{(5-7)}$$
$$= 20.867385 \times 10^{-2}$$
$$= 2.0867385 \times 10^{-1}$$
$$= 0.20867385$$

This last number is written out in plain decimal form because the exponent is between -2 and 2 inclusive.

DIVISION

When numbers are divided in power-of-10 notation, the decimal numbers (to the left of the multiplication symbol) are divided by each other. Then the powers of 10 are subtracted. Finally, the quotient is reduced to standard form. Let's go another round with the same three number pairs we've been using:

$$(3.045 \times 10^5)/(6.853 \times 10^6) = (3.045/6.853) \times (10^5/10^6)$$
$$\approx 0.444331 \times 10^{(5-6)}$$
$$= 0.444331 \times 10^{-1}$$
$$= 0.0444331$$

$$(3.045 \times 10^{-4})/(6.853 \times 10^{-7}) = (3.045/6.853) \times (10^{-4}/10^{-7})$$
$$\approx 0.444331 \times 10^{[-4-(-7)]}$$
$$= 0.444331 \times 10^3$$
$$= 4.44331 \times 10^2$$
$$= 444.331$$

$$(3.045 \times 10^5)/(6.853 \times 10^{-7}) = (3.045/6.853) \times (10^5/10^{-7})$$
$$\approx 0.444331 \times 10^{[5-(-7)]}$$
$$= 0.444331 \times 10^{12}$$
$$= 4.44331 \times 10^{11}$$

Note the "approximately equal to" sign ($\approx$) in the preceding equations. The quotient here doesn't divide out neatly to produce a resultant with a reasonable number of digits. To this, you might naturally ask, "How many digits is reasonable?" The answer lies in the method scientists use to determine significant figures. An explanation of this is coming up soon.

EXPONENTIATION

When a number is raised to a power in scientific notation, both the coefficient and the power of 10 itself must be raised to that power and the result multiplied. Consider this example:

$$(4.33 \times 10^5)^3 - (4.33)^3 \times (10^5)^3$$
$$= 81.182737 \times 10^{(5\times3)}$$
$$= 81.182737 \times 10^{15}$$
$$= 8.1182727 \times 10^{16}$$

If you are a mathematical purist, you will notice gratuitous parentheses in the first and second lines here. From the point of view of a practical scientist, it is more important that the result of a calculation be correct than that the expression be as mathematically lean as possible.

Let's consider another example, in which the exponent is negative:

$$(5.27 \times 10^{-4})^2 = (5.27)^2 \times (10^{-4})^2$$
$$= 27.7729 \times 10^{(-4\times2)}$$
$$= 27.7729 \times 10^{-8}$$
$$= 2.77729 \times 10^{-7}$$

TAKING ROOTS

To find the root of a number in power-of-10 notation, the easiest thing to do is to consider that the root is a fractional exponent. The square root is the same thing as the ½ power; the cube root is the same thing as the ⅓ power. Then you can multiply things out in exactly the same way as you would with whole-number powers. Here is an example:

$$(5.27 \times 10^{-4})^{1/2} = (5.27)^{1/2} \times (10^{-4})^{1/2}$$
$$\approx 2.2956 \times 10^{[(-4\times1/2)]}$$

$$= 2.2956 \times 10^{-2}$$

$$= 0.02956$$

Note the "approximately equal to" sign in the second line. The square root of 5.27 is an irrational number, and the best we can do is to approximate its decimal expansion. Note also that because the exponent in the resultant is within the limits for which we can write the number out in plain decimal form, we have done so, getting rid of the power of 10.

Approximation, Error, and Precedence

In physics, the numbers we work with are not always exact values. In fact, in experimental physics, numbers are rarely the neat, crisp, precise animals familiar to the mathematician. Usually, we must approximate. There are two ways of doing this: *truncation* (simpler but less accurate) and *rounding* (a little more difficult but more accurate).

TRUNCATION

The process of truncation deletes all the numerals to the right of a certain point in the decimal part of an expression. Some electronic calculators use this process to fit numbers within their displays. For example, the number 3.830175692803 can be shortened in steps as follows:

3.830175692803

3.83017569280

3.8301756928

3.830175692

3.83017569

3.8301756

3.830175

3.83017

3.83

3.8

3

ROUNDING

Rounding is the preferred method of rendering numbers in shortened form. In this process, when a given digit (call it r) is deleted at the right-hand extreme of an expression, the digit q to its left (which becomes the new r after the old r is deleted) is not changed if $0 \leq r \leq 4$. If $5 \leq r \leq 9$, then q is increased by 1 ("rounded up"). Some electronic calculators use rounding rather than truncation. If rounding is used, the number 3.830175692803 can be shortened in steps as follows:

3.830175692803

3.83017569280

3.8301756928

3.830175693

3.83017569

3.8301757

3.830176

3.83018

3.8302

3.830

3.83

3.8

4

ERROR

When physical quantities are measured, exactness is impossible. Errors occur because of imperfections in the instruments and in some cases because of human error too. Suppose that x_a represents the actual value of a quantity to be measured. Let x_m represent the measured value of that quantity, in the same units as x_a. Then the *absolute error* D_a (in the same units as x_a) is given by

$$D_a = x_m - x_a$$

The *proportional error* D_p is equal to the absolute error divided by the actual value of the quantity:

$$D_p = (x_m - x_a)/x_a$$

The *percentage error* $D_\%$ is equal to 100 times the proportional error expressed as a ratio:

$$D_\% = 100 \ (x_m - x_a)/x_a$$

Error values and percentages are positive if $x_m > x_a$ and negative if $x_m < x_a$. This means that if the measured value is too large, the error is positive, and if the measured value is too small, the error is negative.

Does something seem strange about the preceding formulas? Are you a little uneasy about them? If you aren't, maybe you should be. Note that the denominators of all three equations contain the value x_a, the actual value of the quantity under scrutiny—the value that we are admitting we do not know exactly because our measurement is imperfect! How can we calculate error based on formulas that contain a quantity subject to the very error in question? The answer is that we can only make a good guess at x_a. This is done by taking several, perhaps even many, measurements, each with its own value x_{m1}, x_{m2}, x_{m3}, and so on, and then averaging them to get a good estimate of x_a. This means that in the imperfect world of physical things, the extent of our uncertainty is uncertain!

The foregoing method of error calculation also can be used to determine the extent to which a single reading x_m varies from a long-term average x_a, where x_a is derived from many readings taken over a period of time.

PRECEDENCE

Mathematicians, scientists, and engineers have all agreed on a certain order in which operations should be performed when they appear together in an expression. This prevents confusion and ambiguity. When various operations such as addition, subtraction, multiplication, division, and exponentiation appear in an expression and you need to simplify that expression, perform the operations in the following sequence:

- Simplify all expressions within parentheses from the inside out.
- Perform all exponential operations, proceeding from left to right.
- Perform all products and quotients, proceeding from left to right.
- Perform all sums and differences, proceeding from left to right.

Here are two examples of expressions simplified according to these rules of precedence. Note that the order of the numerals and operations is the same in each case, but the groupings differ.

$$[(2 + 3)\,(-3 - 1)^2]^2$$

$$[5 \times (-4)^2]^2$$

$$(5 \times 16)^2$$

$$80^2$$

$$6400$$

$$\{[2 + 3 \times (-3) - 1]^2\}^2$$

$$\{[2 + (-9) - 1]^2\}^2$$

$$(-8^2)^2$$

$$64^2$$

$$4096$$

Suppose that you're given a complicated expression and that there are no parentheses, brackets, or braces in it? This does not have to be ambiguous as long as the preceding rules are followed. Consider this example:

$$z = -3x^3 + 4x^2y - 12xy^2 - 5y^3$$

If this were written with parentheses, brackets, and braces to emphasize the rules of precedence, it would look like this:

$$z = [-3\,(x^3)] + \{4\,[(x^2)\,y]\} - \{12\,[x\,(y^2)]\} - [5\,(y^3)]$$

Because we have agreed on the rules of precedence, we can do without the parentheses, brackets, and braces. This will help to keep pure mathematicians happy. It might someday keep one of your research projects from going awry! Nevertheless, if there is any doubt about a crucial equation, you're better off to use a couple of unnecessary parentheses than to make a costly mistake.

Significant Figures

When multiplication or division is done using power-of-10 notation, the number of significant figures in the result cannot legitimately be greater than the number of significant figures in the least-exact expression. You may wonder why, in some of the preceding examples, we come up with answers that have more digits than any of the numbers in the original problem. In pure mathematics, this is not an issue, and up to this point we haven't been concerned with it. In physics, however, things are not so clear-cut.

Consider the two numbers $x = 2.453 \times 10^4$ and $y = 7.2 \times 10^7$. The following is a perfectly valid statement in arithmetic:

$$xy = 2.453 \times 10^4 \times 7.2 \times 10^7$$
$$= 2.453 \times 7.2 \times 10^{11}$$
$$= 17.6616 \times 10^{11}$$
$$= 1.76616 \times 10^{12}$$

However, if x and y represent measured quantities, as they would in experimental physics, the preceding statement needs qualification. We must pay close attention to how much accuracy we claim.

HOW ACCURATE ARE WE?

When you see a product or quotient containing a bunch of numbers in scientific notation, count the number of single digits in the decimal portions of each number. Then take the smallest number of digits. This is the number of significant figures you can claim in the final answer or solution. In the preceding example, there are four single digits in the decimal part of x, and two single digits in the decimal part of y. Thus we must round off the answer, which appears to contain six significant figures, to two. It is important to use rounding and not truncation! We should conclude that

$$xy = 2.453 \times 10^4 \times 7.2 \times 10^7$$
$$= 1.8 \times 10^{12}$$

In situations of this sort, if you insist on being 100 percent rigorous, you should use squiggly equals signs throughout because you are always dealing with approximate values. However, most experimentalists are content to use ordinary equals signs. It is universally understood that physical measurements are inexact, and writing squiggly lines can get tiresome.

Suppose that we want to find the quotient x/y instead of the product xy? Proceed as follows:

$$x/y = (2.453 \times 10^4)/(7.2 \times 10^7)$$

$$= (2.453/7.2) \times 10^{-3}$$

$$= 0.3406944444 \cdots \times 10^{-3}$$

$$= 3.406944444 \cdots \times 10^{-4}$$

$$= 3.4 \times 10^{-4}$$

WHAT ABOUT ZEROS?

Sometimes, when you make a calculation, you'll get an answer that lands on a neat, seemingly whole-number value. Consider $x = 1.41421$ and $y = 1.41422$. Both of these have six significant figures. The product, taking significant figures into account, is

$$xy = 1.41421 \times 1.41422$$

$$= 2.0000040662$$

$$= 2.00000$$

This looks like it's exactly equal to 2. In pure mathematics, $2.00000 = 2$. However, not in physics! (This is the sort of thing that drove the famous mathematician G. H. Hardy to write that mathematicians are in better contact with reality than are physicists.) There is always some error in physics. Those five zeros are important. They indicate how near the exact number 2 we believe the resultant to be. We know that the answer is very close to a mathematician's idea of the number 2, but there is an uncertainty of up to ± 0.000005. If we chop off the zeros and say simply that $xy = 2$, we allow for an uncertainty of up to ± 0.5, and in this case we are entitled to better than this. When we claim a certain number of significant figures, zero gets as much consideration as any other digit.

IN ADDITION AND SUBTRACTION

When measured quantities are added or subtracted, determining the number of significant figures can involve subjective judgment. The best procedure is to expand all the values out to their plain decimal form (if possible), make the calculation as if you were a pure mathematician, and then, at the end of the process, decide how many significant figures you can reasonably claim.

In some cases the outcome of determining significant figures in a sum or difference is similar to what happens with multiplication or division. Take, for example, the sum $x + y$, where $x = 3.778800 \times 10^{-6}$ and $y = 9.22 \times 10^{-7}$. This calculation proceeds as follows:

$$x = 0.000003778800$$

$$y = 0.000000922$$

$$x + y = 0.0000047008$$

$$= 4.7008 \times 10^{-6}$$

$$= 4.70 \times 10^{-6}$$

In other instances, however, one of the values in a sum or difference is insignificant with respect to the other. Let's say that $x = 3.778800 \times 10^{4}$, whereas $y = 9.22 \times 10^{-7}$. The process of finding the sum goes like this:

$$x = 37,788.00$$

$$y = 0.000000922$$

$$x + y = 37,788.000000922$$

$$= 3.7788000000922 \times 10^{4}$$

In this case, y is so much smaller than x that it doesn't significantly affect the value of the sum. Here it is best to regard y, in relation to x or to the sum $x + y$, as the equivalent of a gnat compared with a watermelon. If a gnat lands on a watermelon, the total weight does not change appreciably, nor does the presence or absence of the gnat have any effect on the accuracy of the scales. We can conclude that the "sum" here is the same as the larger number. The value y is akin to a nuisance or a negligible error:

$$x + y = 3.778800 \times 10^{4}$$

G. H. Hardy must be thanking the cosmos that he was not an experimental scientist. However, some people delight in subjectivity and imprecision. A gnat ought to be brushed off a watermelon without giving the matter any thought. A theoretician might derive equations to express the shape of the surface formed by the melon's two-dimensional geometric boundary with surrounding three-space without the gnat and then again with it and marvel at the difference between the resulting two relations. An experimentalist would, after weighing the melon, flick the gnat away, calculate the number of people with whom he could share the melon, slice it up, and have lunch with friends, making sure to spit out the seeds.

 **Quiz**

Refer to the text in this chapter if necessary. A good score is eight correct. Answers are in the back of the book.

1. Two numbers differ in size by exactly six orders of magnitude. This is a factor of
   (a) 6.
   (b) 36.
   (c) 10^6.
   (d) 6^6.

2. Suppose that we invent a new unit called the *flummox* (symbol: *Fx*). What might we logically call 10^{-9} flummox?
   (a) One milliflummox (1 mFx)
   (b) One nanoflummox (1 nFx)
   (c) One picoflummox (1 pFx)
   (d) One kiloflummox (1 kFx)

3. What is the value of $2 \times 4^2 - 6$?
   (a) 26
   (b) 58
   (c) 20
   (d) There is no way to tell; this is an ambiguous expression.

4. What is another way of writing 78,303?
   (a) 7.8303×10^{-4}
   (b) 7.8303E + 04
   (c) 0.78×10^5
   (d) This number is in its most appropriate form already.

5. Suppose that we measure the speed of an Internet connection 100 times and come up with an average of 480 kilobits per second (kbps). Suppose that the speed of the connection is absolutely constant. We go to a test site and obtain a reading of 440 kbps. What is the error of this measurement in percent? Express your answer to three significant figures.

(a) 440/480, or 91.7 percent

(b) (480 − 440)/440, or 9.09 percent

(c) (440 − 480)/480, or −8.33 percent

(d) 440 − 480, or −40.0 percent

6. Suppose that we measure a quantity and get 8.53×10^4 units, accurate to three significant figures. Within what range of whole-number units can we say the actual value is?

(a) 85,250 units to 85,349 units

(b) 85,290 units to 86,310 units

(c) 85,399 units to 86,301 units

(d) We can't say.

7. What is the difference 8.899×10^5 minus 2.02×10^{-12} taking significant figures into account?

(a) 2.02×10^{-12}

(b) 8.9×10^5

(c) 6.88×10^5

(d) 8.899×10^5

8. What is the order in which operations should be performed in an expression containing no parentheses?

(a) Addition, subtraction, multiplication, division, exponentiation

(b) Exponentiation, multiplication and division, addition and subtraction

(c) From left to right

(d) It is impossible to know

9. Suppose that the population of a certain country is found to be 78,790,003 people. What is this expressed to three significant figures?

(a) 7.88×10^7

(b) 7.879×10^7

(c) 78,800,000

(d) 78E + 06

10. What is the product of 8.72×10^5 and 6.554×10^{-5} taking significant figures into account?

(a) 57.15088

(b) 57.151

(c) 57.15

(d) 57.2

CHAPTER 3

Graphing Schemes

Graphs are diagrams of the functions and relations that express phenomena in the physical world. There are all kinds of graphs; the simplest are two-dimensional drawings. The most sophisticated graphs cannot be envisioned even by the most astute human beings, and computers are required to show cross sections of them so that we can get a glimpse of what is going on. In this chapter we will look at the most commonly used methods of graphing. There will be plenty of examples so that you can see what the graphs of various relations and functions look like.

Rectangular Coordinates

The most straightforward two dimensional coordinate system is the *cartesian plane* (Fig. 3-1), also called *rectangular coordinates* or the *xy plane*. The *independent variable* is plotted along the *x* axis, or *abscissa*; the *dependent variable* is plotted along the *y* axis, or *ordinate*. The scales of the abscissa and ordinate are usually (but not always) linear, and they are perpendicular to each other. The divisions of the abscissa need not represent the same increments as the divisions of the ordinate.

SLOPE-INTERCEPT FORM OF LINEAR EQUATION

A linear equation in two variables can be rearranged from standard form to a conveniently graphable form as follows:

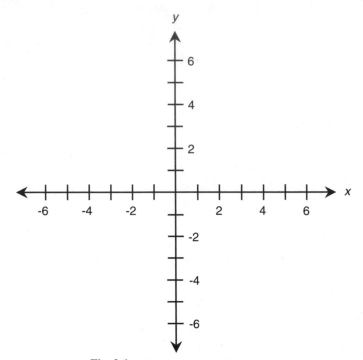

Fig. 3-1.   The cartesian coordinate plane.

$$ax + by + c = 0$$

$$ax + by = -c$$

$$by = -ax - c$$

$$y = (-a/b)x - (c/b)$$

where *a, b,* and *c* are real-number constants, and $b \neq 0$. Such an equation appears as a straight line when graphed on the cartesian plane. Let *dx* represent a small change in the value of *x* on such a graph; let *dy* represent the change in the value of *y* that results from this change in *x*. The ratio *dy/dx* is defined as the *slope* of the line and is commonly symbolized *m*. Let *k* represent the *y* value of the point where the line crosses the ordinate. Then the following equations hold:

$$m = -a/b$$

$$k = -c/b$$

The linear equation can be rewritten in *slope-intercept form* as

$$y = mx + k$$

To plot a graph of a linear equation in cartesian coordinates, proceed as follows:

- Convert the equation to slope-intercept form.
- Plot the point $y = k$.
- Move to the right by n units on the graph.
- Move upward by mn units (or downward by $-mn$ units).
- Plot the resulting point $y = mn + k$.
- Connect the two points with a straight line.

Figures 3-2 and 3-3 illustrate the following linear equations as graphed in slope-intercept form:

$$y = 5x - 3$$
$$y = -x + 2$$

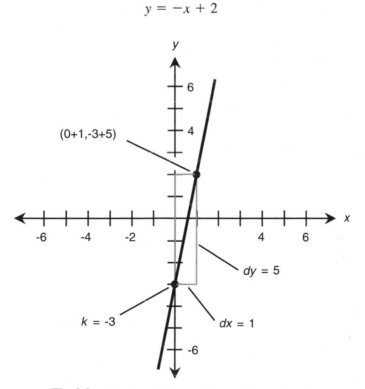

Fig. 3-2.   Slope-intercept plot of the equation $y = 5x - 3$.

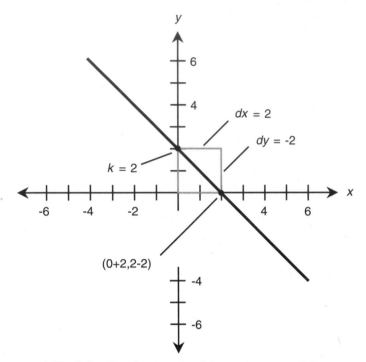

Fig. 3-3. Slope-intercept plot of the equation $y = -x + 2$.

A positive slope indicates that the graph "ramps upward" and a negative slope indicates that the graph "ramps downward" as you move toward the right. A zero slope indicates a horizontal line. The slope of a vertical line is undefined because, in the form shown here, it requires that something be divided by zero.

POINT-SLOPE FORM OF LINEAR EQUATION

It is not always convenient to plot a graph of a line based on the y-intercept point because the part of the graph you are interested in may lie at a great distance from that point. In this situation, the *point-slope form* of a linear equation can be used. This form is based on the slope m of the line and the coordinates of a known point (x_0, y_0):

$$y - y_0 = m (x - x_0)$$

To plot a graph of a linear equation using the point-slope method, you can follow these steps in order:

- Convert the equation to point-slope form.
- Determine a point (x_0, y_0) by "plugging in" values.
- Plot (x_0, y_0) on the plane.
- Move to the right by n units on the graph.
- Move upward by mn units (or downward by $-mn$ units).
- Plot the resulting point (x_1, y_1).
- Connect the points (x_0, y_0) and (x_1, y_1) with a straight line.

   Figures 3-4 and 3-5 illustrate the following linear equations as graphed in point-slope form for regions near points that are a long way from the origin:

$$y - 104 = 3\,(x - 72)$$

$$y \mid 55 = -2\,(x + 85)$$

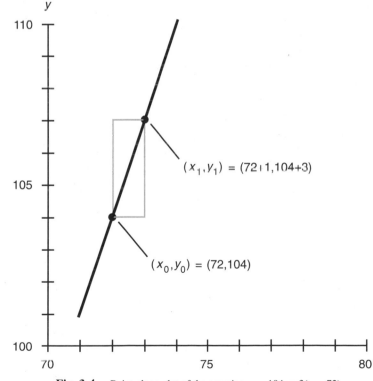

Fig. 3-4.   Point-slope plot of the equation $y - 104 = 3(x - 72)$.

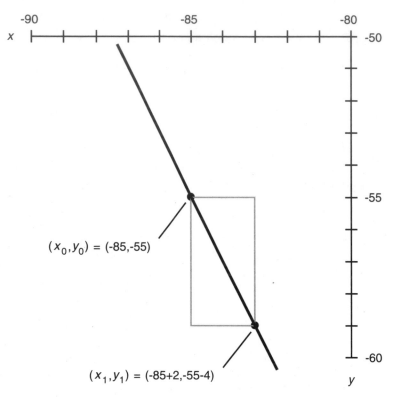

Fig. 3-5.   Point-slope plot of the equation $y + 55 = -2(x + 85)$.

FINDING LINEAR EQUATION BASED ON GRAPH

Now suppose that we are working in rectangular coordinates, and we know the exact values of two points P and Q. These two points define a straight line; this is one of the fundamental rules of geometry. Call the line L. Let's give the coordinates of the points these names:

$$P = (x_p, y_p)$$
$$Q = (x_q, y_q)$$

The slope m of the line L is given by either of the following formulas:

$$m = (y_q - y_p)/(x_q - x_p)$$
$$m = (y_p - y_q)/(x_p - x_q)$$

The point-slope equation of L can be determined based on the known coordinates of P or Q. Therefore, either of the following formulas represent the line L:

$$y - y_p = m(x - x_p)$$
$$y - y_q = m(x - x_q)$$

EQUATION OF PARABOLA

The cartesian-coordinate graph of a quadratic equation where y is substituted for 0 in the standard form (recall this from Chapter 1) is a *parabola:*

$$y = ax^2 + bx + c$$

where $a \neq 0$. (If $a = 0$, then the equation is linear, not quadratic.) To plot a graph of the preceding equation, first determine the coordinates of the point (x_0, y_0) where

$$x_0 = -b/(2a)$$
$$y_0 = c - b^2/(4a)$$

This point represents the *base point* of the parabola; that is, the point at which the curvature is sharpest and at which the slope of a line tangent to the curve is zero. Once this point is known, find four more points by "plugging in" values of x somewhat greater than and less than x_0 and determining the corresponding y values. These x values, call them x_{-2}, x_{-1}, x_1, and x_2, should be equally spaced on either side of x_0, such that

$$x_{-2} < x_{-1} < x_0 < x_1 < x_2$$
$$x_{-1} - x_{-2} = x_0 - x_{-1} = x_1 - x_0 = x_2 - x_1$$

This will give you five points that lie along the parabola and that are symmetrical relative to the axis of the curve. The graph can then be inferred (this means that you can make a good educated guess), provided that the points are chosen judiciously. Some trial and error may be required. If $a > 0$, the parabola will open upward. If $a < 0$, the parabola will open downward.

EXAMPLE A

Consider the following formula:

$$y = x^2 + 2x + 1$$

The base point is

$$x_0 = -2/2 = -1$$
$$y_0 = 1 - 4/4 = 1 - 1 = 0$$

Therefore, $(x_0, y_0) = (-1, 0)$

This point is plotted first. Next, plot the following points:

$$x_{-2} = x_0 - 2 = -3$$
$$y_{-2} = (-3)^2 + 2(-3) + 1 = 9-6 + 1 = 4$$

Therefore, $(x_{-2}, y_{-2}) = (-3, 4)$

$$x_{-1} = x_0 - 1 = -2$$
$$y_{-1} = (-2)^2 + 2(-2) + 1 = 4 - 4 + 1 = 1$$

Therefore, $(x_{-1}, y_{-1}) = (-2, 1)$

$$x_1 = x_0 + 1 = 0$$
$$y_1 = (0)^2 + 2(0) + 1 = 0 + 0 + 1 = 1$$

Therefore, $(x_1, y_1) = (0, 1)$

$$x_2 = x_0 + 2 = 1$$
$$y_2 = (1)^2 + 2(1) + 1 = 1 + 2 + 1 = 4$$

Therefore, $(x_2, y_2) = (1, 4)$

The five known points are plotted as shown in Fig. 3-6. From these, the curve can be inferred.

EXAMPLE B

Consider the following formula:

$$y = -2x^2 + 4x - 5$$

The base point is

$$x_0 = -4/-4 = 1$$
$$y_0 = -5 - 16/-8 = -5 + 2 = -3$$

Therefore, $(x_0, y_0) = (1, -3)$

This point is plotted first. Next, plot the following points:

$$x_{-2} = x_0 - 2 = -1$$
$$y_{-2} = -2(-1)^2 + 4(-1) - 5 = -2 - 4 - 5 = -11$$

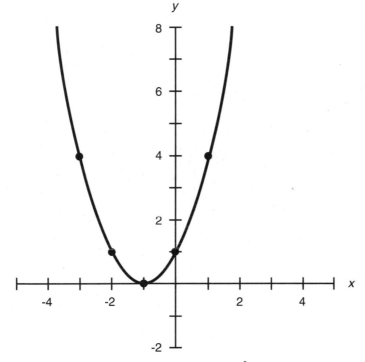

Fig. 3-6.   Plot of the parabola $y = x^2 + 2x + 1$.

Therefore, $(x_{-2}, y_{-2}) = (-1, -11)$

$$x_{-1} = x_0 - 1 = 0$$
$$y_{-1} = -2\,(0)^2 + 4\,(0) - 5 = -5$$

Therefore, $(x_{-1}, y_{-1}) = (0, -5)$

$$x_1 = x_0 + 1 = 2$$
$$y_1 = -2\,(2)^2 + 4\,(2) + 5 = -8 + 8 - 5 = -5$$

Therefore, $(x_1, y_1) = (2, -5)$

$$x_2 = x_0 + 2 = 3$$
$$y_2 = -2\,(3)^2 + 4\,(3) + 5 = -18 + 12 - 5 = -11$$

Therefore, $(x_2, y_2) = (3, -11)$

The five known points are plotted as shown in Fig. 3-7. From these, the curve can be inferred.

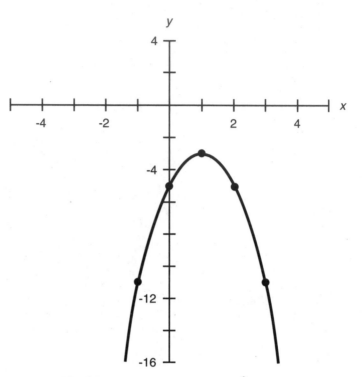

Fig. 3-7.   Plot of the parabola $y = -2x^2 + 4x - 5$.

EQUATION OF CIRCLE

The general form for the equation of a *circle* in the xy plane is given by the following formula:

$$(x - x_0)^2 + (y - y_0)^2 = r^2$$

where (x_0, y_0) represents the coordinates of the center of the circle, and r represents the *radius*. This is illustrated in Fig. 3-8. In the special case where the circle is centered at the origin, the formula becomes

$$x^2 + y^2 = r^2$$

Such a circle intersects the x axis at the points $(r, 0)$ and $(-r, 0)$; it intersects the y axis at the points $(0, r)$ and $(0, -r)$. An even more specific case is the *unit circle:*

$$x^2 + y^2 = 1$$

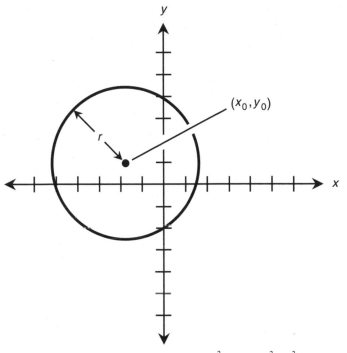

Fig. 3-8.   Plot of the circle $(x - x_0)^2 + (y - y_0)^2 = r^2$.

This curve intersects the x axis at the points $(1, 0)$ and $(-1, 0)$; it also intersects the y axis at the points $(0, 1)$ and $(0, -1)$.

GRAPHIC SOLUTION TO PAIRS OF EQUATIONS

The solutions of pairs of equations can be found by graphing both the equations on the same set of coordinates. Solutions appear as intersection points between the plots.

EXAMPLE A

Suppose that you are given these two equations and are told to solve them for values of x and y that satisfy both at the same time:

$$y = x^2 + 2x + 1$$
$$y = -x + 1$$

These equations are graphed in Fig. 3-9. The line crosses the parabola at two points, indicating that there are two solutions to this set of simultaneous equations. The coordinates of the points corresponding to the solutions are

$$(x_1, y_1) = (-3, 4)$$

$$(x_2, y_2) = (0, 1)$$

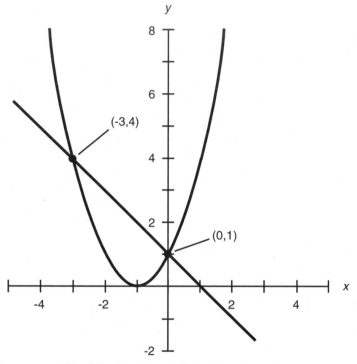

(-3,4)

(0,1)

Fig. 3-9. Graphic method of solving the equations
$y = x^2 + 2x + 1$ and $y = -x + 1$.

EXAMPLE B

Here is another pair of "two by two" equations (two equations in two variables) that can be solved by graphing:

$$y = -2x^2 + 4x - 5$$

$$y = -2x - 5$$

These equations are graphed in Fig. 3-10. The line crosses the parabola at two points, indicating that there are two solutions. The coordinates of the points corresponding to the solutions are

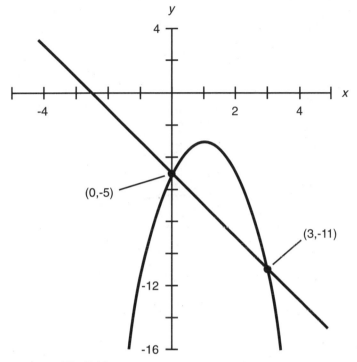

Fig. 3-10.   Graphic method of solving the equations
$y = -2x^2 + 4x - 5$ and $y = -2x - 5$.

$$(x_1, y_1) = (3, -11)$$

$$(x_2, y_2) = (0, -5)$$

Sometimes a graph will reveal that a pair of equations has more than two solutions, or only one solution, or no solutions at all. Solutions to pairs of equations always show up as intersection points on their graphs. If there are n intersection points between the curves representing two equations, then there are n solutions to the pair of simultaneous equations. However, graphing is only good for estimating the values of the solutions. If possible, algebra should be used to find exact solutions to problems of this kind. If the equations are complicated, or if the graphs are the results of experiments, it will be difficult to use algebra to solve them. Then graphs, with the aid of computer programs to accurately locate the points of intersection between graphs, offer a better means of solving pairs of equations.

The Polar Plane

The *polar coordinate plane* is an alternative way of expressing the positions of points and of graphing equations and relations in two dimensions. The independent variable is plotted as the distance or radius r from the origin, and the dependent variable is plotted as an angle q relative to a reference axis. Figure 3-11 shows the polar plane that is used most often by

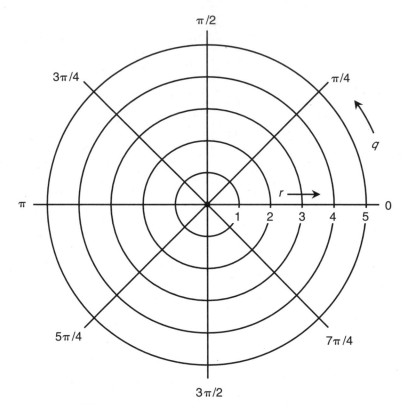

Fig. 3-11. The polar coordinate plane used in physics.

physicists. The angle q is expressed in units called *radians*. One radian is the angle defined by a circular arc whose length is the same as the radius of the circle containing that arc. If this is too complicated to remember, then think of it like this: A radian is a little more than 57°. Or you can remember that there are 2π, or about 6.28, radians in a complete circle. The angle q is plotted counterclockwise from the ray extending to the right.

Figure 3-12 shows the polar system employed by some engineers, especially those in communications. Navigators and astronomers often use this

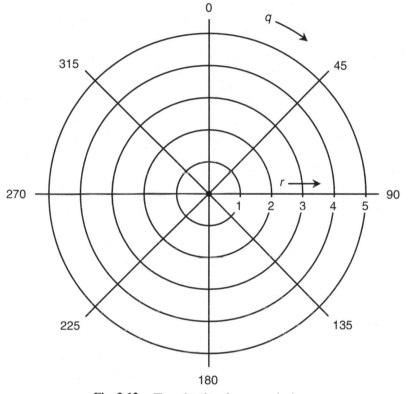

Fig. 3-12.   The polar plane for communications, navigation, and astronomy.

scheme too. The angle q is expressed in degrees here and is plotted clockwise from the ray extending upward (corresponding to geographic north). You have seen this coordinate system in radar pictures of storms. If you're in the military, especially in the Navy or the Air Force, you'll know it as a polar radar display. Sometimes this type of polar display shows the angle referenced clockwise from south rather than from geographic north.

EQUATION OF CIRCLE CENTERED AT THE ORIGIN

The equation of a *circle* centered at the origin in the polar plane is just about as simple as an equation can get. It is given by the following formula:

$$r = a$$

where a is a real number and $a > 0$. This is shown in Fig. 3-13. Certain other graphs, such as cloverleaf patterns, spirals, and cardioids (heart-shaped patterns), also have simple equations in polar coordinates but complicated equations in rectangular coordinates.

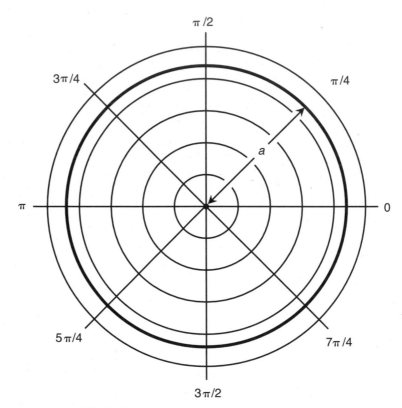

Fig. 3-13. Polar graph of circle centered at the origin.

Other Systems

Here are some other coordinate systems that you are likely to encounter in your journeys through the world of physics. Keep in mind that the technical details are simplified for this presentation. As you gain experience using these systems, you will be introduced to more details, but they would confuse you if we dealt with them now.

LATITUDE AND LONGITUDE

Latitude and *longitude* angles uniquely define the positions of points on the surface of a sphere or in the sky. The scheme for geographic locations on the earth is illustrated in Fig. 3-14*a*. The *polar axis* connects two specified points at antipodes on the sphere. These points are assigned latitude +90

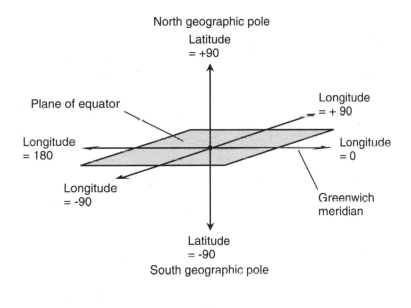

(a)

Fig. 3-14.   (*a*) Latitude and longitude on the earth are measured in degrees.

(north pole) and −90 (south pole). The *equatorial axis* runs outward from the center of the sphere at a right angle to the polar axis. It is assigned longitude 0. Latitude is measured positively (north) and negatively (south) relative to the plane of the equator. Longitude is measured counterclockwise (east) and clockwise (west) relative to the equatorial axis. The angles are restricted as follows:

$$-90° \leq \text{latitude} \leq +90°$$

$$-180° \leq \text{longitude} \leq +180°$$

On the earth's surface, the half-circle connecting the zero-longitude line with the poles passes through Greenwich, England, and is known as the

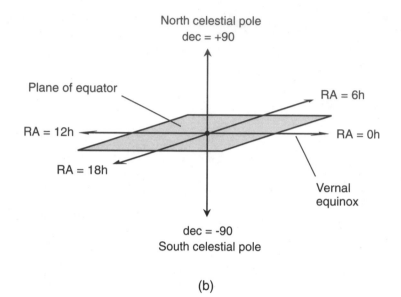

Fig. 3-14.   (*b*) Declination (dec) and right ascension
(RA) are used to find coordinates in the heavens.

Greenwich meridian or the *prime meridian*. Longitude angles are defined
with respect to this meridian.

CELESTIAL COORDINATES

Celestial latitude and *celestial longitude* are extensions of the earth's latitude and longitude into the heavens. An object whose celestial latitude and
longitude coordinates are (*x, y*) appears at the zenith (straight overhead) in
the sky from the point on the earth's surface whose latitude and longitude
coordinates are (*x, y*).

Declination and *right ascension* define the positions of objects in the
sky relative to the stars. Figure 3-14*b* applies to this system. Declination
(abbreviated *dec*) is identical to celestial latitude. Right ascension (abbreviated *RA*) is measured eastward from the *vernal equinox* (the position of
the sun in the heavens at the moment spring begins in the northern hemisphere). Right ascension is measured in hours (symbolized *h*) rather than
degrees, where there are 24h in a 360° circle. The angles are restricted as
follows:

$$-90° \le \text{dec} \le +90°$$

$$0h \le RA < 24h$$

CARTESIAN THREE-SPACE

An extension of rectangular coordinates into three dimensions is *cartesian three-space* (Fig. 3-15), also called *rectangular three-space* or *xyz space*. Independent variables are usually plotted along the x and y axes; the dependent variable is plotted along the z axis. "Graphs" of this sort show up as snakelike curves winding and twisting through space or as surfaces such as spheres, ellipsoids, or those mountain-range-like displays you have seen in the scientific magazines. Usually, the scales are linear; that is, the increments are the same size throughout each scale. However, variations of these schemes can employ nonlinear graduations for one, two, or all three scales.

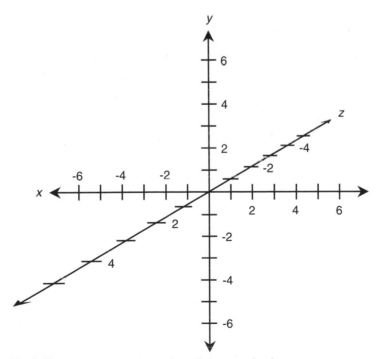

Fig. 3-15.   Cartesian three-space, also called *rectangular three-space* or *xyz space*.

Computers are invaluable in graphing functions in rectangular three-space. Computers can show perspective, and they let you see the true shape of a surface plot. A good three-dimensional (3D) graphics program lets you look at a graph from all possible angles, even rotating it or flipping it over in real time.

CYLINDRICAL COORDINATES

Figure 3-16 shows a system of *cylindrical coordinates* for specifying the positions of points in three-space. Given a set of cartesian coordinates or

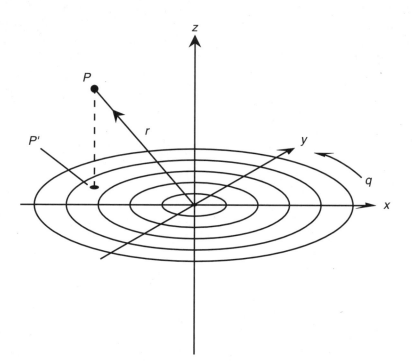

Fig. 3-16.   Cylindrical coordinates for defining points in three-space.

xyz space, an angle *q* is defined in the *xy* plane, measured in radians counterclockwise from the *x* axis. Given a point *P* in space, consider its projection *P'* onto the *xy* plane. The position of *P* is defined by the ordered triple (*q, r, z*) such that

$$q = \text{angle between } P' \text{ and the } x \text{ axis in the } xy \text{ plane}$$

r = distance (radius) from P to the origin

z = distance (altitude) of P above the xy plane

You can think of cylindrical coordinates as a polar plane with the addition of an altitude coordinate to define the third dimension.

SPHERICAL COORDINATES

Figure 3-17 shows a system of *spherical coordinates* for defining points in space. This scheme is similar to the system for longitude and latitude with the addition of a radius r representing the distance of point P from the origin. The location of a point P is defined by the ordered triple (long, lat, r)

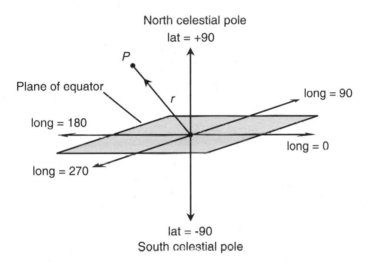

Fig. 3-17. Spherical coordinates for defining points in three-space.

such that

long = longitude of P

lat = latitude of P

r = distance (radius) from P to the origin

In this example, angles are specified in degrees; alternatively, they can be expressed in radians. There are several variations of this system, all of which are commonly called *spherical coordinates*.

SEMILOG (x-LINEAR) COORDINATES

Figure 3-18 shows *semilogarithmic (semilog) coordinates* for defining points in a portion of the *xy* plane. The independent-variable axis is linear, and the dependent-variable axis is logarithmic. The numerical values that can be depicted on the *y* axis are restricted to one sign or the other (positive or negative). In this example, functions can be plotted with domains and ranges as follows:

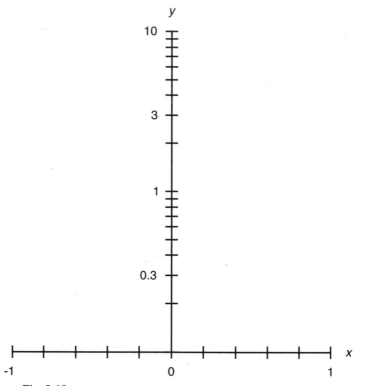

Fig. 3-18.   Semilog *xy* plane with linear *x* axis and logarithmic *y* axis.

$$-1 \leq x \leq 1$$

$$0.1 \leq y \leq 10$$

The *y* axis in Fig. 3-18 spans two orders of magnitude (powers of 10). The span could be larger or smaller than this, but in any case the *y* values cannot extend to zero. In the example shown here, only portions of the first and second quadrants of the xy plane can be depicted. If the y axis were inverted (its values made negative), the resulting plane would cover corresponding parts of the third and fourth quadrants.

SEMILOG (y-LINEAR) COORDINATES

Figure 3-19 shows semilog coordinates for defining points in a portion of the xy plane. The independent-variable axis is logarithmic, and the

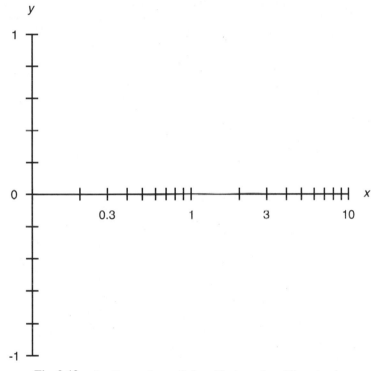

Fig. 3-19.   Semilog xy plane with logarithmic x axis and linear y axis.

dependent-variable axis is linear. The numerical values that can be depicted on the x axis are restricted to one sign or the other (positive or negative). In this example, functions can be plotted with domains and ranges as follows:

$$0.1 \le x \le 10$$
$$-1 \le y \le 1$$

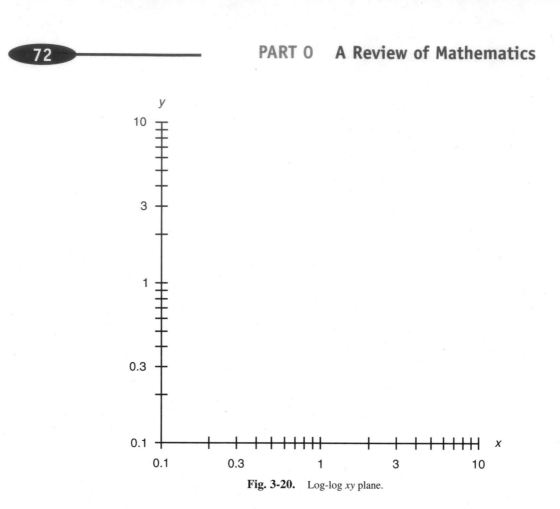

Fig. 3-20.   Log-log xy plane.

The x axis in Fig. 3-19 spans two orders of magnitude (powers of 10). The span could be larger or smaller, but in any case the x values cannot extend to zero. In the example shown here, only portions of the first and fourth quadrants of the xy-plane can be depicted. If the x axis were inverted (its values made negative), the resulting plane would cover corresponding parts of the second and third quadrants.

LOG-LOG COORDINATES

Figure 3-20 shows *log-log coordinates* for defining points in a portion of the xy plane. Both axes are logarithmic. The numerical values that can be depicted on either axis are restricted to one sign or the other (positive or negative). In this example, functions can be plotted with domains and ranges as follows:

$$0.1 \le x \le 10$$

$$0.1 \le y \le 10$$

The axes in Fig. 3-20 span two orders of magnitude (powers of 10). The span of either axis could be larger or smaller, but in any case the values cannot extend to zero. In the example shown here, only a portion of the first quadrant of the xy plane can be depicted. By inverting the signs of one or both axes, corresponding portions of any of the other three quadrants can be covered.

Quiz

Refer to the text in this chapter if necessary. A good score is eight correct. Answers are in the back of the book.

1. A polar plane defines points according to
   (a) two distance coordinates.
   (b) a distance and an angle.
   (c) two angles.
   (d) a distance and two angles.

2. Suppose that you plot the graphs of two equations in the cartesian plane, and the curves meet at a single point. How many solutions are there to this pair of simultaneous equations?
   (a) None
   (b) One
   (c) Two
   (d) There is not enough information to tell.

3. In a semilog coordinate plane,
   (a) both axes are semilogarithmic.
   (b) one axis is semilogarithmic, and the other is logarithmic.
   (c) one axis is linear, and the other is logarithmic.
   (d) both axes are linear.

4. What is the general shape of the graph of the equation $x^2 + y^2 = 16$ as plotted in rectangular coordinates?
   (a) A straight line
   (b) A parabola

(c) A circle

(d) There is not enough information to tell.

5. What is the equation of problem 4 if its graph is plotted in polar coordinates, where r is the radius and q is the angle?

(a) $r = 4$

(b) $q = 4$

(c) $r^2 + q^2 = 16$

(d) There is not enough information to tell.

6. Which three-dimensional coordinate scheme described in this chapter locates a point by defining three different angles relative to a reference axis?

(a) The polar plane

(b) Cylindrical coordinates

(c) Spherical coordinates

(d) None of the above

7. Suppose that two equations are graphed and that their plots are curves A and B in Fig. 3-21. Suppose that both graphs extend infinitely far in both directions. How many solutions are there to this pair of equations?

(a) None

(b) Several

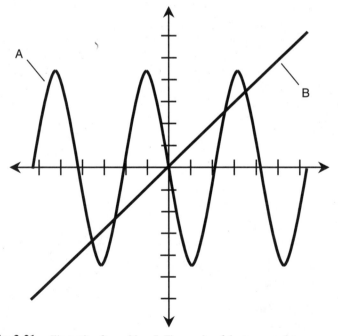

Fig. 3-21.   Illustration for problem 7. The graphs of the two equations are curves A and B. Assume that the graphs extend infinitely in either direction.

(c) Infinitely many

(d) It is impossible to tell.

8. Suppose that two equations are graphed and that their plots are curves A and B in Fig. 3-22. Suppose that both graphs extend infinitely far in both directions. How many solutions are there to this pair of equations?

(a) None

(b) Several

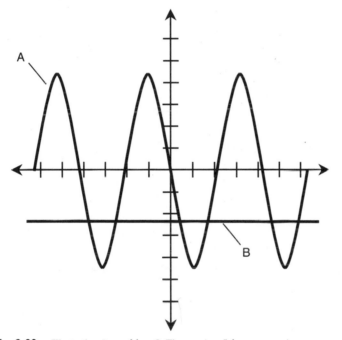

Fig. 3-22.　Illustration for problem 8. The graphs of the two equations are curves A and B. Assume that the graphs extend infinitely in either direction.

(c) Infinitely many

(d) It is impossible to tell.

9. Suppose that two equations are graphed and that their plots are curves A and B in Fig. 3-23. Suppose that both graphs extend infinitely far in both directions. How many solutions are there to this pair of equations?

(a) None

(b) Several

(c) Infinitely many

(d) It is impossible to tell.

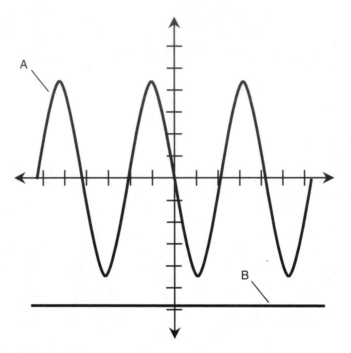

Fig. 3-23.   Illustration for problem 9. The graphs of the two equations are curves
A and *B*. Assume that the graphs extend infinitely in either direction.

10. In which graphing scheme does neither axis extend all the way down to zero?
    (a) Rectangular coordinates
    (b) Cylindrical coordinates
    (c) Spherical coordinates
    (d) Log-log coordinates

CHAPTER 4

Basics of Geometry

If you casually page through the rest of this chapter right now, you might say, "Only a crazy person would expect me to remember all this." Don't worry. You do not have to memorize all these formulas; they're available in books (like this one) and on the Internet. The most often-used formulas, such as the pythagorean theorem for right triangles and the formula for the area of a circle, are worth remembering so that you don't have to run to a reference source every time you need to calculate something. However, it's up to you how much or how little of this stuff you want to burn into your brain.

It's a good idea to be comfortable making calculations with formulas like this before you dive headlong into physics. Therefore, look over these formulas, be sure you can deal with them, and then take the quiz at the end of the chapter. The quiz is "open book," as are all the chapter-ending quizzes in this course. You may check back in the chapter text when taking the quiz so that you can find the formula you need. From there, it's just a matter of punching buttons on a calculator and maybe scratching out a few diagrams to help visualize what is going on.

Fundamental Rules

The fundamental rules of geometry are used widely in physics and engineering. These go all the way back to the time of the ancient Egyptians and Greeks, who used geometry to calculate the diameter of the earth and the distance to the moon. They employed the laws of *euclidean geometry* (named after Euclid, a Greek mathematician who lived thousands of years ago). However much or little Euclid actually had to do with these rules, they're straightforward. So here they are, terse and plain.

TWO-POINT PRINCIPLE

Suppose that P and Q are two distinct geometric points. Then the following statements hold true, as shown in Fig. 4-1:

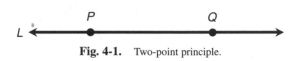

Fig. 4-1.   Two-point principle.

- P and Q lie on a common line L.
- L is the only line on which both points lie.

THREE-POINT PRINCIPLE

Let P, Q, and R be three distinct points, not all of which lie on a straight line. Then the following statements hold true:

- P, Q, and R all lie in a common euclidean plane S.
- S is the only euclidean plane in which all three points lie.

PRINCIPLE OF n POINTS

Let P_1, P_2, P_3,…, and P_n be n distinct points, not all of which lie in the same euclidean space of $n-1$ dimensions. Then the following statements hold true:

- P_1, P_2, P_3,…, and P_n all lie in a common euclidean space U of n dimensions.
- U is the only n-dimensional euclidean space in which all n points lie.

DISTANCE NOTATION

The distance between any two points P and Q, as measured from P toward Q along the straight line connecting them, is symbolized by writing PQ.

MIDPOINT PRINCIPLE

Suppose that there is a line segment connecting two points *P* and *R*. Then there is one and only one point *Q* on the line segment between *P* and *R* such that *PQ = QR*. This is illustrated in Fig. 4-2.

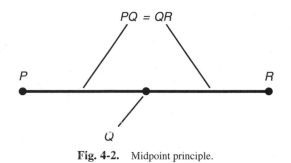

Fig. 4-2.   Midpoint principle.

ANGLE NOTATION

Imagine that *P*, *Q*, and *R* are three distinct points. Let *L* be the line segment connecting *P* and *Q*; let *M* be the line segment connecting *R* and *Q*. Then the angle between *L* and *M*, as measured at point *Q* in the plane defined by the three points, can be written as ∠*PQR* or as ∠*RQP*. If the rotational sense of measurement is specified, then ∠*PQR* indicates the angle as measured from *L* to *M*, and ∠*RQP* indicates the angle as measured from *M* to *L* (Fig. 4-3.) These notations also can stand for the measures of angles, expressed either in degrees or in radians.

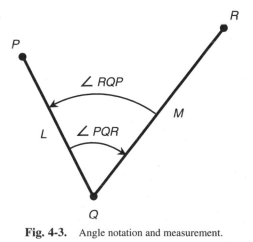

Fig. 4-3.   Angle notation and measurement.

ANGLE BISECTION

Suppose that there is an angle $\angle PQR$ measuring less than 180° and defined by three points P, Q, and R, as shown in Fig. 4-4. Then there is exactly one ray M that bisects the angle $\angle PQR$. If S is any point on M other than the point Q, then $\angle PQS = \angle SQR$. Every angle has one and only one ray that divides the angle in half.

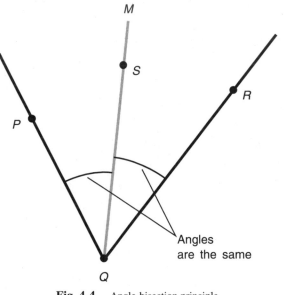

Fig. 4-4.   Angle-bisection principle.

PERPENDICULARITY

Suppose that L is a line through points P and Q. Let R be a point not on L. Then there is one and only one line M through point R intersecting line L at some point S such that M is perpendicular to L. This is shown in Fig. 4-5.

PERPENDICULAR BISECTOR

Suppose that L is a line segment connecting two points P and R. Then there is one and only one perpendicular line M that intersects line segment L in a point Q such that the distance from P to Q is equal to the distance from

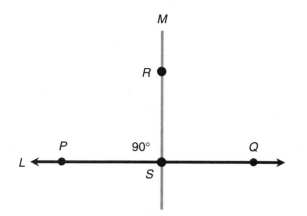

Fig. 4-5.   Perpendicular principle.

Q to R. That is, every line segment has exactly one perpendicular bisector.
This is illustrated in Fig. 4-6.

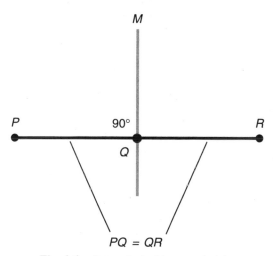

Fig. 4-6.   Perpendicular-bisector principle.

DISTANCE ADDITION AND SUBTRACTION

Let P, Q, and R be points on a line L such that Q is between P and R. Then
the following equations hold concerning distances as measured along L
(Fig. 4-7):

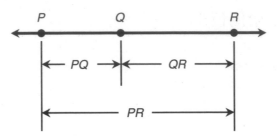

Fig. 4-7.   Distance addition and subtraction.

$$PQ + QR = PR$$

$$PR - PQ = QR$$

$$PR - QR = PQ$$

ANGLE ADDITION AND SUBTRACTION

Let P, Q, R, and S be four points that lie in a common plane. Let Q be the vertex of three angles $\angle PQR$, $\angle PQS$, and $\angle SQR$, as shown in Fig. 4-8. Then the following equations hold concerning the angular measures:

$$\angle PQS + \angle SQR = \angle PQR$$

$$\angle PQR - \angle PQS = \angle SQR$$

$$\angle PQR - \angle SQR = \angle PQS$$

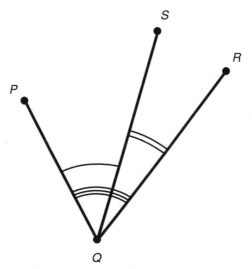

Fig. 4-8.   Angular addition and subtraction.

VERTICAL ANGLES

Suppose that L and M are two lines that intersect at a point P. Opposing pairs of angles, denoted x and y in Fig. 4-9, are known as *vertical angles* and always have equal measure.

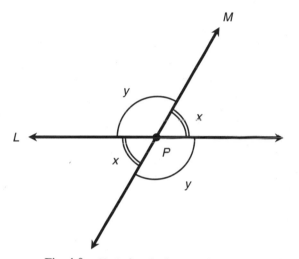

Fig. 4-9.   Vertical angles have equal measure.

ALTERNATE INTERIOR ANGLES

Suppose that L and M are parallel lines. Let N be a line that intersects L and M at points P and Q, respectively. In Fig. 4-10, angles labeled x are *alternate interior angles*; the same holds true for angles labeled y. Alternate interior angles always have equal measure. Line N is perpendicular to lines L and M if and only if $x = y$.

ALTERNATE EXTERIOR ANGLES

Suppose that L and M are parallel lines. Let N be a line that intersects L and M at points P and Q, respectively. In Fig. 4-11, angles labeled x are *alternate exterior angles*; the same holds true for angles labeled y. Alternate exterior angles always have equal measure. The line N is perpendicular to lines L and M if and only if $x = y$.

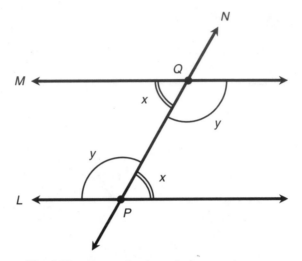

Fig. 4-10   Alternate interior angles have equal measure.

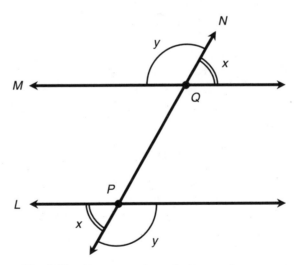

Fig. 4-11   Alternate exterior angles have equal measure.

CORRESPONDING ANGLES

Let *L* and *M* be parallel lines. Let *N* be a transversal line that intersects *L* and *M* at points *P* and *Q*, respectively. In Fig. 4-12, angles labeled *w* are *corresponding angles*; the same holds true for angles labeled *x, y,* and *z*. Corresponding angles always have equal measure. The line *N* is perpendi-

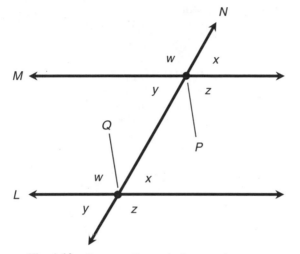

Fig. 4-12   Corresponding angles have equal measure.

cular to lines L and M if and only if $w = x = y = z = 90° = \pi/2$ radians; that is, if and only if all four angles are right angles.

PARALLEL PRINCIPLE

Suppose that L is a line and P is a point not on L. There exists one and only one line M through P such that line M is parallel to line L (Fig. 4-13). This

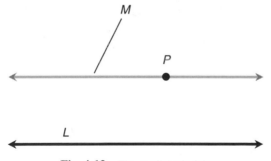

Fig. 4-13   The parallel principle.

is one of the most important postulates in euclidean geometry. Its negation can take two forms: Either there is no such line M, or there exists more than one such line M_1, M_2, M_3,.... Either form of the negation of this principle constitutes a cornerstone of noneuclidean geometry that is important to physicists and astronomers interested in the theories of general relativity and cosmology.

MUTUAL PERPENDICULARITY

Let L and M be lines that lie in the same plane. Suppose that both L and M intersect a third line N and that both L and M are perpendicular to N. Then lines L and M are parallel to each other (Fig. 4-14).

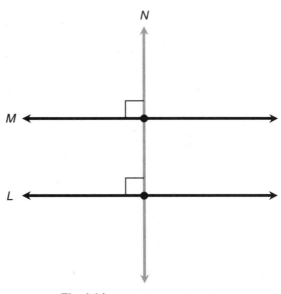

Fig. 4-14　Mutual perpendicularity.

Triangles

If it's been a while since you took a course in plane geometry, perhaps you think of triangles when the subject is brought up. Maybe you recall having to learn all kinds of theoretical proofs concerning triangles using "steps and reasons" tables if your teacher was rigid and less formal methods if your teacher was not so stodgy. Well, here, you don't have to go through the proofs again, but some of the more important facts about triangles are worth stating.

POINT-POINT-POINT

Let P, Q, and R be three distinct points, not all of which lie on the same straight line. Then the following statements are true (Fig. 4-15):

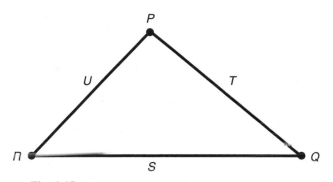

Fig. 4-15   The three-point principle; side-side-side triangles.

- P, Q, and R lie at the vertices of a triangle T.
- T is the only triangle having vertices P, Q, and R.

SIDE-SIDE-SIDE

Let S, T, and U be line segments. Let s, t, and u be the lengths of those three line segments, respectively. Suppose that S, T, and U are joined at their end points P, Q, and R (see Fig. 4-15). Then the following statements hold true:

- Line segments S, T, and U determine a triangle.
- This is the only triangle of its size and shape that has sides S, T, and U.
- All triangles having sides of lengths s, t, and u are congruent (identical in size and shape).

SIDE-ANGLE-SIDE

Let S and T be two distinct line segments. Let P be a point that lies at the ends of both these line segments. Denote the lengths of S and T by their lowercase counterparts s and t, respectively. Suppose that S and T both subtend an angle x at point P (Fig. 4-16). Then the following statements are all true:

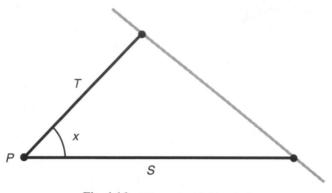

Fig. 4-16   Side-angle-side triangles.

- S, T, and x determine a triangle.
- This is the only triangle having sides S and T that subtend an angle x at point P.
- All triangles containing two sides of lengths s and t that subtend an angle x are congruent.

ANGLE-SIDE-ANGLE

Let S be a line segment having length s and whose end points are P and Q. Let x and y be the angles subtended relative to S by two lines L and M that run through P and Q, respectively (Fig. 4-17). Then the following statements are all true:

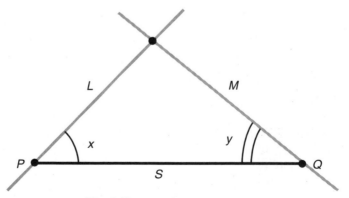

Fig. 4-17   Angle-side-angle triangles.

- *S, x,* and *y* determine a triangle.
- This the only triangle determined by *S, x,* and *y.*
- All triangles containing one side of length *s* and whose other two sides subtend angles of *x* and *y* relative to the side whose length is *s* are congruent.

ANGLE-ANGLE-ANGLE

Let *L, M,* and *N* be lines that lie in a common plane and intersect in three points, as illustrated in Fig. 4-18. Let the angles at these points be *x, y,* and *z.* Then the following statements are true:

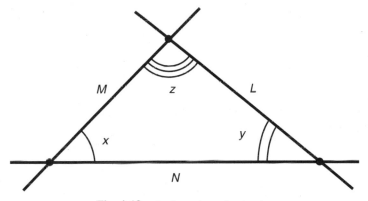

Fig. 4-18   Angle-angle-angle triangles.

- There are infinitely many triangles with interior angles *x, y,* and *z* in the sense shown.
- All triangles with interior angles *x, y,* and *z* in the sense shown are similar (that is, they have the same shape but not necessarily the same size).

ISOSCELES TRIANGLE

Suppose that we have a triangle with sides *S, T,* and *U* having lengths *s, t,* and *u.* Let *x, y,* and *z* be the angles opposite *S, T,* and *U,* respectively (Fig. 4-19). Suppose that any of the following equations hold:

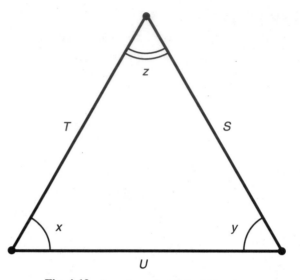

Fig. 4-19   Isosceles and equilateral triangles.

$$s = t$$

$$t = u$$

$$s = u$$

$$x = y$$

$$y = z$$

$$x = z$$

Then the triangle is an *isosceles triangle,* and the following logical statements are valid:

If $s = t$, then $x = y$.

If $t = u$, then $y = z$.

If $s = u$, then $x = z$.

If $x = y$, then $s = t$.

If $y = z$, then $t = u$.

If $x = z$, then $s = u$.

EQUILATERAL TRIANGLE

Suppose that we have a triangle with sides S, T, and U, having lengths s, t, and u. Let x, y, and z be the angles opposite S, T, and U, respectively (see Fig. 4-19). Suppose that either of the following are true:

$$s = t = u \qquad \text{or} \qquad x = y = z$$

Then the triangle is said to be an *equilateral triangle,* and the following logical statements are valid:

$$\text{If } s = t = u, \text{ then } x = y = z.$$

$$\text{If } x = y = z, \text{ then } s = t = u.$$

That is, all equilateral triangles have precisely the same shape; they are all similar.

THEOREM OF PYTHAGORAS

Suppose that we have a right triangle defined by points P, Q, and R whose sides are D, E, and F having lengths d, e, and f, respectively. Let f be the side opposite the right angle (Fig. 4-20). Then the following equation is always true:

$$d^2 + e^2 = f^2$$

The converse of this is also true: If there is a triangle whose sides have lengths d, e, and f and the preceding equation is true, then that triangle is a right triangle.

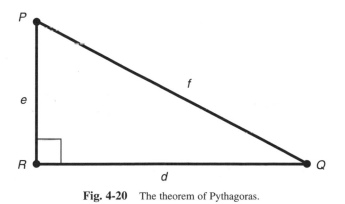

Fig. 4-20   The theorem of Pythagoras.

PERIMETER OF TRIANGLE

Suppose that we have a triangle defined by points P, Q, and R and having sides S, T, and U of lengths s, t, and u, as shown in Fig. 4-21. Let s be the base length, h be the height, and x be the angle between the sides having lengths s and t. Then the perimeter B of the triangle is given by the following formula:

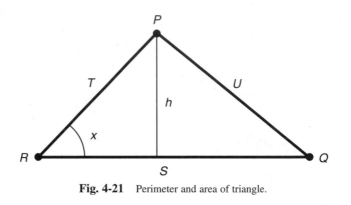

Fig. 4-21   Perimeter and area of triangle.

$$B = s + t + u$$

INTERIOR AREA OF TRIANGLE

Consider the same triangle as defined above; refer again to Fig. 4-21. The interior area A can be found with this formula:

$$A = sh/2$$

Quadrilaterals

A four-sided geometric figure that lies in a single plane is called a *quadrilateral*. There are several classifications and various formulas that apply to each. Here are some of the more common formulas that can be useful in physics.

PARALLELOGRAM DIAGONALS

Suppose that we have a parallelogram defined by four points *P, Q, R,* and *S.* Let *D* be a line segment connecting *P* and *R* as shown in Fig. 4-22a. Then *D* is a *minor diagonal* of the parallelogram, and the triangles defined by *D* are congruent:

$$\Delta PQR \cong \Delta RSP$$

Let *E* be a line segment connecting *Q* and *S* (see Fig. 4-22b). Then *E* is a *major diagonal* of the parallelogram, and the triangles defined by *E* are congruent:

$$\Delta QRS \cong \Delta SPQ$$

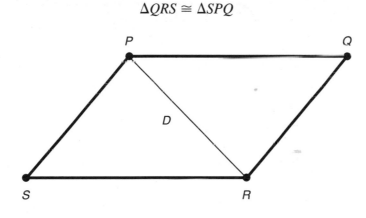

(a)

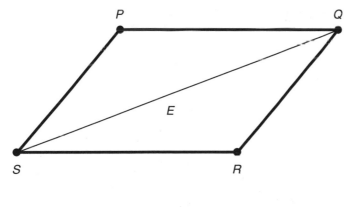

(b)

Fig. 4-22   Triangles defined by the minor diagonal (*a*) or the major diagonal (*b*) of a parallelogram are congruent.

BISECTION OF PARALLELOGRAM DIAGONALS

Suppose that we have a parallelogram defined by four points P, Q, R, and S. Let D be the diagonal connecting P and R; let E be the diagonal connecting Q and S (Fig. 4-23). Then D and E bisect each other at their intersection point T. In addition, the following pairs of triangles are congruent:

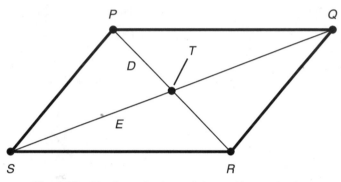

Fig. 4-23   The diagonals of a parallelogram bisect each other.

$$\Delta PQT \cong \Delta RST$$

$$\Delta QRT \cong \Delta SPT$$

The converse of the foregoing is also true: If we have a plane quadrilateral whose diagonals bisect each other, then that quadrilateral is a parallelogram.

RECTANGLE

Suppose that we have a parallelogram defined by four points P, Q, R, and S. Suppose that any of the following statements is true for angles in degrees:

$$\angle PQR = 90° = \pi/2 \text{ radians}$$

$$\angle QRS = 90° = \pi/2 \text{ radians}$$

$$\angle RSP = 90° = \pi/2 \text{ radians}$$

$$\angle SPQ = 90° = \pi/2 \text{ radians}$$

Then all four interior angles measure 90°, and the parallelogram is a *rectangle,* a four-sided plane polygon whose interior angles are all congruent (Fig. 4-24). The converse of this is also true: If a quadrilateral is a rectangle, then any given interior angle has a measure of 90°.

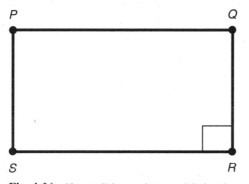

Fig. 4-24   If a parallelogram has one right interior angle, then the parallelogram is a rectangle.

RECTANGLE DIAGONALS

Suppose that we have a parallelogram defined by four points *P, Q, R,* and *S.* Let *D* be the diagonal connecting *P* and *R*; let *E* be the diagonal connecting *Q* and *S.* Let the length of *D* be denoted by *d*; let the length of *E* be denoted by *e* (Fig. 4-25). If *d = e*, then the parallelogram is a rectangle. The converse is also true: If a parallelogram is a rectangle, then *d = e*. Thus a parallelogram is a rectangle if and only if its diagonals have equal lengths.

RHOMBUS DIAGONALS

Suppose that we have a parallelogram defined by four points *P, Q, R,* and *S.* Let *D* be the diagonal connecting *P* and *R*; let *E* be the diagonal connecting *Q* and *S.* If *D* is perpendicular to *E*, then the parallelogram is a *rhombus,* a four-sided plane polygon whose sides are all equally long (Fig. 4-26). The converse is also true: If a parallelogram is a rhombus, then *D* is perpendicular

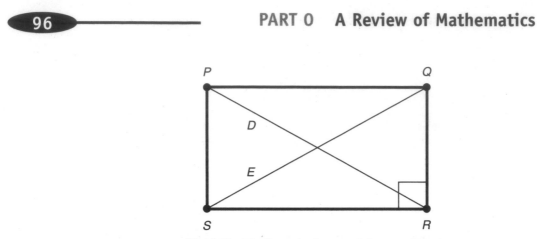

Fig. 4-25   The diagonals of a rectangle have equal length.

to *E*. Thus a parallelogram is a rhombus if and only if its diagonals are perpendicular.

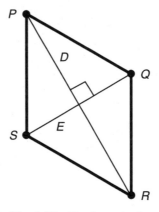

Fig. 4-26   The diagonals of a rhombus are perpendicular.

TRAPEZOID WITHIN TRIANGLE

Suppose that we have a triangle defined by three points *P, Q,* and *R.* Let *S* be the midpoint of side *PR,* and let *T* be the midpoint of side *PQ.* Then line segments *ST* and *RQ* are parallel, and the figure defined by *STQR* is a *trapezoid,* a four-sided plane polygon with one pair of parallel sides (Fig. 4-27). In addition, the length of line segment *ST* is half the length of line segment *RQ.*

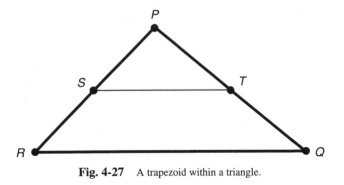

Fig. 4-27 A trapezoid within a triangle.

MEDIAN OF A TRAPEZOID

Suppose that we have a trapezoid defined by four points P, Q, R, and S. Let T be the midpoint of side PS, and let U be the midpoint of side QR. Line segment TU is called the *median* of trapezoid $PQRS$. Let M be the polygon defined by P, Q, U, and T. Let N be the polygon defined by T, U,

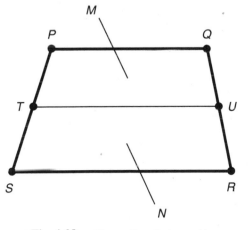

Fig. 4-28 The median of a trapezoid.

R, and S. Then M and N are trapezoids (Fig. 4-28). In addition, the length of line segment TU is half the sum of the lengths of line segments PQ and SR. That is, the length of TU is equal to the average (arithmetic mean) of the lengths of PQ and SR.

SUM OF INTERIOR ANGLES OF PLANE QUADRILATERAL

Suppose that we have a plane quadrilateral and that the interior angles are w, x, y, and z (Fig. 4-29). Then the following equation holds if the angular measures are given in degrees:

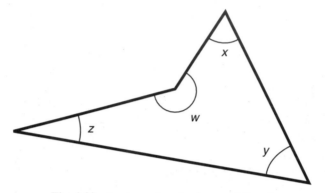

Fig. 4-29   Interior angles of a plane quadrilateral.

$$w + x + y + z = 360°$$

If the angular measures are given in radians, then the following holds:

$$w + x + y + z = 2\pi$$

PERIMETER OF PARALLELOGRAM

Suppose that we have a parallelogram defined by points P, Q, R, and S with sides of lengths d and e, as shown in Fig. 4-30. Let d be the base length, and let h be the height. Then the perimeter B of the parallelogram is given by the following formula:

$$B = 2d + 2e$$

INTERIOR AREA OF PARALLELOGRAM

Suppose that we have a parallelogram as defined above and in Fig. 4-30. The interior area A is given by

$$A = dh$$

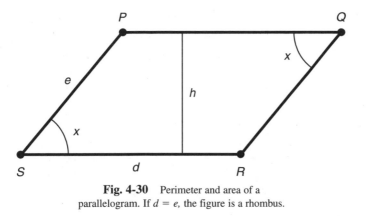

Fig. 4-30   Perimeter and area of a
parallelogram. If $d = e$, the figure is a rhombus.

PERIMETER OF RHOMBUS

Suppose that we have a rhombus defined by points P, Q, R, and S and having sides all of which have the same length. The rhombus is a special case
of the parallelogram (see Fig. 4-30) in which $d = e$. Let the lengths of all
four sides be denoted d. The perimeter B of the rhombus is given by the following formula:

$$B - 4d$$

INTERIOR AREA OF RHOMBUS

Suppose that we have a rhombus as defined above and in Fig. 4-30. The
interior area A of the rhombus is given by

$$A = dh$$

PERIMETER OF RECTANGLE

Suppose that we have a rectangle defined by points P, Q, R, and S and having sides of lengths d and e, as shown in Fig. 4-31. Let d be the base length,
and let e be the height. Then the perimeter B of the rectangle is given by
the following formula:

$$B = 2d + 2e$$

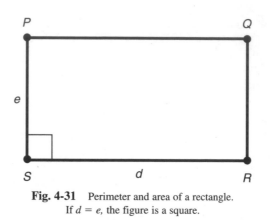

Fig. 4-31   Perimeter and area of a rectangle.
If $d = e$, the figure is a square.

INTERIOR AREA OF RECTANGLE

Suppose that we have a rectangle as defined above and in Fig. 4-31. The interior area A is given by

$$A = de$$

PERIMETER OF SQUARE

Suppose that we have a square defined by points P, Q, R, and S and having sides all of which have the same length. The square is a special case of the rectangle (see Fig. 4-31) in which $d = e$. Let the lengths of all four sides be denoted d. The perimeter B of the square is given by the following formula:

$$B = 4d$$

INTERIOR AREA OF SQUARE

Suppose that we have a square as defined above and in Fig. 4-31. The interior area A is given by

$$A = d^2$$

PERIMETER OF TRAPEZOID

Suppose that we have a trapezoid defined by points *P, Q, R,* and *S* and having sides of lengths *d, e, f,* and *g,* as shown in Fig. 4-32. Let *d* be the base
length, *h* be the height, *x* be the angle between the sides having length *d* and
e, and *y* be the angle between the sides having lengths *d* and *g.* Suppose that
the sides having lengths *d* and *f* (line segments *RS* and *PQ*) are parallel.
Then the perimeter *B* of the trapezoid is

$$B = d + e + f + g$$

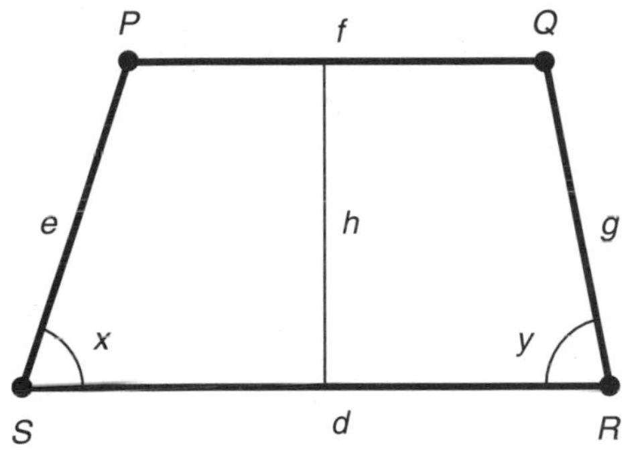

Fig. 4-32   Perimeter and area of a trapezoid.

INTERIOR AREA OF TRAPEZOID

Suppose that we have a trapezoid as defined above and in Fig. 4-32. The
interior area *A* is given by this formula:

$$A = (dh + fh)/2$$

Circles and Ellipses

That's enough of figures with straight lines. Let's get into plane curves. In
some ways, the following formulas are easier for mathematicians to derive
than the ones for figures consisting of lines and angles; in other ways, the
formulas for curves are more troublesome. Fortunately, however, we're
physicists, and the mathematicians have done all the work for us. All we

need to do is take note of the formulas and apply them to situations as the needs arise.

PERIMETER OF CIRCLE

Suppose that we have a circle having radius r, as shown in Fig. 4-33. Then the perimeter B, also called the *circumference*, of the circle is given by the following formula:

$$B = 2\pi r$$

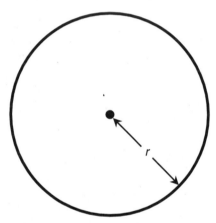

Fig. 4-33   Perimeter and area of a circle.

INTERIOR AREA OF CIRCLE

Suppose that we have a circle as defined above and in Fig. 4-33. The interior area A of the circle can be found using this formula:

$$A = \pi r^2$$

PERIMETER OF ELLIPSE

Suppose that we have an ellipse whose major half-axis measures r and whose minor half-axis measures s, as shown in Fig. 4-34. Then the perimeter B of the ellipse is given approximately by the following formula:

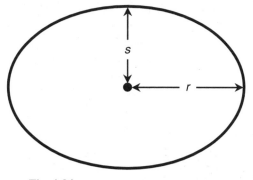

Fig. 4-34   Perimeter and area of an ellipse.

$$B \approx 2\pi \left[(r^2 + s^2)/2 \right]^{1/2}$$

INTERIOR AREA OF ELLIPSE

Suppose that we have an ellipse as defined above and in Fig. 4-34. The interior area A of the ellipse is given by

$$A = \pi rs$$

Surface Area and Volume

Now let's go from two dimensions to three. Here are some formulas for surface areas and volumes of common geometric solids. The three-space involved is *flat*; that is, it obeys the laws of euclidean geometry. These formulas hold in newtonian physics (although in relativistic physics they may not).

VOLUME OF PYRAMID

Suppose that we have a pyramid whose base is a polygon with area A and whose height is h (Fig. 4-35). The volume V of the pyramid is given by

$$V = Ah/3$$

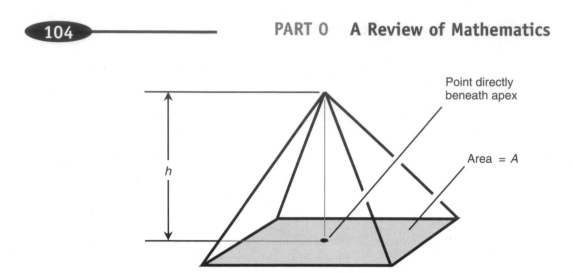

Fig. 4-35   Volume of a pyramid.

SURFACE AREA OF CONE

Suppose that we have a cone whose base is a circle. Let P be the apex of the cone, and let Q be the center of the base (Fig. 4-36). Suppose that line segment PQ is perpendicular to the base so that the object is a *right circular cone*. Let r be the radius of the base, let h be the height of the cone (the length of line segment PQ), and let s be the slant height of the cone as measured from any point on the edge of the circle to the apex P. Then the surface area S of the cone (including the base) is given by either of the following formulas:

$$S = \pi r^2 + \pi r s$$
$$S = \pi r^2 + \pi r (r^2 + h^2)^{1/2}$$

The surface area T of the cone (not including the base) is given by either of the following:

$$T = \pi r s$$
$$T = \pi r (r^2 + h^2)^{1/2}$$

VOLUME OF CONICAL SOLID

Suppose that we have a cone whose base is any enclosed plane curve. Let A be the interior area of the base of the cone. Let P be the apex of the cone, and let Q be a point in the plane X containing the base such that line segment PQ is perpendicular to X (Fig. 4-37). Let h be the height of the cone

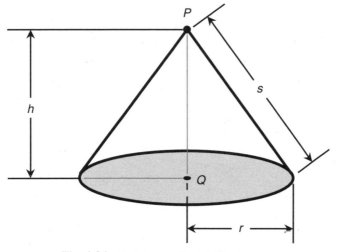

Fig. 4-36   Surface area of a right circular cone.

(the length of line segment PQ). Then the volume V of the corresponding conical solid is given by

$$V = Ah/3$$

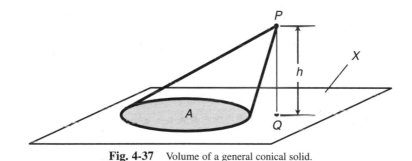

Fig. 4-37   Volume of a general conical solid.

SURFACE AREA OF RIGHT CIRCULAR CYLINDER

Suppose that we have a cylinder whose base is a circle. Let P be the center of the top of the cylinder, and let Q be the center of the base (Fig. 4-38). Suppose that line segment PQ is perpendicular to both the top and the base so that we have a *right circular cylinder.* Let r be the radius of the cylinder, and let h be the height (the length of line segment PQ). Then the surface area S of the cylinder (including the base and the top) is given by

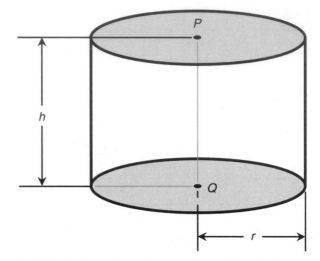

Fig. 4-38　Surface area and volume of a right circular cylinder.

$$S = 2\pi rh + 2\pi r^2 = 2\pi r\,(h + r)$$

The surface area T of the cylinder (not including the base or the top) is

$$T = 2\pi rh$$

VOLUME OF RIGHT CIRCULAR CYLINDRICAL SOLID

Suppose that we have a cylinder as defined above (see Fig. 4-38). The volume V of the corresponding *right circular cylindrical solid* is given by

$$V = \pi r^2 h$$

SURFACE AREA OF GENERAL CYLINDER

Suppose that we have a *general cylinder* whose base is any enclosed plane curve. Let A be the interior area of the base of the cylinder (thus also the interior area of the top). Let B be the perimeter of the base (thus also the perimeter of the top). Let h be the height of the cylinder, or the perpendicular distance separating the planes containing the top and the base. Let x be the angle between the plane containing the base and any line segment PQ connecting corresponding points P and Q in the top and the base, respectively. Let s be the slant height of the cylinder, or the length of line segment PQ (Fig. 4-39). Then the surface area S of the cylinder (including the base and the top) is

$$S = 2A + Bh$$

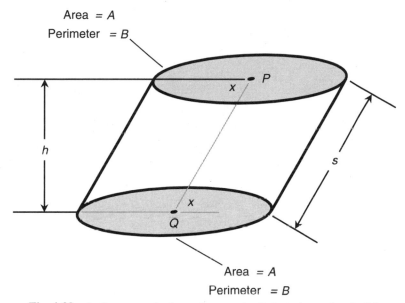

Fig. 4-39   Surface area and volume of a general cylinder and an enclosed solid.

The surface area T of the cylinder (not including the base or the top) is

$$T = Bh$$

VOLUME OF GENERAL CYLINDRICAL SOLID

Suppose that we have a general cylinder as defined above (see Fig. 4-39). The volume V of the corresponding *general cylindrical solid* is

$$V = Ah$$

SURFACE AREA OF SPHERE

Suppose that we have a sphere having radius r, as shown in Fig. 4-40. The surface area A of the sphere is given by

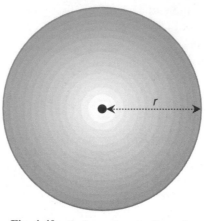

Fig. 4-40   Surface area and volume of a sphere and an enclosed solid.

$$A = 4\pi r^2$$

VOLUME OF SPHERICAL SOLID

Suppose that we have a sphere as defined above and in Fig. 4-40. The volume V of the solid enclosed by the sphere is given by

$$V = 4\pi r^3/3$$

SURFACE AREA OF CUBE

Suppose that we have a cube whose edges each have length s, as shown in Fig. 4-41. The surface area A of the cube is given by

$$A = 6s^2$$

VOLUME OF CUBICAL SOLID

Suppose that we have a cube as defined above and in Fig. 4-41. The volume V of the solid enclosed by the cube is given by

$$V = s^3$$

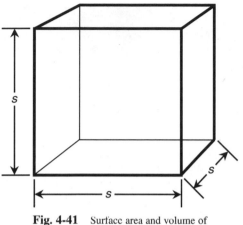

Fig. 4-41   Surface area and volume of
a cube and an enclosed solid.

SURFACE AREA OF RECTANGULAR PRISM

Suppose that we have a rectangular prism whose edges have lengths s, t,
and u, as shown in Fig. 4-42. The surface area A of the prism is given by

$$A = 2st + 2su + 2tu$$

VOLUME OF RECTANGULAR PRISM

Suppose that we have a rectangular prism as defined above and in Fig. 4-42.
The volume V of the enclosed solid is given by

$$V = stu$$

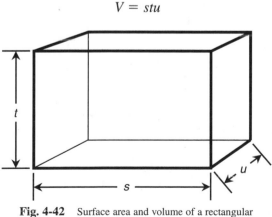

Fig. 4-42   Surface area and volume of a rectangular
prism and an enclosed solid.

Refer to the text in this chapter if necessary. A good score is eight correct. Answers are in the back of the book.

1. A spherical particle has a volume of 8.000×10^{-9} cubic meter. What is the radius of this particle?
    (a) 5.12×10^{-28} millimeter
    (b) 2.000×10^{-3} millimeter
    (c) 1.241 millimeters
    (d) 512 millimeters

2. The earth's radius is approximately 6,400 kilometers (km). What is the earth's surface area in kilometers squared? Go to two significant figures. Assume that the earth is a perfect sphere.
    (a) 1.28×10^{8}
    (b) 5.1×10^{8}
    (c) 1.1×10^{12}
    (d) It cannot be calculated from this information.

3. Imagine that you have a cubical box measuring exactly 1 meter on an edge in its interior. Suppose that you are given a pile of little cubes, each measuring 1 centimeter on an edge, and told to stack the cubes neatly in the box. You are told you will receive 10 cents for each cube you stack in the box. If you complete the task, how much will you have earned?
    (a) $10.00
    (b) $100.00
    (c) $1,000.00
    (d) None of the above

4. Imagine that you are standing on the bottom of a lake whose surface is smooth, without waves. The bottom is flat and level. You shine a laser up toward the surface at an angle of 20° from the horizontal. At what angle, measured relative to the plane of the surface, will the laser beam strike the surface?
    (a) 70°
    (b) 35°
    (c) 20°
    (d) 10°

5. Suppose that you have a cylindrical container whose diameter is 10.00 centimeters and whose height is 20.00 centimeters. What is the volume of this container in centimeters cubed? Give your answer to four significant figures. Assume that $\pi = 3.14159$.
    (a) 1,571

(b) 6,283

(c) 628.3

(d) 1,257

6. If the radius of a sphere is doubled, its surface area increases by a factor of

(a) 2.

(b) 4.

(c) 8.

(d) 16.

7. If the radius of an infinitely thin, flat circular disk is doubled, its surface area increases by a factor of

(a) 2.

(b) 4.

(c) 8.

(d) 16.

8. Suppose that a sample of a certain substance has a mass of 6.000 kilograms and that it is packed into a box measuring 10 centimeters wide by 20 centimeters deep by 30 centimeters high. What is the mass of 1 cubic centimeter of this substance, assuming that its density is uniform?

(a) 0.1000 gram

(b) 1.000 gram

(c) 10.00 grams

(d) 100.0 grams

9. Suppose that a light source is placed at the center of a sphere whose radius is 100 meters. If the sphere's radius is doubled to 200 meters, what will happen to the total light energy striking the interior of the sphere?

(a) It will not change.

(b) It will be cut in half.

(c) It will become ¼ as great.

(d) There is not enough information given here to calculate it.

10. Imagine two triangles. One triangle has a base length of 3 meters, a height of 4 meters, and a slant height of 5 meters. The other triangle has a base length of 15 centimeters, a height of 20 centimeters, and a slant height of 25 centimeters. What can be said about these triangles?

(a) They are both isosceles triangles.

(b) The theorem of Pythagoras applies to both triangles.

(c) The two triangles are congruent.

(d) All of the above are true.

CHAPTER 5

Logarithms, Exponentials, and Trigonometry

This chapter contains formulas similar to those in Chapter 4. Check them out, be sure you can make calculations with them, and then take the "open book" quiz at the end of the chapter. You don't have to memorize these formulas individually, but you should remember where you've seen them. In this way, in case you ever need one of them for reference, you can pull this book off your shelf and look the formula up.

If your calculator cannot deal with logarithms, exponentials, "x to the y power" operations, and the inverses of functions, this is a good time to invest in a good scientific calculator that has these features. Some computer operating systems have calculator programs that are satisfactory.

Logarithms

A *logarithm* (sometimes called a *log*) is an exponent to which a constant is raised to obtain a given number. Suppose that the following relationship exists among three real numbers a, and x, and y:

$$a^y = x$$

Then y is the *base-a logarithm* of x. The expression is written like this:

$$y = \log_a x$$

The two most often-used logarithm bases are 10 and e, where e an irrational number equal to approximately 2.71828.

COMMON LOGS

Base-10 logarithms are also known as *common logarithms* or *common logs*. In equations, common logarithms are written as *log* without a subscript. For example:

$$\log 10 = 1.000$$

Figure 5-1 is an approximate linear-coordinate graph of the function $y = \log x$. Figure 5-2 is the same graph in semilog coordinates. The domain is limited to the positive real numbers. The range of the function encompasses the set of all real numbers.

NATURAL LOGS

Base-e logarithms are also called *natural logs* or *napierian logs*. In equations, the natural-log function is usually denoted *ln* or *log$_e$*. For example:

$$\ln 2.71828 = \log_e 2.71828 \approx 1.00000$$

Figure 5-3 is an approximate linear-coordinate graph of the function $y = \ln x$. Figure 5-4 is the same graph in semilog coordinates. The domain is limited to the positive real numbers, and the range spans the entire set of real numbers.

COMMON LOG IN TERMS OF NATURAL LOG

Suppose that x is a positive real number. The common logarithm of x can be expressed in terms of the natural logarithms of x and 10:

$$\log x = \ln x/\ln 10 \approx 0.434 \ln x$$

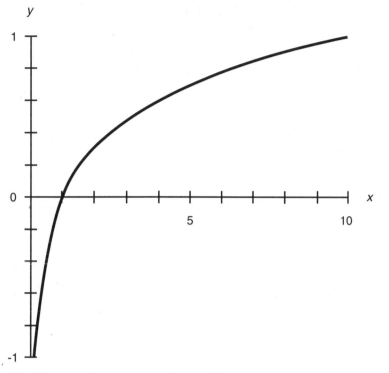

Fig. 5-1   Approximate linear-coordinate graph of the common logarithm function.

NATURAL LOG IN TERMS OF COMMON LOG

Suppose that x is a positive real number. The natural logarithm of x can be expressed in terms of the common logarithms of x and e:

$$\ln x = \log x/\log e \approx 2.303 \log x$$

LOGARITHM OF PRODUCT

Suppose that x and y are both positive real numbers. The common or natural logarithm of the product is equal to the sum of the logarithms of the individual numbers:

$$\log xy = \log x + \log y$$

$$\ln xy = \ln x + \ln y$$

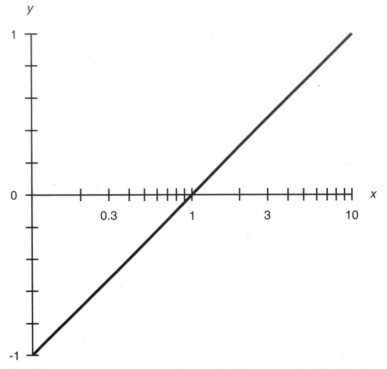

Fig. 5-2   Approximate semilog-coordinate graph of the common logarithm function.

LOGARITHM OF RATIO

Let x and y be positive real numbers. The common or natural logarithm of their ratio, or quotient, is equal to the difference between the logarithms of the individual numbers:

$$\log(x/y) = \log x - \log y$$
$$\ln(x/y) = \ln x - \ln y$$

LOGARITHM OF POWER

Suppose that x is a positive real number; let y be any real number. The common or natural logarithm of x raised to the power y can be reduced to a product as follows:

$$\log x^y = y \log x$$
$$\ln x^y = y \ln x$$

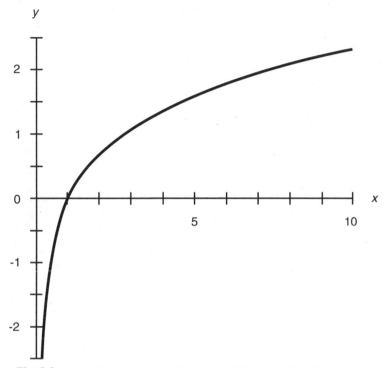

Fig. 5-3   Approximate linear-coordinate graph of the natural logarithm function.

LOGARITHM OF RECIPROCAL

Suppose that x is a positive real number. The common or natural logarithm of the reciprocal (multiplicative inverse) of x is equal to the additive inverse of the logarithm of x:

$$\log(1/x) = -\log x$$
$$\ln(1/x) = -\ln x$$

LOGARITHM OF ROOT

Suppose that x is a positive real number and y is any real number except zero. The common or natural logarithm of the yth root of x (also denoted as x to the $1/y$ power) can be found using the following equations:

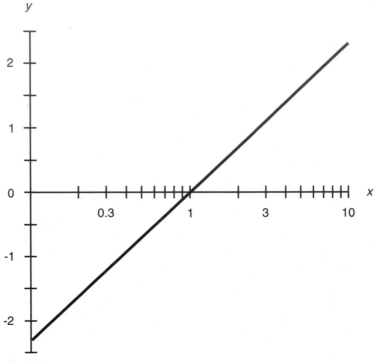

Fig. 5-4   Approximate semilog-coordinate graph of the natural logarithm function.

$$\log(x^{1/y}) = (\log x)/y$$

$$\ln(x^{1/y}) = (\ln x)/y$$

COMMON LOG OF POWER OF 10

The common logarithm of 10 to any real-number power is always equal to that real number:

$$\log(10^x) = x$$

NATURAL LOG OF POWER OF e

The natural logarithm of e to any real-number power is always equal to that real number:

$$\ln(e^x) = x$$

Exponential Functions

An *exponential* is a number that results from the raising of a constant to a given power. Suppose that the following relationship exists among three real numbers *a*, *x*, and *y*:

$$a^x = y$$

Then *y* is the *base-a exponential* of *x*. The two most common exponential-function bases are $a = 10$ and $a = e \approx 2.71828$.

COMMON EXPONENTIALS

Base-10 exponentials are also known as *common exponentials*. For example:

$$10^{-3.000} = 0.001$$

Figure 5-5 is an approximate linear-coordinate graph of the function $y = 10^x$. Figure 5-6 is the same graph in semilog coordinates. The domain encompasses the entire set of real numbers. The range is limited to the positive real numbers.

NATURAL EXPONENTIALS

Base-*e* exponentials are also known as *natural exponentials*. For example:

$$e^{-3.000} \approx 2.71828^{-3.000} \approx 0.04979$$

Figure 5-7 is an approximate linear-coordinate graph of the function $y = e^x$. Figure 5-8 is the same graph in semilog coordinates. The domain encompasses the entire set of real numbers. The range is limited to the positive real numbers.

RECIPROCAL OF COMMON EXPONENTIAL

Let *x* be a real number. The reciprocal of the common exponential of *x* is equal to the common exponential of the additive inverse of *x*:

$$1/(10^x) = 10^{-x}$$

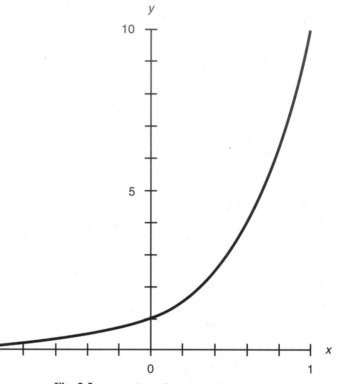

Fig. 5-5   Approximate linear-coordinate graph
of the common exponential function.

RECIPROCAL OF NATURAL EXPONENTIAL

Let x be a real number. The reciprocal of the natural exponential of x is
equal to the natural exponential of the additive inverse of x:

$$1/(e^x) = e^{-x}$$

PRODUCT OF EXPONENTIALS

Let x and y be real numbers. The product of the exponentials of x and y is
equal to the exponential of the sum of x and y. Both these equations hold true:

$$(10^x)(10^y) = 10^{(x+y)}$$

$$(e^x)(e^y) = e^{(x+y)}$$

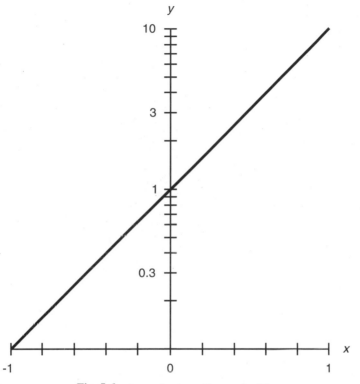

Fig. 5-6   Approximate semilog graph of the
common exponential function.

RATIO OF EXPONENTIALS

Let x and y be real numbers. The ratio (quotient) of the exponentials of x
and y is equal to the exponential of the difference between x and y. Both
these equations hold true:

$$10^x/10^y = 10^{(x-y)}$$

$$e^x/e^y = e^{(x-y)}$$

EXPONENTIAL OF COMMON EXPONENTIAL

Suppose that x and y are real numbers. The yth power of the quantity 10^x is
equal to the common exponential of the product xy:

$$(10^x)^y = 10^{(xy)}$$

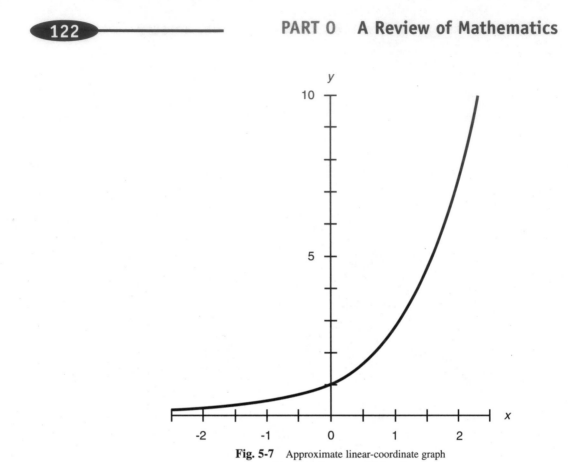

Fig. 5-7　Approximate linear-coordinate graph
of the natural exponential function.

The same situation holds for base *e*. The *y*th power of the quantity e^x is equal to the natural exponential of the product *xy:*

$$(e^x)^y = e^{(xy)}$$

PRODUCT OF COMMON AND NATURAL EXPONENTIALS

Let *x* be a real number. The product of the common and natural exponentials of *x* is equal to the exponential of *x* to the base 10*e*. That is to say:

$$(10^x)(e^x) = (10e)^x \approx (27.1828)^x$$

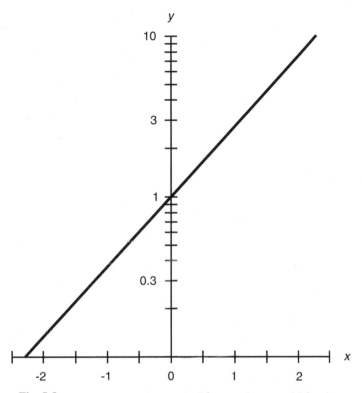

Fig. 5-8   Approximate semilog graph of the natural exponential function.

Now suppose that x is some nonzero real number. The product of the common and natural exponentials of $1/x$ is equal to the exponential of $1/x$ to the base $10e$:

$$(10^{1/x})\,(e^{1/x}) = (10e)^{1/x} \approx (27.1828)^{1/x}$$

RATIO OF COMMON TO NATURAL EXPONENTIAL

Let x be a real number. The ratio (quotient) of the common exponential of x to the natural exponential of x is equal to the exponential of x to the base $10/e$:

$$10^x/e^x = (10/e)^x \approx (3.6788)^x$$

Now suppose that x is some nonzero real number. The ratio (quotient) of the common exponential of $1/x$ to the natural exponential of $1/x$ is equal to the exponential of $1/x$ to the base $10/e$:

$$(10^{1/x})/(e^{1/x}) = (10/e)^{1/x} \approx (3.6788)^{1/x}$$

RATIO OF NATURAL TO COMMON EXPONENTIAL

Let x be a real number. The ratio (quotient) of the natural exponential of x to the common exponential of x is equal to the exponential of x to the base $e/10$. That is to say:

$$e^x/10^x = (e/10)^x \approx (0.271828)^x$$

Now suppose that x is some nonzero real number. The ratio (quotient) of the natural exponential of $1/x$ to the common exponential of $1/x$ is equal to the exponential of $1/x$ to the base $e/10$:

$$(e^{1/x})/(10^{1/x}) = (e/10)^{1/x} \approx (0.271828)^{1/x}$$

COMMON EXPONENTIAL OF RATIO

Let x and y be real numbers, with the restriction that $y \neq 0$. The common exponential of the ratio (quotient) of x to y is equal to the exponential of $1/y$ to the base 10^x:

$$10^{x/y} = (10^x)^{1/y}$$

A similar situation exists for base e. The natural exponential of the ratio (quotient) of x to y is equal to the exponential of $1/y$ to the base e^x:

$$e^{x/y} = (e^x)^{1/y}$$

Trigonometric Functions

There are six basic *trigonometric functions*. They operate on angles to yield real numbers and are known as *sine, cosine, tangent, cosecant, secant,* and *cotangent.* In formulas and equations, they are abbreviated *sin, cos, tan, csc, sec,* and *cot,* respectively.

Until now, angles have been denoted using lowercase italicized English letters from near the end of the alphabet, for example, *w*, *x*, *y*, and *z*. In trigonometry, however, Greek letters are almost always used, particularly θ (italicized lowercase *theta*, pronounced "THAY-tuh") and ϕ (italicized lowercase *phi*, pronounced "FIE" or "FEE"). We will follow this convention here. You should get used to it so that you know how to pronounce the names of the symbols when you see them. This will help you avoid embarrassment when you're around physicists. More important, having a pronunciation in your "mind's ear" may make it easier for you to work with formulas containing such symbols.

BASIC CIRCULAR FUNCTIONS

Consider a circle in rectangular coordinates with the following equation:

$$x^2 + y^2 = 1$$

This is called the *unit circle* because its radius is 1 unit, and it is centered at the origin (0, 0), as shown in Fig. 5-9. Let θ be an angle whose apex is at the origin and that is measured counterclockwise from the abscissa

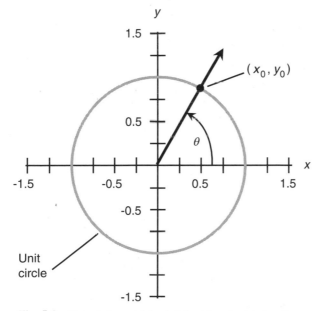

Fig. 5-9 Unit-circle model for defining trigonometric functions.

(x axis). Suppose that this angle corresponds to a ray that intersects the unit circle at some point $P = (x_0, y_0)$. Then

$$y_0 = \sin \theta$$

$$x_0 = \cos \theta$$

$$y_0/x_0 = \tan \theta$$

SECONDARY CIRCULAR FUNCTIONS

Three more circular trigonometric functions are derived from those just defined. They are the *cosecant* function, the *secant* function, and the *cotangent* function. In formulas and equations, they are abbreviated $\csc \theta$, $\sec \theta$, and $\cot \theta$. They are defined as follows:

$$\csc \theta = 1/(\sin \theta) = 1/y_0$$

$$\sec \theta = 1/(\cos \theta) = 1/x_0$$

$$\cot \theta = 1/(\tan \theta) = x_0/y_0$$

RIGHT-TRIANGLE MODEL

Consider a right triangle $\triangle PQR$ such that $\angle PQR$ is the right angle. Let d be the length of line segment RQ, e be the length of line segment QP, and f be the length of line segment RP, as shown in Fig. 5-10. Let θ be the angle between line segments RQ and RP. The six circular trigonometric functions can be defined as ratios between the lengths of the sides as follows:

$$\sin \theta = e/f$$

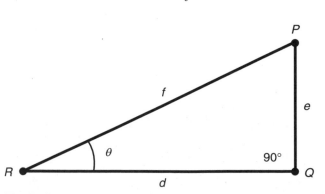

Fig. 5-10   Right-triangle model for defining trigonometric functions.

$$\cos \theta = d/f$$

$$\tan \theta = e/d$$

$$\csc \theta = f/e$$

$$\sec \theta = f/d$$

$$\cot \theta = d/e$$

Trigonometric Identities

The following paragraphs depict common *trigonometric identities* for the circular functions. Unless otherwise specified, these formulas apply to angles θ and ϕ in the standard range as follows:

$$0 \le \theta < 2\pi \text{ (in radians)}$$

$$0 \le \theta < 360 \text{ (in degrees)}$$

$$0 \le \phi < 2\pi \text{ (in radians)}$$

$$0 \le \phi < 360 \text{ (in degrees)}$$

Angles outside the standard range usually are converted to values within the standard range by adding or subtracting the appropriate multiple of 2π radians (360°). You occasionally may hear of an angle with negative measure—that is, measured clockwise rather than counterclockwise—but this can always be converted to some angle with positive measure that is at least zero but less than 360°. The same is true for "angles" greater than 360°. Sometimes physicists will use strange angular expressions (for example, talking about reversed or multiple rotations or revolutions), but it is usually best to reduce angles to values in the standard range. Some of these formulas deal with negative angles, but in these cases, the intent is to let you determine an equivalent value for a trigonometric function of some angle within the standard range.

PYTHAGOREAN THEOREM FOR SINE AND COSINE

The sum of the squares of the sine and cosine of an angle is always equal to 1. The following formula holds:

$$\sin^2 \theta + \cos^2 \theta = 1$$

The expression $\sin^2 \theta$ refers to the sine of the angle squared (not the sine of the square of the angle). That is to say:

$$\sin^2 \theta = (\sin \theta)^2$$

The same holds true for the cosine, tangent, cosecant, secant, cotangent, and all other similar expressions you will see in the rest of this chapter and in physics.

PYTHAGOREAN THEOREM FOR SECANT AND TANGENT

The difference between the squares of the secant and tangent of an angle is always equal to either 1 or −1. The following formulas apply for all angles except $\theta = \pi/2$ radians (90°) and $\theta = 3\pi/2$ radians (270°):

$$\sec^2 \theta - \tan^2 \theta = 1$$
$$\tan^2 \theta - \sec^2 \theta = -1$$

SINE OF NEGATIVE ANGLE

The sine of the negative of an angle (an angle measured in the direction opposite to the normal direction) is equal to the negative (additive inverse) of the sine of the angle. The following formula holds:

$$\sin -\theta = -\sin \theta$$

COSINE OF NEGATIVE ANGLE

The cosine of the negative of an angle is equal to the cosine of the angle. The following formula holds:

$$\cos -\theta = \cos \theta$$

TANGENT OF NEGATIVE ANGLE

The tangent of the negative of an angle is equal to the negative (additive inverse) of the tangent of the angle. The following formula applies for all angles except $\theta = \pi/2$ radians (90°) and $\theta = 3\pi/2$ radians (270°):

$$\tan -\theta = -\tan \theta$$

COSECANT OF NEGATIVE ANGLE

The cosecant of the negative of an angle is equal to the negative (additive inverse) of the cosecant of the angle. The following formula applies for all angles except $\theta = 0$ radians (0°) and $\theta = \pi$ radians (180°):

$$\csc -\theta = -\csc \theta$$

SECANT OF NEGATIVE ANGLE

The secant of the negative of an angle is equal to the secant of the angle. The following formula applies for all angles except $\theta = \pi/2$ radians (90°) and $\theta = 3\pi/2$ radians (270°):

$$\sec -\theta = \sec \theta$$

COTANGENT OF NEGATIVE ANGLE

The cotangent of the negative of an angle is equal to the negative (additive inverse) of the cotangent of the angle. The following formula applies for all angles except $\theta = 0$ radians (0°) and $\theta = \pi$ radians (180°):

$$\cot -\theta = -\cot \theta$$

SINE OF DOUBLE ANGLE

The sine of twice any given angle is equal to twice the sine of the original angle times the cosine of the original angle:

$$\sin 2\theta = 2 \sin \theta \cos \theta$$

COSINE OF DOUBLE ANGLE

The cosine of twice any given angle can be found according to either of the following:

$$\cos 2\theta = 1 - (2 \sin^2 \theta)$$
$$\cos 2\theta = (2 \cos^2 \theta) - 1$$

SINE OF HALF ANGLE

The sine of half of any given angle can be found according to the following formula when $0 \leq \theta < \pi$ radians ($0 \leq \theta < 180°$):

$$\sin(\theta/2) = [(1 - \cos\theta)/2]^{1/2}$$

When $\pi \leq \theta < 2\pi$ radians ($180 \leq \theta < 360°$), the formula is

$$\sin(\theta/2) = -[(1 - \cos\theta)/2]^{1/2}$$

COSINE OF HALF ANGLE

The cosine of half of any given angle can be found according to the following formula when $0 \leq \theta < \pi/2$ radians ($0 \leq \theta < 90°$) or $3\pi/2 \leq \theta < 2\pi$ radians ($270 \leq \theta < 360°$):

$$\cos(\theta/2) = [(1 + \cos\theta)/2]^{1/2}$$

When $\pi/2 \leq \theta < 3\pi/2$ radians ($90 \leq \theta < 270°$) the formula is

$$\cos(\theta/2) = -[(1 + \cos\theta)/2]^{1/2}$$

SINE OF ANGULAR SUM

The sine of the sum of two angles θ and ϕ can be found according to the following formula:

$$\sin(\theta + \phi) = (\sin\theta)(\cos\phi) + (\cos\theta)(\sin\phi)$$

COSINE OF ANGULAR SUM

The cosine of the sum of two angles θ and ϕ can be found according to the following formula:

$$\cos(\theta + \phi) = (\cos\theta)(\cos\phi) - (\sin\theta)(\sin\phi)$$

SINE OF ANGULAR DIFFERENCE

The sine of the difference between two angles θ and ϕ can be found according to the following formula:

$$\sin(\theta - \phi) = (\sin\theta)(\cos\phi) - (\cos\theta)(\sin\phi)$$

COSINE OF ANGULAR DIFFERENCE

The cosine of the difference between two angles θ and ϕ can be found according to the following formula:

$$\cos(\theta - \phi) = (\cos\theta)(\cos\phi) + (\sin\theta)(\sin\phi)$$

 **Quiz**

Refer to the text in this chapter if necessary. A good score is eight correct. Answers are in the back of the book.

1. The range of the common logarithm function extends over the set of
   (a) all real numbers.
   (b) all the positive real numbers.
   (c) all the nonnegative real numbers.
   (d) all real numbers except zero.

2. From a distance of 503 meters, a spherical satellite has an angular diameter (that is, its disk subtends an observed angle) of 2.00 degrees of arc. What is the actual radius of the satellite? Assume that the distance is measured from the center of the satellite.
   (a) 8.78 meters
   (b) 17.6 meters
   (c) 10.6 meters
   (d) 2.79 meters

3. What is sin 45°? Do not use a calculator to determine the answer. Use the pythagorean theorem (as defined in Chapter 4) and simple algebra.
   (a) $2^{1/2}$
   (b) $2^{-1/2}$
   (c) 1
   (d) It cannot be determined from this information.

4. The natural logarithm of -5.670, to four significant figures, is equal to
   (a) 1.735.
   (b) -1.735.
   (c) 0.7536.
   (d) Nothing; the value is not defined.

5. What is the value of the quotient $(10^{(4.553)}/10^{(3.553)})$? Parentheses have been added to make the meaning of this expression completely clear.
   (a) 10

(b) 1

(c) 4.553

(d) 3.553

6. Suppose that you are given the equation $e^x = -5$ and told to solve it. What can you say right away about the value of x?

(a) It is a large positive real number.

(b) It is between 0 and 1.

(c) It is a real number and is large negatively.

(d) It is not a real number.

7. What is ln e to three significant figures? Use a calculator if you need it.

(a) 0.434

(b) 2.718

(c) 1.000

(d) It can't be figured out without more information.

8. Suppose that the cosine of a small angle is 0.950. What is the cosine of the negative of that angle—that is, the cosine of the same angle measured clockwise rather than counterclockwise?

(a) 0.950

(b) −0.950

(c) 0.050

(d) −0.050

9. Given that the length of a day on earth is 24 hours (as measured with respect to the sun), how many degrees of arc does the earth rotate through in 1 minute of time?

(a) 1/60

(b) 15

(c) 1/3,600

(d) 0.25

10. Suppose that two points on the earth's equator are separated by one second of arc (that is, 1/3,600 of an angular degree). If the circumference of the earth at the equator is given as 4.00×10^7 meters, how far apart are these two points?

(a) 1.11×10^4 meters

(b) 463 meters

(c) 30.9 meters

(d) It cannot be calculated from this information.

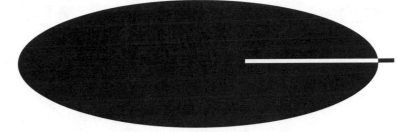

Test: Part Zero

Do not refer to the text when taking this test. A good score is at least 37 correct. Answers are in the back of the book. It is best to have a friend check your score the first time so that you won't memorize the answers if you want to take the test again.

1. A constant that is a real number but is not expressed in units is known as
   (a) a euclidean constant.
   (b) a cartesian constant.
   (c) a dimensionless constant.
   (d) an irrational constant.
   (e) a rational constant.

2. If someone talks about a *gigameter* in casual conversation, how many kilometers could you logically assume this is?
   (a) 1,000
   (b) 10,000
   (c) 100,000
   (d) 1 million
   (e) 1 billion

3. In log-log coordinates,
   (a) one axis is linear, and the other is determined according to an angle.
   (b) both axes are logarithmic.
   (c) all possible real-number ordered pairs can be shown in a finite area.
   (d) all three values are determined according to angles.
   (e) points are defined according to right ascension and declination.

4. Consider this sequence of numbers: 7.899797, 7.89979, 7.8997, 7.899, 7.89, ... Each number in this sequence has been modified to get the next one. This process is an example of
   (a) truncation.
   (b) vector multiplication.

(c) rounding.

(d) extraction of roots.

(e) scientific notation.

5. The expression 3_x (read "three sub x") is another way of writing
    (a) 3 raised to the xth power.
    (b) the product of 3 and x.
    (c) 3 divided by x.
    (d) The xth root of 3.
    (e) nothing; this is a nonstandard expression.

6. Which of the following statements is true?
    (a) A quadrilateral can be uniquely determined according to the lengths of its four sides.
    (b) The diagonals of a parallelogram always bisect each other.
    (c) Any given four points always lie in a single plane.
    (d) If one of the interior angles in a triangle measures 90°, then all the interior angles in that triangle measure 90°.
    (e) All of the above statements are true.

7. If you see the lowercase italic letter c in an equation or formula describing the physical properties of a system, it would most likely represent
    (a) the exponential base, roughly equal to 2.71828.
    (b) the ratio of a circle's diameter to its radius.
    (c) the square root of -1.
    (d) the speed of light in free space.
    (e) the 90° angle in a right triangle.

8. Suppose that an airplane is flying on a level course over an absolutely flat plain. At a certain moment in time you measure the angle x at which the airplane appears to be above the horizon. At the same moment the pilot of the aircraft sees you and measures the angle y at which you appear to be below the horizon. Which of the following statements is true?
    (a) $x < y$.
    (b) $x = y$.
    (c) $x > y$.
    (d) The relationship between x and y depends on the plane's altitude.
    (e) The relationship betwen x and y depends on the plane's speed.

9. Suppose that you are told that the diameter of the sun is 1.4×10^6 kilometers and you measure its angular diameter in the sky as 0.50°. Based on this information, approximately how far away is the sun, to two significant figures?
    (a) 1.6×10^8 kilometers
    (b) 6.2×10^8 kilometers
    (c) 1.6×10^7 kilometers
    (d) 6.2×10^7 kilometers
    (e) 6.2×10^6 kilometers

10. What is the diameter of a sphere whose volume is 100 cubic meters? (The formula for the volume V of a sphere in cubic meters, in terms of its radius R in meters, is $V = 4\pi r^3/3$.)
    (a) 2.88 meters
    (b) 4.19×10^6 meters
    (c) 5.76 meters
    (d) 8.28×10^6 meters
    (e) There is not enough information to determine this.

11. What is the difference, from the point of view of an experimental physicist, between 2.0000000×10^5 and 2.000×10^5?
    (a) One expression has eight significant figures, and the other has four significant figures.
    (b) Four orders of magnitude
    (c) One part in 10,000
    (d) One number is rounded, and the other is truncated.
    (e) There is no difference whatsoever between these two expressions.

12. Refer to Fig. T-0-1. What is the domain of this function?
    (a) All the real numbers between and including 0 and 1
    (b) All real numbers greater than 0 but less than or equal to 1
    (c) All real numbers greater than or equal to 0 but less than 1

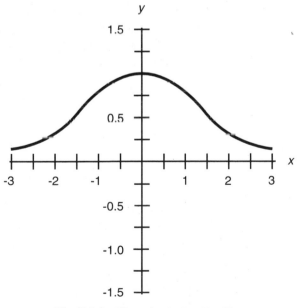

Fig. T-0-1   Illustration for Part Zero Test
Questions 12 and 13.

(d)　All real numbers between but not including 0 and 1

(e)　All real numbers

13.　Refer again to Fig. T-0-1. What is the range of this function?

(a)　All the real numbers between and including 0 and 1

(b)　All real numbers greater than 0 but less than or equal to 1

(c)　All real numbers greater than or equal to 0 but less than 1

(d)　All real numbers between but not including 0 and 1

(e)　All real numbers

14.　Suppose that a car travels down a straight road at a constant speed. Then the distance traveled in a certain amount of time is equal to

(a)　the product of the speed and the elapsed time.

(b)　the speed divided by the elapsed time.

(c)　the elapsed time divided by the speed.

(d)　the sum of the speed and the elapsed time.

(e)　the difference between the speed and the elapsed time.

15.　A two-dimensional coordinate system that locates points based on an angle and a radial distance, similar to circular radar displays, is called

(a)　the cartesian plane.

(b)　semilog coordinates.

(c)　cylindrical coordinates.

(d)　circular coordinates.

(e)　polar coordinates.

16.　The expression 6! is equivalent to

(a)　the common logarithm of 6.

(b)　the natural logarithm of 6.

(c)　⅙.

(d)　21.

(e)　720.

17.　Suppose that you come across a general single-variable equation written in the following form:

$$(x - q)\,(x - r)\,(x - s)\,(x - t) = 0$$

This can be classified as a

(a)　quadratic equation.

(b)　cubic equation.

(c)　quartic equation.

(d)　quintic equation.

(e)　linear equation.

18.　Suppose that you have a pair of equations in two variables. What is the least number of common solutions this pair of equations can have?

(a)　None

(b) One
(c) Two
(d) Three
(e) Four

19. Suppose that you have a brick wall 1.5 meters high and you need to build a ramp to the top of this wall from some point 3.2 meters away on level ground. Which of the following plank lengths is sufficient to make such a ramp without being excessively long?
    (a) 4.7 meters
    (b) 4.8 meters
    (c) 3.6 meters
    (d) 1.7 meters
    (e) There is not enough information given here to answer that question.

20. In a cylindrical coordinate system, a point is determined relative to the origin and a reference ray according to
    (a) angle, radius, and elevation.
    (b) three radii.
    (c) three angles.
    (d) height, width, and depth.
    (e) celestial latitude and longitude.

21. What is the standard quadratic form of $(x + 2)(x - 5)$?
    (a) $2x - 3 = 0$
    (b) $x^2 - 10 = 0$
    (c) $x^2 - 3x - 10 = 0$
    (d) $x^2 + 7x + 10 = 0$
    (e) There is no such form because this is not a quadratic equation.

22. Suppose that a piston has the shape of a cylinder with a circular cross section. If the area of the circular cross section (the end of the cylinder) is 10 square centimeters and the cylinder itself is 10 centimeters long, what, approximately, is the volume of the cylinder?
    (a) 10 square centimeters
    (b) 100 square centimeters
    (c) 62.8 cubic centimeters
    (d) 100 cubic centimeters
    (e) More information is needed to determine this.

23. Suppose that you see this equation in a physics paper: $z_0 = 3h + \sin q$. What does $\sin q$ mean?
    (a) The logarithm of the quantity q
    (b) The inverse sine (arcsine) of the quantity q
    (c) The sine of the quantity q
    (d) The exponential of the quantity q
    (e) None of the above

24. The expression $-5.44E + 04$ is another way of writing
    (a) -5.4404
    (b) $-544,004$
    (c) -5.44×10^{-4}
    (d) $-54,400$
    (e) nothing; this is a meaningless expression.

25. When two equations in two variables are graphed, their common approximate solutions, if any, appear as
    (a) points where the curves cross the x axis.
    (b) points where the curves cross the y axis.
    (c) points where the curves intersect each other.
    (d) points where the curves intersect the origin $(0, 0)$.
    (e) nothing special; the graphs give no indication of the solutions.

26. What is the product of 5.8995×10^{-8} and 1.03×10^{6}? Take significant figures into account.
    (a) 6.0764845×10^{-2}
    (b) 6.076485×10^{-2}
    (c) 6.07648×10^{-2}
    (d) 6.076×10^{-2}
    (e) 6.08×10^{-2}

27. Suppose that you see the following expression in a physics thesis:

$$\mathrm{sech}^{-1} x = \ln [x^{-1} + (x^{-2} - 1)^{1/2}]$$

What does the expression *ln* mean in this context?
    (a) Some real number multiplied by 1
    (b) The common logarithm
    (c) The natural logarithm
    (d) The inverse secant
    (e) The square root

28. Suppose that there are two vectors **a** and **b,** represented in the cartesian plane as follows:

$$\mathbf{a} = (3, 5)$$
$$\mathbf{b} = (-3, -5)$$

What is the sum of these vectors in the cartesian plane?
    (a) $\mathbf{a} + \mathbf{b} = -34$
    (b) $\mathbf{a} + \mathbf{b} = (0, 0)$
    (c) $\mathbf{a} + \mathbf{b} = (6, 10)$
    (d) $\mathbf{a} + \mathbf{b} = (-9, -25)$
    (e) There is no such sum, because the sum of these vectors is not defined.

29. How many points does it take to uniquely define a geometric plane?
    (a) One
    (b) Two
    (c) Three
    (d) Four
    (e) Five

30. Refer to Fig. T-0-2. What does this graph represent?
    (a) The sine function
    (b) The cosine function
    (c) A quadratic equation
    (d) A linear equation
    (e) A logarithmic function

31. Refer again to Fig. T-0-2. The coordinate system in this illustration is
    (a) polar.
    (b) spherical.
    (c) semilog.

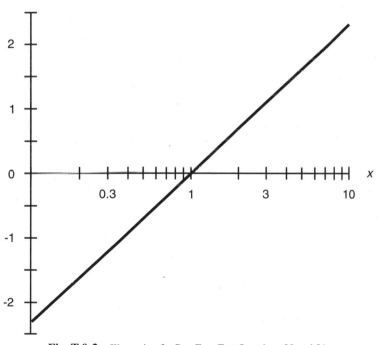

Fig. T-0-2   Illustration for Part Zero Test Questions 30 and 31

(d) log-log.

(e) trigonometric.

32. Suppose that there are two vectors. Vector **a** points straight up with a magnitude of 3, and vector **b** points directly toward the western horizon with a magnitude of 4. The cross product **a** × **b** has the following characteristics:

(a) It is a scalar with a value of 12.

(b) It is a vector pointing toward the southern horizon with a magnitude of 12.

(c) It is a vector pointing upward and toward the west with a magnitude of 5.

(d) It is a vector pointing straight down with a magnitude of 5.

(e) We need more information to answer this.

33. Using a calculator, you determine the ⅔ power of 2 (that is, $2^{2/3}$) to four significant figures. The result is

(a) 1.587.

(b) 2.828.

(c) 4.000.

(d) 8.000.

(e) The expression $2^{2/3}$ is not defined and cannot be determined by any means.

34. The numbers 34 and 34,000 differ by

(a) a factor of 10.

(b) three orders of magnitude.

(c) five orders of magnitude.

(d) seven orders of magnitude.

(e) the same ratio as a foot to a mile.

35. Right ascension is measured in

(a) degrees.

(b) radians.

(c) linear units.

(d) logarithmic units.

(e) hours.

36. Consider the function $y = 2x$ with the domain restricted to $0 < x < 2$. What is the range?

(a) $0 < y < 1/2$

(b) $0 < y < 1$

(c) $0 < y < 2$

(d) $0 < y < 4$

(e) There is not enough information given to answer this question.

37. The fifth root of 12 can be written as

(a) $12^{1/5}$.

(b) 12/5.

(c) 12^5.

(d) 5^{12}.

(e) $5^{1/12}$.

38. Suppose that you are given the equation $x^2 + y^2 = 10$. What does this look like when graphed in rectangular coordinates?
    (a) A straight line
    (b) A parabola
    (c) An elongated ellipse
    (d) A hyperbola
    (e) A circle

39. Suppose that an experimenter takes 10,000 measurements of the voltage on a household utility line over a period of several days and comes up with an average figure of 115.85 volts. This is considered the nominal voltage on the line. Suppose that a second experimenter takes a single measurement and gets a value of 112.20 volts. The percentage departure of the single observer's measurement from the nominal voltage is approximately
    (a) -0.03 percent.
    (b) $+0.03$ percent.
    (c) $+3$ percent.
    (d) -3 percent.
    (e) impossible to determine from the data given.

40. Suppose that there is a four-sided geometric figure that lies in a single plane and whose sides all have the same length. Then the perimeter of that figure is
    (a) the product of the base length and the height.
    (b) the square of the length of any given side.
    (c) the sum of the lengths of all four sides.
    (d) half the sum of the lengths of all four sides.
    (e) impossible to define without knowing more information.

41. Suppose that you view a radio tower on a perfectly flat plain and find that it appears to extend up to a height of $2.2°$ above the horizon. How far away is the base of the tower from where you stand, expressed to two significant figures?
    (a) 0.5 kilometer
    (b) 1.0 kilometer
    (c) 1.5 kilometers
    (d) 2.2 kilometers
    (e) More information is needed to determine this.

42. The product of 3.88×10^7 and 1.32×10^{-7} is
    (a) 5.12.
    (b) 5.12×10^{14}.
    (c) 5.12×10^{-14}.
    (d) 5.12×10^{49}.
    (d) 5.12×10^{-49}.

43. The cosine of a negative angle is the same as the cosine of the angle. Knowing this and the fact that the cosine of $60°$ is equal to 0.5, what can be said about the cosine of $300°$ without making any calculations?

    (a) Nothing; we need more information to know.

    (b) The cosine of 300° is equal to 0.5.

    (c) The cosine of 300° is equal to −0.5.

    (d) The cosine of 300° can be equal to either 0.5 or −0.5.

    (e) The cosine of 300° is equal to zero.

44. The slope of a *vertical line* (a line parallel to the ordinate) in cartesian rectangular coordinates is

    (a) undefined.

    (b) equal to 0.

    (c) equal to 1.

    (d) variable, depending on how far you go from the origin.

    (e) imaginary.

45. Which of the following statements is false?

    (a) A triangle can be uniquely determined according to the lengths of its sides.

    (b) A triangle can be uniquely determined according to the length of one side and the measures of the two angles at either end of that side.

    (c) A triangle can be uniquely determined according to the measures of its three interior angles.

    (d) All equilateral triangles are similar to each other.

    (e) An isosceles triangle has two sides whose lengths are the same.

46. The equation $-4x^2 + 17x = 7$ is an example of

    (a) a two-variable equation.

    (b) a linear equation.

    (c) a quadratic equation.

    (d) an exponential function.

    (e) none of the above.

47. The sum of a real number and an imaginary number is

    (a) undefined.

    (b) an irrational number.

    (c) a rational number.

    (d) a transcendental number.

    (e) a complex number.

48. In astronomy, the equivalent of celestial longitude, based on the vernal equinox and measured relative to the stars, is called

    (a) longitude.

    (b) azimuth.

    (c) right ascension.

    (d) arc span.

    (e) meridian.

49. Consider a plane that contains the axis of a true parabolic dish antenna or mirror. The dish or mirror intersects this plane along a curve that can be defined by
    (a) imaginary numbers.
    (b) a linear equation.
    (c) a quadratic equation.
    (d) a cubic equation.
    (e) no particular equation.

50. Suppose that you are given two positive numbers, one of them 25 orders of magnitude larger than the other and both expressed to four significant figures. If you add these numbers and express the sum to four significant figures,
    (a) the smaller number vanishes into insignificance.
    (b) you must write both numbers out in full.
    (c) you must have the aid of a computer.
    (d) the sum is 25 orders of magnitude bigger than the larger number.
    (c) you must subtract the numbers and then take the negative of the result.

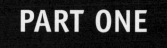

Classical Physics

CHAPTER 6

Units and Constants

Units are devices that scientists use to indicate, estimate, and calculate aspects of the world and the universe. Numbers by themselves are abstract. Try to envision the number 5 in your mind. You think of a set or object: five objects, five dots, a line five meters long, a five-pointed star, or a pentagon. However, these are sets or objects, not the actual number. It is still more difficult to directly envision the square root of two ($2^{1/2}$), pi (π), or the natural logarithm base (e), which aren't whole numbers.

Most people think of numbers as points on a line that are certain distances from the *origin,* or zero point. Displacement might be $2^{1/2}$ units or π meters. You might think of a specific length of time, such as e seconds. Maybe you think of mass in kilograms or even something more exotic, such as the intensity of an electric current in *amperes,* or the brilliance of a light bulb in *candelas.*

Systems of Units

There are various schemes, or systems, of physical units in use throughout the world. The *meter-kilogram-second (mks) system,* also called the *metric system* or the *International System,* is favored by most physicists. The *centimeter-gram-second (cgs) system* is used less often, and the *foot-pound-second (fps) system,* also called the *English system,* is used rarely by scientists but is popular among nonscientists. Each system has several fundamental, or *base,* units from which all the others are derived.

THE INTERNATIONAL SYSTEM (SI)

The International System is often abbreviated SI, which stands for *Système International* in French. This scheme in its earlier form, mks, has existed since the 1800s, but more recently it has been defined in a rigorous fashion by the General Conference on Weights and Measures.

The base units in SI quantify *displacement, mass, time, temperature, electric current, brightness of light,* and *amount of matter* (in terms of the number of atoms or molecules in a sample). Respectively, the units in SI are known as the *meter,* the *kilogram,* the *second,* the *kelvin* (or *degree kelvin*), the *ampere,* the *candela,* and the *mole.* We'll define these in detail shortly.

THE CGS SYSTEM

In the centimeter-gram-second (cgs) system, the base units are the *centimeter* (exactly 0.01 meter), the *gram* (exactly 0.001 kilogram), the *second,* the *degree Celsius* (approximately the number of Kelvins minus 273), the *ampere,* the *candela,* and the *mole.* The second, the ampere, the candela, and the mole are the same in cgs as they are in SI.

THE ENGLISH SYSTEM

In the English or foot-pound-second (fps) system, the base units are the *foot* (approximately 30.5 centimeters), the *pound* (equivalent to about 2.2 kilograms in the gravitational field at the Earth's surface), the *second,* the *degree Fahrenheit* (where water freezes at 32 degrees and boils at 212 degrees at standard sea-level atmospheric pressure), the *ampere,* the *candela,* and the *mole.* The second, the ampere, the candela, and the mole are the same in fps as they are in SI.

Base Units in SI

In all systems of measurement, the base units are those from which all the others can be derived. Base units represent some of the most elementary properties or phenomena we observe in nature.

THE METER

The fundamental unit of distance, length, linear dimension, or displacement (all different terms meaning essentially the same thing) is the meter, symbolized by the lowercase nonitalicized English letter m. Originally, the meter was designated as the distance between two scratches on a platinum bar put on display in Paris, France. The original idea was that there ought to be 10 million (10^7) meters along a great circle between the north pole and the equator of Earth, as it would be measured if the route passed through Paris (Fig. 6-1). Mountains, bodies of water, and other barriers were ignored; the Earth was imagined to be a perfectly round, smooth ball. The circumference of the Earth is about 40 million (4.0×10^7) m, give or take a little depending on which great circle around the globe you choose.

Nowadays, the meter is defined more precisely as the distance a beam of light travels through a perfect vacuum in 3.33564095 billionths of a second, that is, $3.33564095 \times 10^{-9}$ second. This is approximately the length of an adult's full stride when walking at a brisk pace.

THE KILOGRAM

The base SI unit of mass is the kilogram, symbolized by the lowercase non-italicized pair of English letters kg. Originally, the kilogram was defined as the mass of 0.001 cubic meter (or 1 liter) of pure liquid water (Fig. 6-2).

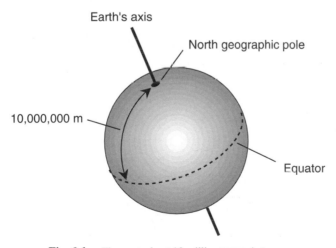

Fig. 6-1.   There are about 10 million meters between the Earth's north pole and the equator.

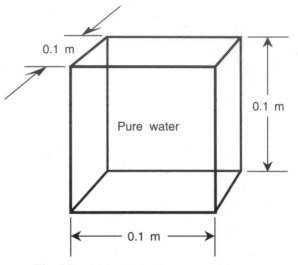

Fig. 6-2.   Originally, the kilogram was defined as the mass of 0.001 cubic meter of pure liquid water.

This is still an excellent definition, but these days scientists have come up with something more absolute. A kilogram is the mass of a sample of platinum-iridium alloy that is kept under lock and key at the International Bureau of Weights and Measures.

It is important to realize that mass is not the same thing as *weight*. A mass of 1 kg maintains this same mass no matter where it is located. That standard platinum-iridium ingot would mass 1 kg on the Moon, on Mars, or in intergalactic space. Weight, in contrast, is a force exerted by gravitation or acceleration on a given mass. On the surface of the Earth, a 1-kg mass happens to weigh about 2.2 pounds. In interplanetary space, the same mass weighs 0 pounds; it is *weightless.*

THE SECOND

The SI unit of time is the second, symbolized by the lowercase nonitalicized English letter s (or sometimes abbreviated as sec). It was defined originally as 1/60 of a minute, which is 1/60 of an hour, which in turn is 1/24 of a *mean solar day*. A second was thus thought of as 1/86,400 of a mean solar day, and this is still an excellent definition (Fig. 6-3). However, formally, these days, 1 s is defined as the amount of time taken for a certain cesium atom to oscillate through 9.192631770×10^9 complete cycles.

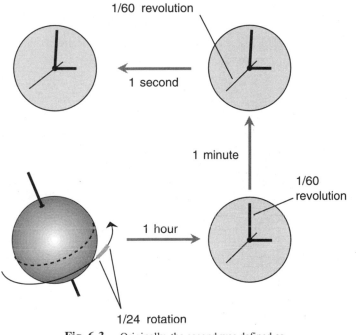

1/60 revolution

1 second

1 minute

1/60 revolution

1 hour

1/24 rotation

Fig. 6-3. Originally, the second was defined as (1/60)(1/60)(1/24), or 1/86,400, of a mean solar day.

One second also happens to be the time it takes for a ray of light to travel 2.99792458×10^8 m through space. This is about three-quarters of the way to the Moon. You may have heard of the Moon being a little more than one *light-second* away from Earth. If you are old enough to remember the conversations Earth-based personnel carried on with Apollo astronauts as the astronauts walked around on the Moon, you will recall the delay between comments or questions from earthlings and the replies from the moonwalkers. The astronauts were not hesitating; it took more than 2 seconds for radio signals to make a round trip between Earth and the Moon. In a certain manner of thinking, time is a manifestation or expression of linear dimension, and vice versa. Both of these aspects of nature are intimately related by the speed of light, which Albert Einstein hypothesized is an absolute.

THE KELVIN

The SI unit of temperature is the kelvin, symbolized K (uppercase and nonitalicized). It is a measure of how much heat exists relative to absolute zero, which represents the absence of all heat and which is therefore the coldest possible temperature. A temperature of 0 K represents absolute zero. Formally, the kelvin is defined as a temperature increment (an increase or decrease) of 0.003661 part of the thermodynamic temperature of the triple point of pure water. Pure water at sea level freezes (or melts) at +273.15 K and boils (or condenses) at +373.15 K.

What, you might ask, is the meaning of *triple point?* In the case of water, it's almost exactly the same as the freezing point. For water, it is the temperature and pressure at which it can exist as vapor, liquid, and ice in equilibrium. For practical purposes, you can think of it as freezing.

THE AMPERE

The ampere, symbolized by the uppercase nonitalicized English letter A (or abbreviated as amp), is the unit of electric current. A flow of approximately 6.241506×10^{18} electrons per second past a given fixed point in an electrical conductor produces an electrical current of 1 A.

Various units smaller than the ampere are often employed to measure or define current. A *milliampere* (mA) is one-thousandth of an ampere, or a flow of 6.241506×10^{15} electrons per second past a given fixed point. A *microampere* (μA) is one-millionth or 10^{-6} of an ampere, or a flow of 6.241506×10^{12} electrons per second. A *nanoampere* (nA) is 10^{-9} of an ampere; it is the smallest unit of electric current you are likely to hear about or use. It represents a flow of 6.241506×10^{9} electrons per second past a given fixed point.

The formal definition of the ampere is highly theoretical: 1 A is the amount of constant charge-carrier flow through two straight, parallel, infinitely thin, perfectly conducting media placed 1 m apart in a vacuum that results in a force between the conductors of 2×10^{-7} newton per linear meter. There are two problems with this definition. First, we haven't defined the term *newton* yet; second, this definition asks you to imagine some theoretically ideal objects that cannot exist in the real world. Nevertheless, there you have it: the physicist venturing into the mathematician's back yard again. It has been said that mathematicians and physicists can't live with each other and they can't live without each other.

THE CANDELA

The candela, symbolized by the lowercase nonitalicized pair of English letters cd, is the unit of luminous intensity. It is equivalent to 1/683 of a watt of radiant energy emitted at a frequency of 5.4×10^{14} hertz (cycles per second) in a solid angle of one steradian. (The steradian will be defined shortly.) This is a sentence full of arcane terms! However, there is a simpler, albeit crude, definition: 1 cd is roughly the amount of light emitted by an ordinary candle.

Another definition, more precise than the candle reference, does not rely on the use of derived units, a practice to which purists legitimately can object. According to this definition, 1 cd represents the radiation from a surface area of 1.667×10^{-6} square meter of a perfectly radiating object called a *blackbody* at the solidification temperature of pure platinum.

THE MOLE

The mole, symbolized or abbreviated by the lowercase nonitalicized English letters mol, is the standard unit of material quantity. It is also known as *Avogadro's number* and is a huge number, approximately 6.022169×10^{23}. This is the number of atoms in precisely 0.012 kg of carbon-12, the most common isotope of elemental carbon with six protons and six neutrons in the nucleus.

The mole arises naturally in the physical world, especially in chemistry. It is one of those strange numbers for which nature seems to have reserved a special place. Otherwise, scientists surely would have chosen a round number such as 1,000, or maybe even 12 (one dozen).

A NOTE ABOUT SYMBOLOGY

Up to this point we've been rigorous about mentioning that symbols and abbreviations consist of lowercase or uppercase nonitalicized letters or strings of letters. This is important because if this distinction is not made, especially relating to the use of italics, the symbols or abbreviations for physical units can be confused with the constants, variables, or coefficients that appear in equations. When a letter is italicized, it almost always represents a constant, a variable, or a coefficient. When it is nonitalicized, it often represents a physical unit. A good example is s, which represents second, versus s, which is often used to represent linear dimension or displacement.

From now on we won't belabor this issue every time a unit symbol or abbreviation comes up. But don't forget it. Like the business about significant figures, this seemingly trivial thing can matter a lot!

Other Units

The preceding seven units can be combined in various ways, usually as products and ratios, to generate many other units. Sometimes these *derived units* are expressed in terms of the base units, although such expressions can be confusing (for example, seconds cubed or kilograms to the −1 power). If you see combinations of units in a physics book, article, or paper that don't seem to make sense, don't be alarmed. You are looking at a derived unit that has been put down in terms of base units.

THE RADIAN

The standard unit of *plane angular measure* is the *radian* (rad). It is the angle subtended by an arc on a circle whose length, as measured on the circle, is equal to the radius of the circle as measured on a flat geometric plane containing the circle. Imagine taking a string and running it out from the center of a circle to some point on the edge and then laying that string down around the periphery of the circle. The resulting angle is 1 rad. Another definition goes like this: One radian is the angle between the two straight edges of a slice of pie whose straight and curved edges all have the same length *r* (Fig. 6-4). It is equal to about 57.2958 *angular degrees*.

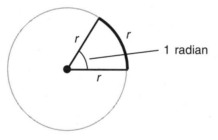

Fig. 6-4.   One radian is the angle at the apex of a slice of pie whose straight and curved edges all have the same length *r*.

THE ANGULAR DEGREE

The angular degree, symbolized by a little elevated circle (°) or by the three-letter abbreviation deg, is equal to 1/360 of a complete circle. The history of the degree is uncertain, although one theory says that ancient mathematicians chose it because it represents approximately the number of days in the year. One angular degree is equal to approximately 0.0174533 radians.

THE STERADIAN

The standard unit of *solid angular measure* is the *steradian,* symbolized sr. A solid angle of 1 sr is represented by a cone with its apex at the center of a sphere and intersecting the surface of the sphere in a circle such that, within the circle, the enclosed area on the sphere is equal to the square of the radius of the sphere. There are 4π, or approximately 12.56636, steradians in a complete sphere.

THE NEWTON

The standard unit of *mechanical force* is the *newton,* symbolized N. One newton is the amount of force that it takes to make a mass of 1 kg accelerate at a rate of one meter per second squared (1 m/s^2). Jet or rocket engine propulsion is measured in newtons. Force is equal to the product of mass and acceleration; reduced to base units in SI, newtons are equivalent to kilogram-meters per second squared (kg · m/s^2).

THE JOULE

The standard unit of *energy* is the *joule,* symbolized J. This is a fairly small unit in real-world terms. One joule is the equivalent of a newton-meter (n · m). If reduced to base units in SI, the joule can be expressed in terms of unit mass multiplied by unit distance squared per unit time squared:

$$1\ J = 1\ kg \cdot m^2/s^2$$

THE WATT

The standard unit of *power* is the *watt,* symbolized W. One watt is equivalent to one joule of energy expended for one second of time (1 J/s). In fact, power is a measure of the rate at which energy is produced, radiated, or consumed. The expression of watts in terms of SI base units begins to get esoteric, as you have been warned:

$$1\ W = 1\ kg \cdot m^2/s^3$$

THE COULOMB

The standard unit of *electric charge quantity* is the *coulomb,* symbolized C. This is the electric charge that exists in a congregation of approximately 6.241506×10^{18} electrons. It also happens to be the electric charge contained in that number of protons, antiprotons, or positrons (antielectrons). When you walk along a carpet with hard-soled shoes in the winter or anywhere the humidity is very low, your body builds up a static electric charge that can be expressed in coulombs (or more likely a fraction of one coulomb). Reduced to base units in SI, one coulomb is equal to one ampere-second (1 A · s).

THE VOLT

The standard unit of *electrical potential* or *potential difference,* also called *electromotive force* (emf), is the *volt,* symbolized V. One volt is equivalent to one joule per coulomb (1 J/C). The volt is, in real-world terms, a moderately small unit of electrical potential. A standard dry cell of the sort you find in a flashlight (often erroneously called a *battery*), produces about 1.5 V. Most automotive batteries in the United States produce between 12 and 13.5 V.

THE OHM

The standard unit of *electrical resistance* is the *ohm,* symbolized by the upper-case Greek letter omega (Ω). When one volt is applied across a resistance of ohm, the result is one ampere of current flow. The ohm is thus equivalent to one volt per ampere (V/A).

THE SIEMENS

The standard unit of *electrical conductance* is the *siemens,* symbolized S. It was formerly called the *mho,* and in some papers and texts you'll still see this term. Conductance is the reciprocal of resistance. One siemens can be considered the equivalent of one ampere per volt (A/V). If R is the resistance of a component in ohms and G is the conductance of the component in siemens, then

$$G = 1/R$$
$$R = 1/G$$

THE HERTZ

The standard unit of *frequency* is the *hertz,* symbolized Hz. It was formerly called the *cycle per second* or simply the *cycle.* The hertz is a small unit in the real world, and 1 Hz represents an extremely low frequency. Usually, frequency is measured in thousands, millions, billions, or trillions of hertz. These units are called *kilohertz* (kHz), *megahertz* (MHz), *gigahertz* (GHz), and *terahertz* (THz), respectively. In terms of SI units, the hertz is mathematically simple, but the concept is esoteric for some people to grasp: It is an inverse second (s^{-1}) or per second (/s).

THE FARAD

The standard unit of *capacitance* is the *farad,* which is symbolized F. The farad is equivalent to one coulomb per volt (1 C/V). This is a large unit in real-world applications. Most values of capacitance that you will find in electrical and electronic circuits are on the order of millionths, billionths, or trillionths of a farad. These units are called *microfarads* (μF), *nanofarads* (nF), and *picofarads* (pF).

THE HENRY

The standard unit of *inductance* is the *henry,* symbolized H. One henry is equivalent to one volt-second per ampere (V · s/A or V · s · A^{-1}). This is a large unit in practice but not quite as gigantic as the farad. In electrical and electronic circuits, most values of inductance are on the order of thousandths or millionths of a henry. These units are called *millihenrys* (mH) and *microhenrys* (μH).

THE WEBER

The standard unit of *magnetic flux* is the *weber,* symbolized Wb. This is a large unit in practical applications. One weber is equal to one ampere-henry

(1 A · H). This is represented in the real world as the amount of magnetism produced by a constant direct current of 1 A flowing through a coil having an inductance of 1 H.

THE TESLA

The standard unit of *magnetic flux density* is the *tesla,* symbolized T. One tesla is equivalent to one weber per meter squared (1 Wb/m^2 or Wb · m^{-2}) when the flux is perpendicular to the surface under consideration. Sometimes magnetic flux density is spoken of in terms of the number of "lines of flux" per unit cross-sectional area; this is an imprecise terminology unless we are told exactly how much magnetic flux is represented by a line.

Prefix Multipliers

Sometimes the use of standard units is inconvenient or unwieldy because a particular unit is very large or small compared with the magnitudes of phenomena commonly encountered in real life. We've already seen some good examples: the hertz, the farad, and the henry. Scientists use *prefix multipliers,* which can be attached in front of the words representing units, to express power-of-10 multiples of those units.

In general, the prefix multipliers range in increments of 10^3, or 3 orders of magnitude, all the way down to 10^{-24} (septillionths) and all the way up to 10^{24} (septillions). This is a range of 48 orders of magnitude! It's not easy to think of an illustrative example to demonstrate the hugeness of this ratio. Table 6-1 outlines these prefix multipliers and what they stand for.

PROBLEM 6-1
Suppose that you are told that a computer's microprocessor has a clock frequency of 5 GHz. What is this frequency in hertz?

SOLUTION 6-1
From Table 6-1, observe that the gigahertz (GHz) represents 10^9 Hz. Thus 5 GHz is equal to 5×10^9 Hz, or 5 billion Hz.

PROBLEM 6-2
A capacitor is specified as having a value of 0.001 μF. What is this value in farads?

SOLUTION 6-2
From Table 6-1, note that μ stands for micro-, or a unit of 10^{-6}. Therefore, 0.001 μF is 0.001 microfarad, equivalent to 0.001×10^{-6} F $= 10^{-3} \times 10^{-6}$ $= 10^{-9}$ F. This could be called a nanofarad (1 nF), but for some reason, engineers rarely use the nano- multiplier when speaking or writing about capacitances. Instead, they likely would stick with the 0.001 μF notation; alternatively, they might talk about 10,000 picofarads (pF).

PROBLEM 6-3
An inductor has a value of 0.1 mH. What is this in microhenrys?

SOLUTION 6-3
From Table 6-1, you can see that the prefix multiplier m stands for *milli-* or 10^{-3}. Therefore, 0.1 mH $= 0.1 \times 10^{-3}$ H $= 10^{-4}$ H $= 10^{2} \times 10^{-6}$ H $= 100$ μH.

Table 6-1   Prefix Multipliers and Their Abbreviations

Designator	Symbol	Multiplier
yocto-	y	10^{-24}
zepto-	z	10^{-21}
atto-	a	10^{-18}
femto-	f	10^{-15}
pico-	p	10^{-12}
nano-	n	10^{-9}
micro-	μ or mm	10^{-6}
milli-	m	10^{-3}
centi-	c	10^{-2}
deci-	d	10^{-1}
(none)	—	10^{0}
deka-	da or D	10^{1}
hecto-	h	10^{2}
kilo-	K or k	10^{3}
mega-	M	10^{6}
giga-	G	10^{9}
tera-	T	10^{12}
peta-	P	10^{15}
exa-	E	10^{18}
zetta-	Z	10^{21}
yotta-	Y	10^{24}

Constants

Constants are characteristics of the physical and mathematical world that can be "taken for granted." They don't change, at least not within an ordinary human lifetime, unless certain other factors change too.

MATH VERSUS PHYSICS

In pure mathematics, constants are usually presented all by themselves as plain numbers without any units associated. These are called *dimensionless constants* and include π, the circumference-to-diameter ratio of a circle, and *e,* the natural logarithm base. In physics, there is almost always a unit equivalent attached to a constant. An example is *c,* the speed of light in free space, expressed in meters per second.

Table 6-2 is a list of constants you'll encounter in physics. This is by no means a complete list. Do you not know what most of the constants in this table mean? Are they unfamiliar or even arcane to you? Don't worry about this now. As you keep on reading this book, you'll learn about most of them. This table can serve as a reference long after you're done with this course.

Here are a few examples of constants from the table and how they relate to the physical universe and the physicist's modes of thought.

MASS OF THE SUN

It should come as no surprise to you that the Sun is a massive object. But just how massive, really, is it? How can we express the mass of the Sun in terms that can be comprehended? Scientific notation is generally used; we come up with the figure 1.989×10^{30} kg if we go to four significant figures. This is just a little less than 2 nonillion kilograms or 2 octillion metric tons. (This doesn't help much, does it?)

How big is 2 octillion? It's represented numerically as a 2 with 27 zeros after it. In scientific notation it's 2×10^{27}. We can split this up into $2 \times 10^9 \times 10^9 \times 10^9$. Now imagine a huge box 2,000 kilometers (km) tall by 1,000 km wide by 1,000 km deep. [A thousand kilometers is about 620 miles (mi); 2000 km is about 1240 mi.] Suppose now that you are called on to stack this box neatly full of little cubes measuring 1 millimeter (1 mm) on an edge. These cubes are comparable in size to grains of coarse sand.

You begin stacking these little cubes with the help of tweezers and a magnifying glass. You gaze up at the box towering high above the Earth's

Table 6-2   Some Physical Constants

Quantity or Phenomenon	Value	Symbol
Mass of Sun	1.989×10^{30} kg	m_{sun}
Mass of Earth	5.974×10^{24} kg	m_{earth}
Avogadro's number	6.022169×10^{23} mol^{-1}	N or N_A
Mass of Moon	7.348×10^{22} kg	m_{moon}
Mean radius of Sun	6.970×10^{8} m	r_{sun}
Speed of electromagnetic-field propagation in free space	2.99792×10^{8} m/s	c
Faraday constant	9.64867×10^{4} C/mol	F
Mean radius of Earth	6.371×10^{6} m	r_{earth}
Mean orbital speed of Earth	2.978×10^{4} m/s	
Base of natural logarithms	2.718282	e or ε
Ratio of circle circumference to radius	3.14159	π
Mean radius of Moon	1.738×10^{6} m	r_{moon}
Characteristic impedance of free space	376.7 Ω	Z_0
Speed of sound in dry air at standard atmospheric temperature and pressure	344 m/s	
Gravitational acceleration at sea level	9.8067 m/s^2	g
Gas constant	8.31434 J/K/mol	R or R_0
Fine structure constant	7.2974×10^{-3}	α
Wien's constant	0.0029 m · K	σ_W
Second radiation constant	0.0143883 m · K	c_2
Permeability of free space	1.257×10^{-6} H/m	μ_0
Stefan-Boltzmann constant	5.66961×10^{-8} W/m^2/K^4	σ
Gravitational constant	6.6732×10^{-11} N · m^2/kg^2	G
Permittivity of free space	8.85×10^{-12} F/m	ε_0
Boltzmann's constant	1.380622×10^{23} J/K	k
First radiation constant	4.99258×10^{-24} J · m	c_1
Atomic mass unit (amu)	1.66053×10^{-27} kg	u
Bohr magneton	9.2741×10^{-24} J/T	μ_B
Bohr radius	5.2918×10^{-11} m	α_0
Nuclear magneton	5.0510×10^{-27} J/T	μ_n
Mass of alpha particle	6.64×10^{-27} kg	m_α
Mass of neutron at rest	1.67492×10^{-27} kg	m_n
Mass of proton at rest	1.67261×10^{-27} kg	m_p
Compton wavelength of proton	1.3214×10^{-15} m	λ_{cp}
Mass of electron at rest	9.10956×10^{-31} kg	m_e
Radius of electron	2.81794×10^{-15} m	r_e

Table 6-2 Some Physical Constants (*Continued*)

Quantity or Phenomenon	Value	Symbol
Elementary charge	1.60219×10^{-19} C	e
Charge-to-mass ratio of electron	1.7588×10^{11} C/kg	e/m_e
Compton wavelength of electron	2.4263×10^{-12} m	λ_c
Planck's constant	6.6262×10^{-34} J · s	h
Quantum-charge ratio	4.1357×10^{-15} J · s/C	h/e
Rydberg constant	1.0974×10^7 m^{-1}	R_∞
Euler's constant	0.577216	γ

atmosphere and spanning several states or provinces (or even whole countries) over the Earth's surface. You can imagine it might take you quite a while to finish this job. If you could live long enough to complete the task, you would have stacked up 2 octillion little cubes, which is the number of *metric tons* in the mass of our Sun. A metric ton is slightly more than an English ton.

The Sun is obviously a massive chunk of matter. But it is small as stars go. There are plenty of stars that are many times larger than our Sun.

MASS OF THE EARTH

The Earth, too, is massive, but it is a mere speck compared with the Sun. Expressed to four significant figures, the Earth masses 5.974×10^{24} kg. This works out to approximately 6 hexillion metric tons.

How large a number is 6 hexillion? Let's use a similar three-dimensional analogy. Suppose that you have a cubical box measuring 2.45×10^5 meters, or 245 kilometers, on an edge. This is a cube about 152 mi tall by 152 mi wide by 152 mi deep. Now imagine an endless supply of little cubes measuring 1 centimeter (1 cm) on an edge. This is about the size of a gambling die or a sugar cube. Now suppose that you are given the task of—you guessed it—stacking up all the little cubes in the huge box. When you are finished, you will have placed approximately 6 hexillion little cubes in the box. This is the number of metric tons in the mass of our planet Earth.

SPEED OF ELECTROMAGNETIC (EM) FIELD PROPAGATION

The so-called speed of light is about 2.99792×10^8 m/s. This works out to approximately 186,282 miles per second (mi/s). Radio waves, infrared, visible light, ultraviolet, x-rays, and gamma rays all propagate at this speed, which Albert Einstein postulated to be the same no matter from what point of view it is measured.

How fast, exactly, is this? One way to grasp this is to calculate how long it would take a ray of light to travel from home plate to the center-field fence in a major league baseball stadium. Most ballparks are about 122 m deep to center field; this is pretty close to 400 feet (ft). To calculate the time t it takes a ray of light to travel that far, we must divide 122 m by 2.99792 $\times 10^8$ m/s:

$$t = 122/(2.99792 \times 10^8)$$
$$= 4.07 \times 10^{-7}$$

That is just a little more than four-tenths of a microsecond (0.4 μs), an imperceptibly short interval of time.

Two things should be noted at this time. First, remember the principles of significant figures. We are justified in going to only three significant figures in our answer here. Second, the units must be consistent with each other to get a meaningful answer. Mixing units is a no-no in any calculation. It almost always leads to trouble.

If we were to take the preceding problem and calculate in terms of units without using any numbers at all, this is what we would get:

$$\text{seconds} = \text{meters}/(\text{meters per second})$$
$$s = m/(m/s) = m \times s/m$$

In this calculation, meters cancel out, leaving only seconds. Suppose, however, that we were to try to make this calculation using feet as the figure for the distance from home plate to the center-field fence? We would then obtain some value in undefined units; call them *fubars* (fb):

$$\text{fubars} = \text{feet}/(\text{meters per second})$$
$$fb = ft/(m/s) = ft \times s/m$$

Feet do not cancel out meters. Thus we have invented a new unit, the fubar, that is equivalent to a foot-second per meter. This unit is basically useless, as would be our numerical answer. (As an aside, *fubar* is an acronym for "fouled up beyond all recognition.")

Always remember to be consistent with units when making calculations! When in doubt, reduce all the "givens" in a problem to SI units before starting to make calculations. You will then be certain to get an answer in derived SI units and not fubars or gobbledygooks or any other non-sensical things.

GRAVITATIONAL ACCELERATION AT SEA LEVEL

The term *acceleration* can be somewhat confusing to the uninitiated when it is used in reference to gravitation. Isn't gravity just a force that pulls on things? The answer to this question is both yes and no.

Obviously, gravitation pulls things downward toward the center of the Earth. If you were on another planet, you would find gravitation there, too, but it would not pull on you with the same amount of force. If you weigh 150 pounds here on Earth, for example, you would weigh only about 56 pounds on Mars. (Your mass, 68 kilograms, would be the same on Mars as on Earth.) Physicists measure the intensity of a gravitational field according to the rate at which an object accelerates when it is dropped in a vacuum so that there is no atmospheric resistance. On the surface of the Earth, this rate of acceleration is approximately 9.8067 meters per second per second, or 9.8067 m/s^2. This means that if you drop something, say, a brick, from a great height, it will be falling at a speed of 9.8067 m/s after 1 s, (9.8067 $\times$ 2) m/s after 2 s, (9.8067 $\times$ 3) m/s after 3 s, and so on. The speed becomes 9.8067 m/s greater with every second of time that passes. On Mars, this rate of increase would be less. On Jupiter, if Jupiter had a definable surface, it would be more. On the surface of a hugely dense object such as a neutron star, it would be many times more than it is on the surface of the Earth.

The rate of gravitational acceleration does not depend on the mass of the object being "pulled on" by gravity. You might think that heavier objects fall faster than slower ones. This is sometimes true in a practical sense if you drop, say, a Ping-Pong ball next to a golf ball. However, the reason the golf ball falls faster is that its greater density lets it overcome air resistance more effectively than the Ping-Pong ball can. If both were dropped in a vacuum, they would fall at the same speed. Astronomer and physicist Galileo Galilei is said to have proven this fact several centuries ago by dropping two heavy objects, one more massive than the other, from the Leaning Tower of Pisa in Italy. He let go of the objects at the same time, and they hit the ground at the same time. This upset people who believed that heavier objects fall faster than lighter ones. Galileo appeared to have shown that an ancient law of physics, which had become ingrained as an article of religious faith, was false. People had a word for folks of that sort: *heretic*. In those days, being branded as a heretic was like being accused of a felony.

Unit Conversions

With all the different systems of units in use throughout the world, the business of conversion from one system to another has become the subject matter for whole books. Web sites are devoted to this task; at the time of this writing, a good site could be found at Test and Measurement World (*www.tmworld.com*). Click on the "Software" link, and then go to the page called "Calculator Programs."

A SIMPLE TABLE

Table 6-3 shows conversions for quantities in base SI units to other common schemes. In this table and in any expression of quantity in any units, the coefficient is the number by which the power of 10 is multiplied.

This is by no means a complete table. It is amazing how many different units exist; for example, you might someday want to know how many bushels there are in a cubic kilometer! Some units seem to have been devised out of whimsy, as if the inventors knew the confusion and consternation their later use would cause.

DIMENSIONS

When converting from one unit system to another, always be sure you're talking about the same quantity or phenomenon. For example, you cannot convert meters squared to centimeters cubed or candela to meters per second. You must keep in mind what you're trying to express and be sure that you are not, in effect, trying to change an apple into an orange.

The particular thing that a unit quantifies is called the *dimension* of the quantity or phenomenon. Thus meters per second, feet per hour, and furlongs per fortnight represent expressions of the speed dimension; seconds, minutes, hours, and days are expressions of the time dimension. Units are always associated with dimensions. So are most constants, although there are a few constants that stand by themselves (π and e are two well-known examples).

> **PROBLEM 6-4**
> You step on a scale, and it tells you that you mass 63 kilograms. How many pounds does this represent?
>
> **SOLUTION 6-4**
> Assume that you are on the planet Earth, so your mass-to-weight conversion can be defined in a meaningful way. (Remember, mass is not the same thing

Table 6-3 Conversions for Base Units in the International System (SI) to Units in Other Systems (When no coefficient is given, it is exactly equal to 1.)

To convert:	To:	Multiply by:	Conversely, multiply by:
meters (m)	Angstroms	10^{10}	10^{-10}
meters (m)	nanometers (nm)	10^9	10^{-9}
meters (m)	microns (μ)	10^6	10^{-6}
meters (m)	millimeters (mm)	10^3	10^{-3}
meters (m)	centimeters (cm)	10^2	10^{-2}
meters (m)	inches (in)	39.37	0.02540
meters (m)	feet (ft)	3.281	0.3048
meters (m)	yards (yd)	1.094	0.9144
meters (m)	kilometers (km)	10^{-3}	10^3
meters (m)	statute miles (mi)	6.214×10^{-4}	1.609×10^3
meters (m)	nautical miles	5.397×10^{-4}	1.853×10^3
meters (m)	light-seconds	3.336×10^{-9}	2.998×10^8
meters (m)	astronomical units (AU)	6.685×10^{-12}	1.496×10^{11}
meters (m)	light-years	1.057×10^{-16}	9.461×10^{15}
meters (m)	parsecs (pc)	3.241×10^{-17}	3.085×10^{16}
kilograms (kg)	atomic mass units (amu)	6.022×10^{26}	1.661×10^{-27}
kilograms (kg)	nanograms (ng)	10^{12}	10^{-12}
kilograms (kg)	micrograms (μg)	10^9	10^{-9}
kilograms (kg)	milligrams (mg)	10^6	10^{-6}
kilograms (kg)	grams (g)	10^3	10^{-3}
kilograms (kg)	ounces (oz)	35.28	0.02834
kilograms (kg)	pounds (lb)	2.205	0.4535
kilograms (kg)	English tons	1.103×10^{-3}	907.0
seconds (s)	minutes (min)	0.01667	60.00
seconds (s)	hours (h)	2.778×10^{-4}	3.600×10^3
seconds (s)	days (dy)	1.157×10^{-5}	8.640×10^4
seconds (s)	years (yr)	3.169×10^{-8}	3.156×10^7
seconds (s)	centuries	3.169×10^{-10}	3.156×10^9
seconds (s)	millennia	3.169×10^{-11}	3.156×10^{10}
degrees Kelvin (K)	degrees Celsius (°C)	Subtract 273	Add 273
degrees Kelvin (K)	degrees Fahrenheit (°F)	Multiply by 1.80, then subtract 459	Multiply by 0.556, then add 255
degrees Kelvin (K)	degrees Rankine (°R)	1.80	0.556
amperes (A)	carriers per second	6.24×10^{18}	1.60×10^{-19}
amperes (A)	statamperes (statA)	2.998×10^9	3.336×10^{-10}

Table 6-3   Conversions for Base Units in the International System (SI) to Units in Other Systems (When no coefficient is given, it is exactly equal to 1.) (*Continued*)

To convert:	To:	Multiply by:	Conversely, multiply by:
amperes (A)	nanoamperes (nA)	10^9	10^{-9}
amperes (A)	microamperes (μA)	10^6	10^{-6}
amperes (A)	abamperes (abA)	0.10000	10.000
amperes (A)	milliamperes (mA)	10^3	10^{-3}
candela (cd)	microwatts per steradian (μW/sr)	1.464×10^3	6.831×10^{-4}
candela (cd)	milliwatts per steradian (mW/sr)	1.464	0.6831
candela (cd)	lumens per steradian (lum/sr)	Identical; no conversion	Identical; no conversion
candela (cd)	watts per steradian (W/sr)	1.464×10^{-3}	683.1
moles (mol)	coulombs (C)	9.65×10^4	1.04×10^{-5}

as weight.) Use Table 6-3. Multiply 63 by 2.205 to get 139 pounds. Because you are given your mass to only two significant figures, you must round this off to 140 pounds to be purely scientific.

PROBLEM 6-5
You are driving in Europe and you see that the posted speed limit is 90 kilometers per hour (km/h). How many miles per hour (mi/h) is this?

SOLUTION 6-5
In this case, you only need to worry about miles versus kilometers; the "per hour" part doesn't change. Thus you convert kilometers to miles. First remember that 1 km − 1,000 m; then 90 km = 90,000 m = 9.0×10^4 m. The conversion of meters to statute miles (these are the miles used on land) requires that you multiply by 6.214×10^{-4}. Therefore, you multiply 9.0×10^4 by 6.214×10^{-4} to get 55.926. This must be rounded off to 56, or two significant figures, because the posted speed limit quantity, 90, only goes that far.

PROBLEM 6-6
How many feet per second is the speed limit in Problem 6-5?

SOLUTION 6-6
This is a two-step problem. You're given the speed in kilometers per hour. You must convert kilometers to feet, and you also must convert hours to seconds. These two steps should be done separately. It does not matter in which order you do them, but you must do both conversions independently if you want to avoid getting confused. (Some of the Web-based calculator programs will do it all for you in a flash, but here, all we have is Table 6-3.)

Let's convert kilometers per hour to kilometers per second first. This requires division by 3,600, the number of seconds in an hour. Thus 90 km/h = 90/3600 km/s = 0.025 km/s. Now convert kilometers to meters; multiply by 1,000 to obtain 25 m/s as the posted speed limit. Finally, convert meters to feet; multiply 25 by 3.281 to get 82.025. This must be rounded off to 82 ft/s, again because the posted speed limit is expressed to only two significant figures.

Quiz

Refer to the text in this chapter if necessary. A good score is eight correct. Answers are in the back of the book.

1. The mole is a unit that expresses the
   (a) number of electrons in an ampere.
   (b) number of particles in a sample.
   (c) distance from the Sun to a planet.
   (d) time required for an electron to orbit an atomic nucleus.

2. A joule is the equivalent of a
   (a) foot-pound.
   (b) meter per second.
   (c) kilogram per meter.
   (d) watt-second.

3. A direct current of 3 A flows through a coil whose inductance is 1 H. The magnetic flux caused by this current is
   (a) 3 Wb.
   (b) 3 H.
   (c) 3 T.
   (d) impossible to determine from this information.

4. A light source generates the equivalent of 4.392 mW/sr of energy at the peak visible wavelength. This is approximately equal to
   (a) 6.4×10^{-6} cd.
   (b) 3.0 cd.
   (c) 6.4 cd.
   (d) 3.0×10^{-6} cd.

5. A temperature of 0 K represents
   (a) the freezing point of pure water at sea level.
   (b) the boiling point of pure water at sea level.
   (c) the absence of all heat.
   (d) nothing. This is a meaningless expression.

6. A newton is equivalent to a
   (a) kilogram-meter.
   (b) kilogram-meter per second.
   (c) kilogram-meter per second squared.
   (d) kilogram-meter per second cubed.

7. Kilograms can be converted to pounds only if you also know the
   (a) temperature.
   (b) mass of the object in question.
   (c) gravitational-field intensity.
   (d) material quantity.

8. The SI system is an expanded form of the
   (a) English system.
   (b) metric system.
   (c) European system.
   (d) American system.

9. The radian is a unit of
   (a) visible-light intensity.
   (b) temperature.
   (c) solid angular measure.
   (d) plane angular measure.

10. The pound is a unit of
    (a) mass.
    (b) substance.
    (c) material quantity.
    (d) none of the above.

CHAPTER 7

Mass, Force, and Motion

The earliest physicists were curious about the way matter behaves: What happens to pieces of it when they move or are acted on by forces. Scientists set about doing experiments and then tried to develop mathematical *models* (theories) to explain what happened and that would predict what would occur in future situations. This chapter involves *classical mechanics,* the study of mass, force, and motion.

Mass

The term *mass,* as used by physicists, refers to quantity of matter in terms of its ability to resist motion when acted on by a *force.* A good synonym for mass is *heft.* Every material object has a specific, definable mass. The Sun has a certain mass; Earth has a much smaller mass. A lead shot has a far smaller mass still. Even subatomic particles, such as protons and neutrons, have mass. Visible-light particles, known as *photons,* act in some ways as if they have mass. A ray of light puts pressure on any surface it strikes. The pressure is tiny, but it exists and sometimes can be measured.

MASS IS A SCALAR

The mass of an object or particle has magnitude (size or extent) but not direction. It can be represented as a certain number of kilograms, such as the mass of the Sun or the mass of the Earth. Mass is customarily denoted by the lowercase italicized letter m.

You might think that mass can have direction. When you stand somewhere, your body presses downward on the floor, the pavement, or the ground. If someone is more massive than you, his or her body presses downward too, but harder. If you get in a car and accelerate, your body presses backward in the seat as well as downward toward the center of the Earth. But this is force, not mass. The force you feel is caused in part by your mass and in part by gravity or acceleration. Mass itself has no direction. If you go into outer space and become weightless, you will have the same mass as you do on Earth (assuming that you do not lose or gain mass in between times). There won't be any force in any direction unless the space vessel begins to accelerate.

HOW MASS IS DETERMINED

The simplest way to determine the mass of an object is to measure it with a scale. However, this isn't the best way. When you put something on a scale, you are measuring that object's *weight* in the gravitational field of the Earth. The intensity of this field is, for most practical purposes, the same wherever you go on the planet. If you want to get picayune about it, though, there is a slight variation of weight for a given mass with changes in the geographic location. A scale with sufficient accuracy will show a specific object, such as a lead shot, as being a tiny bit heavier at the equator than at the north pole. The weight changes, but the mass does not.

Suppose that you are on an interplanetary journey, coasting along on your way to Mars or in orbit around Earth, and everything in your space vessel is weightless. How can you measure the mass of a lead shot under these conditions? It floats around in the cabin along with your body, the pencils you write with, and everything else that is not tied down. You are aware that the lead shot is more massive than, say, a pea, but how can you measure it to be certain?

One way to measure mass, independently of gravity, involves using a pair of springs set in a frame with the object placed in the middle (Fig. 7-1). If you put something between the springs and pull it to one side, the object oscillates. You try this with a pea, and the springs oscillate rapidly. You try it again with a lead shot, and the springs oscillate slowly. This "mass meter" is anchored to a desk in the space ship's cabin, which is in turn anchored to the "floor" (however you might define this in a weightless environment). Anchoring the scale keeps the whole apparatus from wagging back and forth in midair after you start the object oscillating.

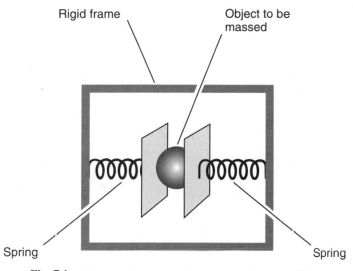

Rigid frame   Object to be
                     massed

Spring   Spring

Fig. 7.1.   Mass can be measured by setting an object to oscillate
between a pair of springs in a weightless environment.

A scale of this type must be calibrated in advance before it can render meaningful figures for masses of objects. The calibration will result in a graph that shows oscillation period or frequency as a function of the mass. Once this calibration is done in a weightless environment and the graph has been drawn, you can use it to measure the mass of anything within reason. The readings will be thrown off if you try to use the "mass meter" on Earth, the Moon, or Mars because there is an outside force, gravity, acting on the mass. The same problem will occur if you try to use the scale when the space ship is accelerating rather than merely coasting or orbiting through space.

PROBLEM 7-1
Suppose that you place an object in a "mass meter" similar to the one shown in Fig. 7-1. Also suppose that the mass-versus-frequency calibration curve for this device has been determined and looks like the graph of Fig. 7-2. The object oscillates with a frequency of 5 complete cycles per second (that is, 5 hertz or 5 Hz).What is the approximate mass of this object?

SOLUTION 7-1
Locate the frequency on the horizontal scale. Draw a vertical line (or place a ruler) parallel to the vertical (mass) axis. Note where this straight line intersects the curve. Draw a horizontal line from this point toward the left until it intersects the mass scale. Read the mass off the scale. It is approximately 0.8 kg, as shown in Fig. 7-3.

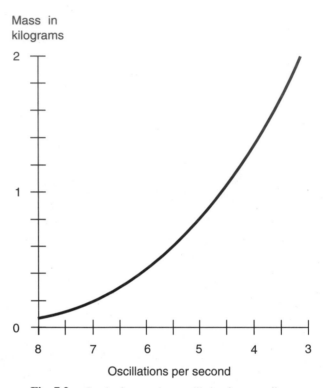

Mass in kilograms

Oscillations per second

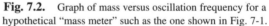

Fig. 7.2.   Graph of mass versus oscillation frequency for a hypothetical "mass meter" such as the one shown in Fig. 7-1.

PROBLEM 7-2
What will the "mass meter" shown in Fig. 7-1, and whose mass-versus-frequency function is graphed in Figs. 7-2 and 7-3, do if a mass of only 0.000001 kg (that is, 1 milligram or 1 mg) is placed in between the springs?

SOLUTION 7-2
The scale will oscillate at essentially the frequency corresponding to zero mass. This is off the graph scale in this example. You might be tempted at first to suppose that the oscillation frequency would be extremely high, but in fact, any practical "mass meter" will oscillate at a certain maximum frequency even with no mass placed in between the springs. This happens because the springs and the clamps themselves have mass.

PROBLEM 7-3
Wouldn't it be easier and more accurate in real life to program the mass-ver-sus-frequency function into a computer instead of using graphs like the ones shown here? In this way, we could simply input frequency data into the computer and read the mass on the computer display.

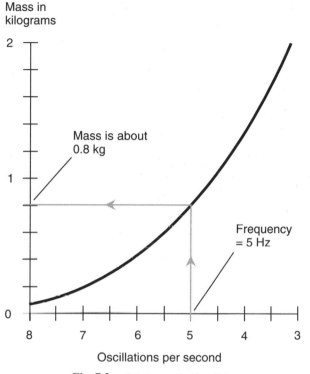

Mass in
kilograms

Mass is about
0.8 kg

Frequency
= 5 Hz

Oscillations per second

Fig. 7.3. Solution to Problem 7-1.

SOLUTION 7-3

Yes, such a method would be easier, and in a real-life situation, this is exactly what a physicist would do. In fact, we might expect the scale to have its own built-in microcomputer and a numerical display to tell us the mass directly.

Force

Imagine again that you are in a spacecraft orbiting the Earth, so everything in the cabin is weightless. Two objects float in front of you: a brick and a marble. You know that the brick is more massive than the marble. However, either the brick or the marble can be made to move across the cabin if you give it a push.

Suppose that you flick your finger against the marble. It flies across the cabin and bounces off the wall. Then you flick your finger just as hard (no more, no less) against the brick. The brick takes several minutes to float across the cabin and bump into the opposite wall. The flicking of your finger imparts a force to the marble or the brick for a moment, but that force has a different effect on the brick than on the marble.

FORCE AS A VECTOR

Force is a vector quantity. It can have any magnitude, from the flick of a finger to a swift leg kick, the explosion of powder in a cannon, or the thrust of a rocket engine. Force always has a defined direction as well. You can fire a pop-gun in any direction you want (and bear the consequences if you make a bad choice). Vectors are commonly symbolized using boldface letters of the alphabet. A force vector, for example, can be denoted by the uppercase boldface letter **F.**

Sometimes the direction of a force is not important. In such instances, we can speak of the magnitude of a force vector and denote it as an uppercase italicized letter *F.* The standard international unit of force magnitude is the *newton* (N), which is the equivalent of a kilogram-meter per second squared ($kg \cdot m/s^2$). Suppose that the brick in your spacecraft has a mass of 1 kg and that you push against it with a force of 1 N for 1 s and then let go. The brick will then be traveling at a speed of 1 m/s. It will have gone from stationary (with respect to its surroundings) to a speed of 1 m/s, which might seem rather slow unless it hits someone.

HOW FORCE IS DETERMINED

Force can be measured by the effect it has on an object with mass. It also can be measured by the amount of deflection or distortion it produces in an elastic object such as a spring. The "mass meter" described earlier for determining mass can be modified to make a "force meter" if one-half of it is taken away and a calibrated scale is placed alongside (Fig. 7-4). This scale must be calibrated in advance in a laboratory environment.

Displacement

Displacement is also known as *distance.* Unless otherwise specified, displacement is defined along a straight line. We might say that Minneapolis,

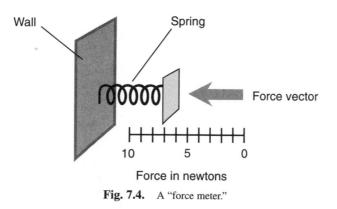

Fig. 7.4. A "force meter."

Minnesota, is 100 km from Rochester, Minnesota, "as the crow flies," or along a straight line. If you were to drive along U.S. Route 52, however, the displacement would turn out to be closer to 120 km because this highway does not follow a straight path from Rochester to Minneapolis.

DISPLACEMENT AS A VECTOR

When displacement is defined in a straight line, it is a vector quantity because it has both magnitude (expressed in meters, kilometers, or other distance units) and direction (which can be defined in various ways). Displacement magnitude is denoted by a lowercase italicized letter; let's call it q. A displacement vector is denoted by a lowercase boldface letter. In this discussion, let's use **q**.

The displacement vector $\mathbf{q}_{rm}$ of Minneapolis relative to Rochester would be approximately 100 km in a northwesterly direction "as the crow flies." As an azimuth bearing, it would be around 320 degrees, measured clockwise from true north. However, if we speak about driving along Route 52, we can no longer define the displacement as a vector because the direction changes as the road bends, goes over hills, and dips into valleys. In this case, we must denote displacement as a scalar, usually in lowercase italics. In this discussion let's use q. We write $q_{rm} \approx 120$ km.

HOW DISPLACEMENT IS DETERMINED

Displacement magnitude is determined by mechanically measuring distance or by inferring it with observations and mathematical calculations. In the case of a car or truck driving along Route 52, displacement is measured with an *odometer* that counts the number of wheel rotations and multiplies this by the circumference of the wheel. In a laboratory environment, displacement magnitude can be measured with a *meter stick*, by *triangulation*, or by measuring the time it takes for a ray of light to travel between two points given the constancy of the speed of light ($c \approx 2.99792 \times 10^8$ m/s).

The direction component of a displacement vector is determined by measuring one or more angles or coordinates relative to a reference axis. In the case of a local region on the Earth's surface, direction can be found by specifying the *azimuth,* which is the angle clockwise relative to true north. This is the scheme used by hikers and backpackers. In three-dimensional space, *direction angles* are used. A reference axis, for example, a vector pointing toward Polaris, the North Star, is defined. Then two angles are specified in a coordinate system based on this axis. The most common system used by astronomers and space scientists involves angles called *celestial latitude* and *celestial longitude* or, alternatively, *right ascension* and *declination.* Both these schemes are defined in Chapter 3. (If this is not familiar to you, and if you didn't see fit to study Part Zero, this might be a good time to reconsider that decision!)

Speed

Speed is an expression of the rate at which an object moves relative to some defined reference point of view. The reference frame is considered *stationary,* even though this is a relative term. A person standing still on the surface of the Earth considers himself or herself to be stationary, but this is not true with respect to the distant stars, the Sun, the Moon, or most other celestial objects.

SPEED IS A SCALAR

The standard unit of speed is the meter per second (m/s). A car driving along Route 52 might have a cruise control device that you can set at, say, 25 m/s. Then, assuming that the cruise control works properly, you will be traveling, relative to the pavement, at a constant speed of 25 m/s. This will be true whether you are on a level straightaway, rounding a curve, cresting a hill, or passing the bottom of a valley. Speed can be expressed as a simple number, and the direction is not important. Thus speed is a scalar quantity. In this discussion, let's symbolize speed by the lowercase italic letter *v*.

Speed can, of course, change with time. If you hit the brakes to avoid a deer crossing the road, your speed will decrease suddenly. As you pass the deer, relieved to see it bounding off into a field unharmed, you pick up speed again.

Speed can be considered as an average over time or as an instantaneous quantity. In the foregoing example, suppose that you are moving along at 25 m/s and then see the deer, put on the brakes, slow down to a minimum of 10 m/s, watch the deer run away, and then speed up to 25 m/s again, all in a time span of 1 minute. Your average speed over that minute might be 17 m/s. However, your instantaneous speed varies from instant to instant and is 17 m/s for only two instants (one as you slow down, the other as you speed back up).

HOW SPEED IS DETERMINED

In an automobile or truck, speed is determined by the same odometer that measures distance. However, instead of simply counting up the number of wheel rotations from a given starting point, a *speedometer* counts the number of wheel rotations in a given period of time. Knowing the wheel circumference, the number of wheel rotations in a certain time interval can be translated directly into meters per second.

You know, of course, that most speedometers respond almost immediately to a change in speed. These instruments measure the rotation rate of a car or truck axle by another method, similar to that used by the engine's *tachometer* (a device that measures revolutions per minute, or rpm). A real-life car or truck speedometer measures instantaneous speed, not average speed. In fact, if you want to know the average speed you have traveled during a certain period of time, you must measure the distance on the odometer and then divide by the time elapsed.

In a given period of time t, if an object travels over a displacement of magnitude q at an average speed v_{avg}, then the following formulas apply. These are all arrangements of the same relationship among the three quantities.

$$q = v_{avg}t$$

$$v_{avg} = q/t$$

$$t = q/v_{avg}$$

PROBLEM 7-4
Look at the graph of Fig. 7-5. Curve *A* is a straight line. What is the instantaneous speed v_{inst} at $t = 5$ seconds?

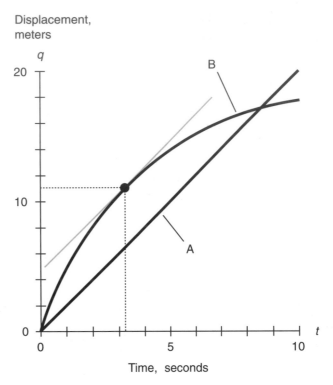

Fig. 7.5.   Illustration for Problems 7-4 through 7-8.

SOLUTION 7-4
The speed depicted by curve A is constant; you can tell because the curve is a straight line. The number of meters per second does not change throughout the time span shown. In 10 seconds, the object travels 20 meters; that's 20 m/10 s = 2 m/s. Therefore, the speed at $t = 5$ s is $v_{inst} = 2$ m/s.

PROBLEM 7-5
What is the average speed v_{avg} of the object denoted by curve A in Fig. 7-5 during the time span from $t = 3$ s to $t = 7$ s?

SOLUTION 7-5
Because the curve is a straight line, the speed is constant; we already know that it is 2 m/s. Therefore, $v_{avg} = 2$ m/s between any two points in time shown in the graph.

PROBLEM 7-6
Examine curve B in Fig. 7-5. What can be said about the instantaneous speed of the object whose motion is described by this curve?

SOLUTION 7-6
The object starts out moving relatively fast, and the instantaneous speed decreases with the passage of time.

PROBLEM 7-7

Use visual approximation in the graph of Fig. 7-5. At what time t is the instantaneous speed v_{inst} of the object described by curve B equal to 2 m/s?

SOLUTION 7-7

Take a ruler and find a straight line tangent to curve B whose slope is the same as that of curve A. That is, find the straight line parallel to line A that is tangent to curve B. Then locate the point on curve B where the line touches curve B. Finally, draw a line straight down, parallel to the displacement (q) axis, until it intersects the time (t) axis. Read the value off the t axis. In this example, it appears to be approximately $t = 3.2$ s.

PROBLEM 7-8

Use visual approximation in the graph of Fig. 7-5. Consider the object whose motion is described by curve B. At the point in time t where the instantaneous speed v_{inst} is 2 m/s, how far has the object traveled?

SOLUTION 7-8

Locate the same point that you found in Problem 7-7, corresponding to the tangent point of curve B and the line parallel to curve A. Draw a horizontal line to the left until it intersects the displacement (q) axis. Read the value off the q axis. In this example, it looks like it's about $q = 11$ m.

Velocity

Velocity consists of two independent components: *speed* and *direction*. The direction can be defined in one dimension (either way along a straight line), in two dimensions (within a plane), or in three dimensions (in space). Some physicists get involved with expressions of velocity in more than three spatial dimensions; that is a realm beyond the scope of this book.

VELOCITY IS A VECTOR

Because velocity has both magnitude and direction components, it is a vector quantity. You can't express velocity without defining both these components. In the earlier example of a car driving along a highway from one town to another, its speed might be constant, but the velocity changes nevertheless. If you're moving along at 25 m/s and then you come to a bend in the road, your velocity changes because your direction changes.

Vectors can be illustrated graphically as line segments with arrowheads. The speed component of a velocity vector is denoted by the length of the line segment, and the direction is denoted by the orientation of the arrow.

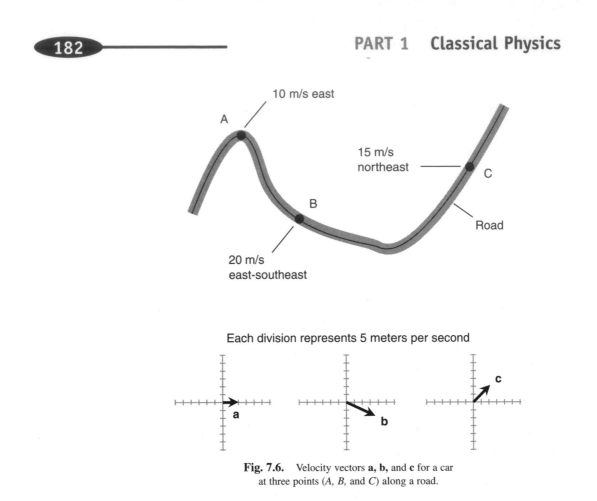

Fig. 7.6. Velocity vectors **a, b,** and **c** for a car
at three points (*A, B,* and *C*) along a road.

In Fig. 7-6, three velocity vectors are shown for a car traveling along a curving road. Three points are shown, called *A, B,* and *C.* The corresponding vectors are **a, b,** and **c.** Both the speed and the direction of the car change with time.

HOW VELOCITY IS DETERMINED

Velocity can be measured by using a speedometer in combination with some sort of device that indicates the instantaneous direction of travel. In a car, this might be a magnetic compass. In a strict sense, however, even a speedometer and a compass don't tell the whole story unless you're driving on a flat plain or prairie. In midstate South Dakota, a speedometer and compass can define the instantaneous velocity of your car, but when you get into the Black Hills, you'll have to include a *clinometer* (a device for measuring the steepness of the grade you're ascending or descending).

Two-dimensional direction components can be denoted either as compass (azimuth) bearings or as angles measured counterclockwise with respect to the axis pointing "east." The former system is preferred by hikers and navigators, whereas the latter scheme is preferred by theoretical physicists and mathematicians. In Fig. 7-6, the azimuth bearings of vectors **a, b,** and **c** are approximately 90, 120, and 45 degrees, respectively. In the mathematical model, they are about 0, −30 (or 330), and 45 degrees, respectively.

A three-dimensional velocity vector consists of a magnitude component and two direction angles. Celestial latitude and longitude or right ascension and declination are used commonly to denote the directions of velocity vectors.

Acceleration

Acceleration is an expression of the change in the velocity of an object. This can occur as a change in speed, a change in direction, or both. Acceleration can be defined in one dimension (along a straight line), in two dimensions (within a flat plane), or in three dimensions (in space), just as can velocity. Acceleration sometimes takes place in the same direction as an object's velocity vector, but this is not necessarily the case.

ACCELERATION IS A VECTOR

Acceleration is a vector quantity. Sometimes the magnitude of the acceleration vector is called *acceleration,* and is symbolized by lowercase italic letters such as *a.* Technically, however, the vector expression should be used; it is symbolized by lowercase boldface letters such as **a.**

In the example of a car driving along a highway, suppose that the speed is constant at 25 m/s (Fig. 7-7). The velocity changes when the car goes around curves and also if the car crests a hilltop or bottoms out in a ravine or valley (although these can't be shown in this two-dimensional drawing). If the car is going along a straightaway and its speed is increasing, then the acceleration vector points in the same direction that the car is traveling. If the car puts on the brakes, still moving along a straight path, then the acceleration vector points exactly opposite the direction of the car's motion.

Acceleration vectors can be illustrated graphically as arrows. In Fig. 7-7, three acceleration vectors are shown approximately for a car traveling

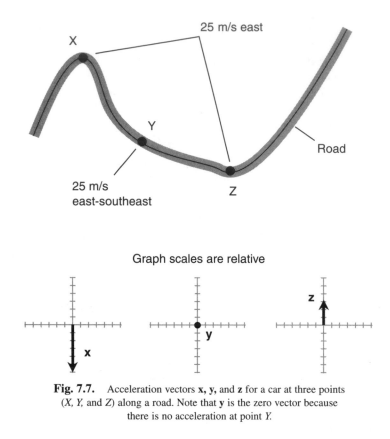

Fig. 7.7. Acceleration vectors **x, y,** and **z** for a car at three points (*X, Y,* and *Z*) along a road. Note that **y** is the zero vector because there is no acceleration at point *Y.*

along a curving road at a constant speed of 25 m/s. Three points are shown, called *X, Y,* and *Z.* The corresponding acceleration vectors are **x, y,** and **z.** Acceleration only takes place where the car is following a bend in the road. At point *Y,* the road is essentially straight, so the acceleration is zero. This is shown as a point at the origin of a vector graph.

HOW ACCELERATION IS DETERMINED

Acceleration magnitude is expressed in meters per second per second, also called *meters per second squared* (m/s^2). This might seem esoteric at first. What is a *second squared?* Think of it in terms of a concrete example. Suppose that you have a car that can go from 0 to 60 miles per hour (0.00 to 60.0 mi/h) in 5 seconds (5.00 s). A speed of 60.0 mi/h is roughly equivalent to 26.8 m/s. Suppose that the acceleration rate is constant from the moment you first hit the gas pedal until you have attained a speed of 26.8 m/s on a level straightaway. Then you can calculate the acceleration magnitude:

$$a = (26.8 \text{ m/s})/(5 \text{ s}) = 5.36 \text{ m/s}^2$$

Of course, the instantaneous acceleration will not be constant in a real-life test of a car's get-up-and-go power. However, the average acceleration magnitude will still be 5.36 m/s²—a speed increase of 5 meters per second with each passing second—assuming that the vehicle's speed goes from 0.00 to 60.0 mi/h in 5.00 s.

Instantaneous acceleration magnitude can be measured in terms of the force it exerts on a known mass. This can be determined according to the amount of distortion in an elastic object such as a spring. The "force meter" shown in Fig. 7-4 can be adapted to make an "acceleration meter," technically known as an *accelerometer*, for measuring acceleration magnitude. A fixed, known mass is placed in the device, and the deflection scale is calibrated in a laboratory environment. For the accelerometer to work, the direction of the acceleration vector must be in line with the spring axis, and the acceleration vector must point outward from the fixed anchor toward the mass. This will produce a force on the mass directly against the spring. A functional diagram of the basic arrangement is shown in Fig. 7-8.

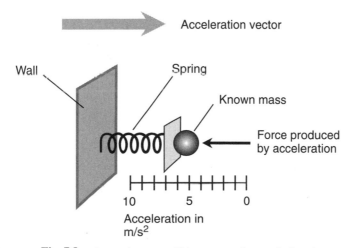

Fig. 7.8.   An accelerometer. This measures the magnitude only and must be oriented properly to provide an accurate reading.

A common spring scale can be used to measure acceleration indirectly. When you stand on the scale, you compress a spring or balance a set of masses on a lever. This measures the downward force that the mass of your body exerts as a result of a phenomenon called the *acceleration of gravity*. The effect of gravitation on a mass is the same as that of an upward acceleration of approximately 9.8 m/s^2. Force, mass, and acceleration are intimately related, as we shall soon see.

Suppose that an object starts from a dead stop and accelerates at an average magnitude of a_{avg} in a straight line for a period of time t. Suppose, after this length of time, that the magnitude of the displacement from the starting point is q. Then this formula applies:

$$q = a_{avg}\, t^2/2$$

In this example, suppose that the acceleration magnitude is constant; call it a. Let the instantaneous speed be called v_{inst} at time t. Then the instantaneous speed is related to the acceleration magnitude as follows:

$$v_{inst} = at$$

PROBLEM 7-9
Suppose that two objects, denoted by curves A and B in Fig. 7-9, accelerate along straight-line paths. What is the instantaneous acceleration a_{inst} at $t = 4$ seconds for object A?

SOLUTION 7-9
The acceleration depicted by curve A is constant because the speed increases at a constant rate with time. (This is why the graph is a straight line.) The number of meters per second squared does not change throughout the time span shown. In 10 seconds, the object accelerates from 0 to 10 m/s; this is a rate of speed increase of 1 meter per second per second (1 m/s^2). Therefore, the acceleration at $t = 4$ s is $a_{inst} = 1$ m/s^2.

PROBLEM 7-10
What is the average acceleration a_{avg} of the object denoted by curve A in Fig. 7-9 during the time span from $t = 2$ s to $t = 8$ s?

SOLUTION 7-10
Because the curve is a straight line, the acceleration is constant; we already know that it is 1 m/s^2. Therefore, $a_{avg} = 1$ m/s^2 between any two points in time shown in the graph.

PROBLEM 7-11
Examine curve B in Fig. 7-9. What can be said about the instantaneous acceleration of the object whose motion is described by this curve?

Speed,
meters per second

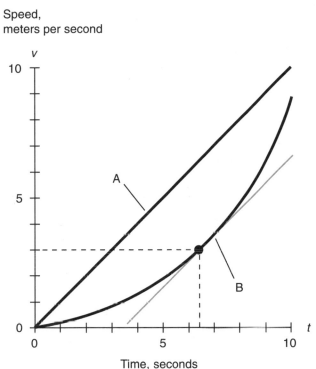

Fig. 7.9.   Illustration for Problems 7-9 through 7-13.

SOLUTION 7-11
The object starts out accelerating slowly, and as time passes, its instantaneous rate of acceleration increases.

PROBLEM 7-12
Use visual approximation in the graph of Fig. 7-9. At what time t is the instantaneous acceleration a_{inst} of the object described by curve B equal to 1 m/s²?

SOLUTION 7-12
Take a ruler and find a straight line tangent to curve B whose slope is the same as that of curve A. Then locate the point on curve B where the line touches curve B. Finally, draw a line straight down, parallel to the speed (v) axis, until it intersects the time (t) axis. Read the value off the t axis. Here it appears to be about $t = 6.3$ s.

PROBLEM 7-13
Use visual approximation in the graph of Fig. 7-9. Consider the object whose motion is described by curve B. At the point in time t where the instantaneous acceleration a_{inst} is 1 m/s, what is the instantaneous speed v_{inst} of the object?

SOLUTION 7-13
Locate the same point that you found in Problem 7-12 corresponding to the tangent point of curve *B* and the line parallel to curve *A*. Draw a horizontal line to the left until it intersects the speed (*v*) axis. Read the value off the *v* axis. In this example, it looks like it's about $v_{inst} = 3.0$ m/s.

Newton's Laws of Motion

Three laws, credited to the physicist, astronomer, and mathematician Isaac Newton, apply to the motions of objects in classical physics. These laws do not take into account the relativistic effects that become significant when speeds approach the speed of light or when extreme gravitational fields exist.

NEWTON'S FIRST LAW

This law is twofold: (1) Unless acted on by an outside force, an object at rest, stays at rest; and (2) unless acted on by an outside force, an object moving with uniform velocity continues to move at that velocity.

NEWTON'S SECOND LAW

If an object of mass *m* (in kilograms) is acted on by a force of magnitude *F* (in newtons), then the magnitude of the acceleration *a* (in meters per second squared) can be found according to the following formula:

$$a = F/m$$

The more familiar version of this formula is

$$F = ma$$

When force and acceleration are defined as vector quantities, the formula becomes

$$\mathbf{F} = \mathbf{ma}$$

NEWTON'S THIRD LAW

Every action is attended by an equal and opposite reaction. In other words, if an object *A* exerts a force vector **F** on an object *B*, then object *B* exerts a force vector $-\mathbf{F}$ (the negative of **F**) on object *A*.

PROBLEM 7-14

A spacecraft of mass $m = 10{,}500$ (1.0500×10^4) kg in interplanetary space is acted on by a force vector $\mathbf{F} = 100{,}000$ (1.0000×10^5) N in the direction of Polaris, the North Star. Determine the magnitude and direction of the acceleration vector.

SOLUTION 7-14

Use the first formula stated earlier in Newton's second law. Plugging in the numbers for force magnitude F and mass m yields the acceleration magnitude a:

$$a = F/m$$
$$= 1.0000 \times 10^5 / 1.0500 \times 10^4$$
$$= 9.5238 \text{ m/s}^2$$

The direction of the acceleration vector $\mathbf{a}$ is the same as the direction of the force vector $\mathbf{F}$ in this case, that is, toward the North Star. As an interesting aside, you might notice that this acceleration is just a little less than the acceleration of gravity at the Earth's surface, 9.8 m/s^2. Therefore, a person inside this spacecraft would feel quite at home; there would be an artificial gravitational field produced that would be just about the same strength as the gravity on Earth.

PROBLEM 7-15

According to Newton's first law, shouldn't the Moon fly off in a straight line into interstellar space? Why does it orbit the Earth?

SOLUTION 7-15

The Moon is acted on constantly by a force vector that tries to pull it down to Earth. This force is exactly counterbalanced by the inertia of the Moon, which tries to get it to fly away in a straight line. The *speed* of the Moon around the Earth is nearly constant, but its *velocity* is always changing because of the force imposed by the gravitational attraction between the Moon and the Earth.

 **Quiz**

Refer to the text in this chapter if necessary. A good score is eight correct. Answers are in the back of the book.

1. One newton is equivalent to
   (a) one kilogram meter.
   (b) one kilogram meter per second.
   (c) one kilogram meter per second squared.
   (d) one meter per second squared.

2. A mass of 19 kg moves at a constant speed of 1.0 m/s relative to an observer. From the point of view of the observer, what is the magnitude of the force vector on that mass? Assume the mass travels in a straight line.
   (a) 19 newtons.
   (b) 0.053 newtons.
   (c) 1 newton.
   (d) 0 newtons.

3. A velocity vector has components of
   (a) magnitude and direction.
   (b) speed and mass.
   (c) time and mass.
   (a) speed and time.

4. The gravitational acceleration of the Earth, near the surface, is 9.81 m/s^2. If a brick of mass 3.00 kg is dropped from a great height, how far will the brick fall in 2.00 s?
   (a) 6.00 m.
   (b) 29.4 m.
   (c) 19.6 m.
   (d) 58.8 m.

5. Suppose that the mass of the brick in quiz item 4 has a mass of only 1.00 kg. How far will this brick fall in 2.00 s?
   (a) 2.00 m.
   (b) 19.6 m.
   (c) 29.4 m.
   (d) 9.80 m.

6. What would happen if Earth's gravitational pull on the Moon suddenly stopped?
   (a) Nothing.
   (b) The Moon would fly out of Earth orbit.
   (c) The Moon would fall into the Earth.
   (d) The Moon would fall into the Sun.

7. A mass vector consists of
   (a) weight and direction.
   (b) weight and speed.
   (c) weight and time.
   (d) There is no such thing; mass is not a vector.

8. You are docking a small boat. As you approach the dock, you leap from the boat. You fall short of the dock and land in the water because the boat was thrust backward when you jumped forward. This is a manifestation of
   (a) Newton's first law.
   (b) Newton's second law.
   (c) Newton's third law.
   (d) the fact that weight is not the same thing as mass.

9. In three-dimensional space, the direction of a vector might be described in
terms of
(a) right ascension and declination.
(b) distance and speed.
(c) time and distance.
(d) its length.

10. While driving on a level straightaway, you hit the brakes. The acceleration vector
(a) points in the direction the car is traveling.
(b) points opposite to the direction the car is traveling.
(c) points at a right angle to the direction the car is traveling.
(d) does not exist; it is zero.

CHAPTER 8

Momentum, Work, Energy, and Power

Classical mechanics describes the behavior of objects in motion. Any moving mass has *momentum* and *energy*. When two objects collide, the momentum and energy contained in each object changes. In this chapter we will continue our study of basic newtonian physics.

Momentum

Momentum is the product of an object's mass and its velocity. The standard unit of mass is the kilogram (kg), and the standard unit of speed is the meter per second (m/s). Momentum magnitude is expressed in kilogram-meters per second (kg · m/s). If the mass of an object moving at a certain speed increases by a factor of 5, then the momentum increases by a factor of 5, assuming that the speed remains constant. If the speed increases by a factor of 5 but the mass remains constant, then again, the momentum increases by a factor of 5.

MOMENTUM AS A VECTOR

Suppose that the speed of an object (in meters per second) is v, and that the mass of the object (in kilograms) is m. Then the magnitude of the momentum p is their product:

$$p = mv$$

This is not the whole story. To fully describe momentum, the direction as well as the magnitude must be defined. This means that we must consider the velocity of the mass in terms of its speed and direction. (A 2-kg brick flying through your east window is not the same as a 2-kg brick flying through your north window.) If we let **v** represent the velocity vector and **p** represent the momentum vector, then we can say

$$\mathbf{p} = m\mathbf{v}$$

IMPULSE

The momentum of a moving object can change in any of three different ways:

- A change in the mass of the object
- A change in the speed of the object
- A change in the direction of the object's motion

Let's consider the second and third of these possibilities together; then this constitutes a change in the velocity.

Imagine a mass, such as a space ship, coasting along a straight-line path in interstellar space. Consider a point of view, or *reference frame,* such that the velocity of the ship can be expressed as a nonzero vector pointing in a certain direction. A force **F** may be applied to this vessel by firing a rocket engine. Imagine that there are several engines on this space ship, one intended for driving the vessel forward at increased speed and others capable of changing the vessel's direction. If any engine is fired for t seconds with force vector of **F** newtons (as shown by the three examples in Fig. 8-1), then the product **F**t is called the *impulse.* Impulse is a vector, symbolized by the uppercase boldface letter **I,** and is expressed in kilogram meters per second (kg · m/s):

$$\mathbf{I} = \mathbf{F}t$$

Impulse produces a change in velocity. This is clear enough; this is the purpose of rocket engines in a space ship! Recall the formula from Chapter 7 concerning mass $m,$ force **F,** and acceleration **a:**

$$\mathbf{F} = m\mathbf{a}$$

Substitute $m\mathbf{a}$ for **F** in the equation for impulse. Then we get this:

$$\mathbf{I} = (m\mathbf{a})t$$

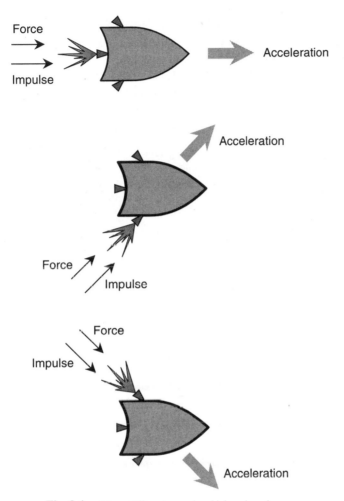

Fig. 8-1.   Three different ways in which an impulse can cause an object (in this case, a space ship) to accelerate.

Now remember that acceleration is a change in velocity per unit time. Suppose that the velocity of the space ship is v_1 before the rocket is fired and v_2 afterwards. Then, assuming that the rocket engine produces a constant force while it is fired,

$$\mathbf{a} = (\mathbf{v}_2 - \mathbf{v}_1)/t$$

We can substitute in the preceding equation to get

$$\mathbf{I} = m[(\mathbf{v}_2 - \mathbf{v}_1)/t]t = m\mathbf{v}_2 - m\mathbf{v}_1$$

This means the impulse is equal to the change in the momentum.

We have just derived an important law of newtonian physics. Reduced to base units in SI, impulse is expressed in kilogram-meters per second (kg · m/s), just as is momentum. You might think of impulse as momentum in another form. When an object is subjected to an impulse, the object's momentum vector **p** changes. The vector **p** can grow larger or smaller in magnitude, it can change direction, or both these things can happen.

PROBLEM 8-1

Suppose that an object of mass 2.0 kg moves at a constant speed of 50 m/s in a northerly direction. An impulse, acting in a southerly direction, slows this mass down to 25 m/s, but it still moves in a northerly direction. What is the impulse responsible for this change in momentum?

SOLUTION 8-1

The original momentum $\mathbf{p}_1$ is the product of the mass and the initial velocity:

$$\mathbf{p}_1 = 2.0 \text{ kg} \times 50 \text{ m/s} = 100 \text{ kg} \cdot \text{m/s}$$

in a northerly direction. The final momentum $\mathbf{p}_2$ is the product of the mass and the final velocity:

$$\mathbf{p}_2 = 2.0 \text{ kg} \times 25 \text{ m/s} = 50 \text{ kg} \cdot \text{m/s}$$

in a northerly direction. Thus, the change in momentum is $\mathbf{p}_2 - \mathbf{p}_1$:

$$\mathbf{p}_2 - \mathbf{p}_1 = 50 \text{ kg} \cdot \text{m/s} - 100 \text{ kg} \cdot \text{m/s} = -50 \text{ kg} \cdot \text{m/s}$$

in a northerly direction. This is the same as 50 kg · m/s in a southerly direction. Because impulse is the same thing as the change in momentum, the impulse is 50 kg · m/s in a southerly direction.

Don't let this result confuse you. A vector with a magnitude $-x$ in a certain direction is the same as a vector with magnitude x in the exact opposite direction. Problems sometimes will work out to yield vectors with negative magnitude. When this happens, just reverse the direction and then take the absolute value of the magnitude.

Collisions

When two objects strike each other because they are in relative motion and their paths cross at exactly the right time, a *collision* is said to occur.

CONSERVATION OF MOMENTUM

According to the *law of conservation of momentum*, the total momentum contained in two objects is the same after a collision as before. The characteristics of the collision do not matter as long as it is an *ideal system*. In an ideal system, there is no friction or other real-world imperfection, and the total system momentum never changes unless a new mass or force is introduced.

The law of conservation of momentum applies not only to systems having two objects or particles but also to systems having any number of objects or particles. However, the law holds only in a *closed system,* that is, a system in which the total mass remains constant, and no forces are introduced from the outside.

This is a good time to make an important announcement. From now on in this book, if specific units are not given for quantities, assume that the units are intended to be expressed in the International System (SI). Therefore, in the following examples, masses are in kilograms, velocity magnitudes are in meters per second, and momentums are in kilogram-meters per second. Get into the habit of making this assumption, whether the units end up being important in the discussion or not. Of course, if other units are specified, then use those. But beware when making calculations. Units always must agree throughout a calculation, or you run the risk of getting a nonsensical or inaccurate result.

STICKY OBJECTS

Look at Fig. 8-2. The two objects have masses m_1 and m_2, and they are moving at speeds v_1 and v_2, respectively. The velocity vectors $\mathbf{v}_1$ and $\mathbf{v}_2$ are not specifically shown here, but they point in the directions shown by the arrows. At A in this illustration, the two objects are on a collision course. The momentum of the object with mass m_1 is equal to $\mathbf{p}_1 = m_1\mathbf{v}_1$; the momentum of the object with mass m_2 is equal to $\mathbf{p}_2 = m_2\mathbf{v}_2$.

At B, the objects have just hit each other and stuck together. After the collision, the composite object cruises along at a new velocity $\mathbf{v}$ that is different from either of the initial velocities. The new momentum, call it $\mathbf{p}$, is equal to the sum of the original momentums. Therefore:

$$\mathbf{p} = m_1\mathbf{v}_1 + m_2\mathbf{v}_2$$

The final velocity $\mathbf{v}$ can be determined by noting that the final mass is $m_1 + m_2$. Therefore:

$$\mathbf{p} = (m_1 + m_2)\mathbf{v}$$
$$\mathbf{v} = \mathbf{p}/(m_1 + m_2)$$

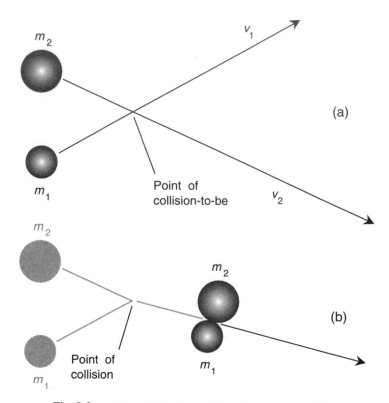

Fig. 8-2. (*a*) Two sticky objects, both with constant but different velocities, approach each other. (*b*) The objects after the collision.

BOUNCY OBJECTS

Now examine Fig. 8-3. The two objects have masses m_1 and m_2, and they are moving at speeds v_1 and v_2, respectively. The velocity vectors $\mathbf{v}_1$ and $\mathbf{v}_2$ are not specifically shown here, but they point in the directions shown by the arrows. At *A* in this illustration, the two objects are on a collision course. The momentum of the object with mass m_1 is equal to $\mathbf{p}_1 = m_1\mathbf{v}_1$; the momentum of the object with mass m_2 is equal to $\mathbf{p}_2 = m_2\mathbf{v}_2$. Thus far the situation is just the same as that in Fig. 8-2. But here the objects are made of different stuff. They bounce off of each other when they collide.

At *B,* the objects have just hit each other and bounced. Of course, their masses have not changed, but their velocities have, so their individual momentums have changed. However, the total momentum of the system has not changed, according to the law of conservation of momentum.

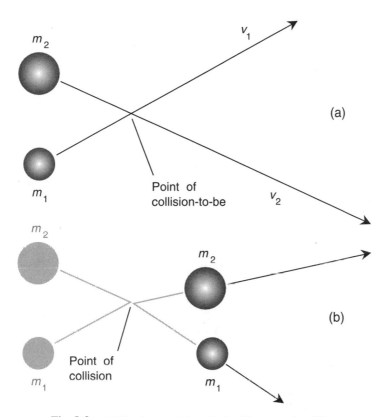

m_2

v_1

(a)

m_1

Point of
collision-to-be

v_2

m_2

m_2

(b)

Point of
collision

m_1

m_1

Fig. 8-3.   (a) Two bouncy objects, both with constant but different
velocities, approach each other. (b) The objects after the collision.

Suppose that the new velocity of m_1 is $\mathbf{v}_{1a}$ and that the new velocity of m_2
is $\mathbf{v}_{2a}$. The new momentums of the objects are therefore

$$\mathbf{P}_{1a} = m_1 \mathbf{v}_{1a}$$

$$\mathbf{P}_{2a} = m_2 \mathbf{v}_{2a}$$

According to the law of conservation of momentum,

$$\mathbf{P}_1 + \mathbf{P}_2 = \mathbf{P}_{1a} + \mathbf{P}_{2a}$$

and therefore:

$$m_1\mathbf{v}_1 + m_2\mathbf{v}_2 = m_1\mathbf{v}_{1a} + m_2\mathbf{v}_{2a}$$

The examples shown in Figs. 8-2 and 8-3 represent idealized situations.
In the real world, there would be complications that we are ignoring here

for the sake of demonstrating basic principles. For example, you might already be wondering whether or not the collisions shown in these drawings would impart spin to the composite mass (in Fig. 8-2) or to either or both masses (in Fig. 8-3). In the real world, this would usually happen, and it would make our calculations vastly more complicated. In these idealized examples, we assume that no spin is produced by the collisions.

PROBLEM 8-2

Suppose that you have two toy electric trains set up on a long, straight track running east and west. Train A has a mass 1.60 kg and travels east at 0.250 m/s. Train B has a mass of 2.50 kg and travels west at 0.500 m/s. The trains have stick pads on the fronts of their engines so that if they crash, they will not bounce off each other. The trains are set up so that they will crash. Suppose that the friction in the wheel bearings is zero, and suppose that the instant the trains hit each other, you shut off the power to the engines. How fast and in what direction will the composite train be moving after the crash? Assume that neither train derails.

SOLUTION 8-2

First, calculate the momentum of each train. Call the masses of the trains m_a and m_b, respectively. Let us assign the directions east as positive and west as negative. (We can do this because they're exactly opposite each other along a straight line.) Let the velocity vector of train A be represented as $\mathbf{v}_a$ and the velocity vector of train B be represented as $\mathbf{v}_b$. Then $m_a = 1.60$ kg, $m_b = 2.50$ kg, $\mathbf{v}_a = +0.250$ m/s, and $\mathbf{v}_b = -0.500$ m/s. Their momentums, respectively, are

$$\mathbf{p}_a = m_a\mathbf{v}_a = (1.60 \text{ kg})(+ 0.250 \text{ m/s}) = + 0.400 \text{ kg} \cdot \text{m/s}$$

$$\mathbf{p}_b = m_b\mathbf{v}_b = (2.50 \text{ kg})(-0.500 \text{ m/s}) = -1.25 \text{ kg} \cdot \text{m/s}$$

The sum total of their momentums is therefore

$$\mathbf{p} = \mathbf{p}_a + \mathbf{p}_b = + 0.400 \text{ kg} \cdot \text{m/s} + (-1.25 \text{ kg} \cdot \text{m/s})$$

$$= -0.850 \text{ kg} \cdot \text{m/s}$$

The mass m of the composite is simply the sum of the masses of trains A and B, which remain the same throughout this violent process:

$$m = m_a + m_b = 1.60 \text{ kg} + 2.50 \text{ kg} = 4.10 \text{ kg}$$

Let the final velocity, for which we are trying to solve, be denoted **v**. We know that momentum is conserved in this collision, as it is in all ideal collisions. Therefore, the final velocity must be equal to the final momentum **p** divided by the final mass m:

$$\mathbf{v} = \mathbf{p}/m = (-0.850 \text{ kg} \cdot \text{m/s})/(4.10 \text{ kg})$$

$$\approx -0.207 \text{ m/s}$$

This means the composite "train," after the crash, will move west at 0.207 m/s.

TWO MORE NOTEWORTHY ITEMS

Two things ought to be mentioned before we continue on. First, until now, units have been included throughout calculations for illustrative purposes. Units can be multiplied and divided, just as can numbers. For example, 0.850 kg · m/s divided by 4.10 kg causes kilograms to cancel out in the final result, yielding meters per second (m/s). It's a good idea to keep the units in calculations, at least until you get comfortable with them, so that you can be sure that the units in the final result make sense. If we had come up with, say, kilogram-meters (kg · m) in the final result for Problem 8-2, we would know that something was wrong because kilogram-meters are not units of speed or velocity magnitude.

The second thing you should know is that it's perfectly all right to multiply and divide vector quantities, such as velocity or momentum, by scalar quantities, such as mass. This always yields another vector. For example, in the solution of Problem 8-2, we divided momentum (a vector) by mass (a scalar). However, we cannot add a vector to a scalar so easily or subtract a scalar from a vector. Nor can we multiply two vectors and expect to get a meaningful answer unless we define whether we are to use the dot product or the cross product. You should be familiar with this from your high school mathematics courses. If not, go to Part Zero of this book and review the material on vectors. It can be found in the latter part of Chapter 1.

PROBLEM 8-3

Suppose that you have two toy electric trains set up as in Problem 8-2. Train *A* has a mass 2.00 kg and travels east at 0.250 m/s. Train *B* has a mass of 1.00 kg and travels west at 0.500 m/s. How fast and in what direction will the composite train be moving after the crash? Assume that neither train derails.

SOLUTION 8-3

Call the masses of the trains m_a and m_b, respectively. Assign the directions east as positive and west as negative. Let the velocity vectors be represented as $\mathbf{v}_a$ and $\mathbf{v}_b$. Then $m_a = 2.00$ kg, $m_b = 1.00$ kg, $\mathbf{v}_a = +0.250$ m/s, and $\mathbf{v}_b = -0.500$ m/s. Their momentums, respectively, are

$$\mathbf{p}_a = m_a\mathbf{v}_a = (2.00 \text{ kg})(+ 0.250 \text{ m/s}) = + 0.500 \text{ kg} \cdot \text{m/s}$$

$$\mathbf{p}_b = m_b\mathbf{v}_b = (1.00 \text{ kg})(-0.500 \text{ m/s}) = -0.500 \text{ kg} \cdot \text{m/s}$$

The sum total of their momentums is therefore

$$\mathbf{p} = \mathbf{p}_a + \mathbf{p}_b = + 0.500 \text{ kg} \cdot \text{m/s} + (-0.500 \text{ kg} \cdot \text{m/s})$$

$$= 0 \text{ kg} \cdot \text{m/s}$$

The mass m of the composite is simply the sum of the masses of trains A and B, neither of which change:

$$m = m_a + m_b = 2.00 \text{ kg} + 1.00 \text{ kg} = 3.00 \text{ kg}$$

The final velocity $\mathbf{v}$ is equal to the final momentum $\mathbf{p}$ divided by the final mass m:

$$\mathbf{v} = \mathbf{p}/m = (0 \text{ kg} \cdot \text{m/s})/(3.00 \text{ kg})$$

$$= 0 \text{ m/s}$$

This means that the composite "train," after the crash, is at rest. This might at first seem impossible. If momentum is conserved, how can it be zero after the crash? Where does it all go? The answer to this question is that the total momentum of this system is zero before the crash as well as after. Remember that momentum is a vector quantity. Look at the preceding equations again:

$$\mathbf{p}_a = m_a \mathbf{v}_a = (2.00 \text{ kg})(+0.250 \text{ m/s}) = +0.500 \text{ kg} \cdot \text{m/s}$$

$$\mathbf{p}_b = m_b \mathbf{v}_b = (1.00 \text{ kg})(-0.500 \text{ m/s}) = -0.500 \text{ kg} \cdot \text{m/s}$$

The momentums of the trains have equal magnitude but opposite direction. Thus their vector sum is zero before the crash.

Work

In physics, *work* refers to a specific force applied over a specific distance. The most common examples are provided by lifting objects having significant mass ("weights" or "masses") directly against the force of gravity. The amount of work w done by the application over a displacement q of a force whose magnitude is F is given by

$$w = Fq$$

The standard unit of work is the newton-meter (N $\cdot$ m), equivalent to a kilogram-meter squared per second squared (kg $\cdot$ m^2/s^2).

WORK AS A DOT PRODUCT OF VECTORS

The preceding formula is not quite complete because, as you should know by now, both force and displacement are vector quantities. How can we multiply two vectors? Fortunately, in this case it is easy because the force and displacement vectors generally point in the same direction when work is done. It turns out that the dot product provides the answer we need. Work is a scalar, therefore, and is equivalent to

$$w = \mathbf{F} \cdot \mathbf{q}$$

where $\mathbf{F}$ is the force vector, represented as newtons in a certain direction, and $\mathbf{q}$ is the displacement vector, represented as meters in a certain direction. The directions of $\mathbf{F}$ and $\mathbf{q}$ are almost always the same. Note the dot symbol here ($\cdot$), which is a heavy dot so that the dot product of vectors can be distinguished from the ordinary scalar product of variables, units, or numbers, as in $kg \cdot m^2/s^2$.

As long as the force and displacement vectors point in the same direction, we can simply multiply their magnitudes and get a correct result for work done. Just remember that work is a scalar, not a vector.

LIFTING AN OBJECT

Imagine a 1.0-kg object lifted upward against the Earth's gravity. The easiest way to picture this is with a rope-and-pulley system. (Suppose that the pulley is frictionless and that the rope doesn't stretch.) You stand on the floor, holding the rope, and pull downward. You must exert a certain force over a certain distance. The force and displacement vectors through which your hands move point in the same direction. You can wag your arms back and forth while you pull, but in practice this won't make any difference in the amount of work required to lift the object a certain distance, so let's keep things simple and suppose that you pull in a straight line.

The force of your pulling downward is translated to an equal force vector $\mathbf{F}$ upward on the object (Fig. 8-4 on p. 206). The object moves upward as far as you pull the rope, that is, by a distance $\mathbf{q}$. What is the force with which you pull? It is the force required to exactly counteract the force of gravitation on the mass. The force of gravity $\mathbf{F}_g$ on the object is the product of the object's mass m and the acceleration vector $\mathbf{a}_g$ of gravity. The value of $\mathbf{a}_g$ is approximately 9.8 m/s^2 directly downward. To lift the object, you must exert a force $\mathbf{F} = m\mathbf{a}_g = (9.8 \text{ m/s}^2)(1.0 \text{ kg}) = 9.8 \text{ kg} \cdot \text{m/s}^2 = 9.8 \text{ N}$ directly upward.

PROBLEM 8-4
Consider the example described earlier and in Fig. 8-4. Suppose that you lift the object 1.5 m. How much work have you done?

SOLUTION 8-4

The force **F**, applied straight up, to move the object is equal to the product of the mass m = 1.0 kg and the acceleration of gravity $\mathbf{a}_g$ = 9.8 m/s^2 directed straight up:

$$\mathbf{F} = m\mathbf{a}_g = (1.0 \text{ kg})(9.8 \text{ m/s}^2) = 9.8 \text{ kg} \cdot \text{m/s}^2$$

This force is to be applied over a distance **q** = 1.5 m straight up, so the work w is equal to the dot product **F** $\cdot$ **q**. Because **F** and **q** point in the same direction, we can simply multiply their magnitudes:

$$w = Fq = (9.8 \text{ kg} \cdot \text{m/s}^2)\,(1.5 \text{ m})$$

$$= 14.7 \text{ kg} \cdot \text{m}^2/\text{s}^2$$

We should round this off to 15 kg $\cdot$ m^2/s^2 because our input data are given only to two significant digits.

This unit, the kilogram-meter squared per second squared, seems arcane, doesn't it? Thinking of it as a newton-meter might help a little. Fortunately, however, there is another name for this unit, the *joule,* symbolized J. In lifting the object in the example of Problem 8-4, you did approximately 15 J of work. The joule is a significant unit in physics, chemistry, electricity, and electronics. It will turn up again and again if you do much study in any of these fields.

Energy

Energy exists in many forms. From time to time, we hear news about an "energy crisis." Usually newscasters are talking about shortages of the energy available from burning fossil fuels, such as oil and natural gas. You might find a barrel of oil and sit down in front of it. Where is the energy in it? It doesn't seem to be doing anything; it's just a big container of dark, thick liquid. However, if you light it on fire (don't!), the energy it contains becomes vividly apparent. Energy is measured in joules, just as is work. In fact, one definition of energy is "the capacity to do work."

POTENTIAL ENERGY

Look again at the situation shown by Fig. 8-4. When the object with mass m is raised through a displacement **q**, a force **F** is applied to it. Imagine what would happen if you let go of the rope and the object were allowed to fall.

Suppose that $m = 5$ kg. This is about 11 pounds in Earth's gravitational field. Suppose that the object is hard and solid, such as a brick. If you raise the brick a couple of millimeters, it will strike the floor without much fanfare. If you raise it 2 m, it will crack or dent a linoleum floor, and the brick itself might break apart. If you raise it 4 m, there will certainly be trouble when it hits. The landing of a heavy object can be put to some useful task, such as pounding a stake into the ground. It also can do a lot of damage.

There is something about lifting up an object that gives it the ability to do work. This "something" is *potential energy*. Potential energy is the same thing as work, in a mechanical sense. If a force vector of magnitude F is applied to an object against Earth's gravitation and that object is lifted by a displacement vector of magnitude q, then the potential energy E_p is given by this formula:

$$E_p = Fq$$

This is a simplistic view of potential energy. As we just discussed, potential energy can exist in a barrel of oil even if it is not lifted. Potential energy also exists in electrochemical cells, such as the battery in your car. It exists in gasoline, natural gas, and rocket fuel. It is not as easy to quantify in those forms as it is in the mechanical example of Fig. 8-4, but it exists nevertheless.

PROBLEM 8-5

Refer again to Fig. 8-4. If the object has a mass of 5.004 kg and it is lifted 3.000 m, how much potential energy will it attain? Take the value of the magnitude of Earth's gravitational acceleration as $a_g - 9.8067$ m/s^2. We can neglect vectors here because everything takes place along a single straight line.

SOLUTION 8-5

First, we must determine the force required to lift a 5.004-kg object in Earth's gravitational field:

$$F = ma_g = (5.004 \text{ kg}) (9.8067 \text{ m/s}^2) = 49.0727268 \text{ N}$$

The potential energy is the product of this force and the displacement:

$$Ep = Fq = (49.0727268 \text{ N}) (3.000 \text{ m}) = 147.2181804 \text{ J}$$

We are entitled to go to four significant figures here because the least accurate input data are given to four significant figures. Therefore, $E_p = 147.2$ J.

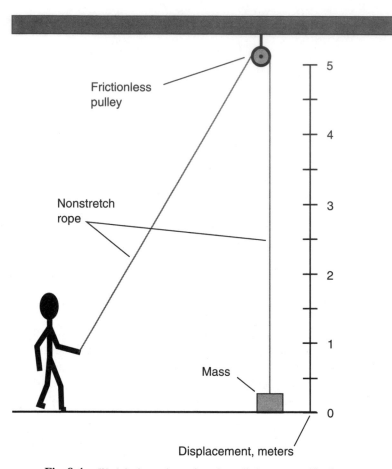

Fig. 8-4. Work is done when a force is applied over a specific distance. In this case, the force is applied upward to an object against Earth's gravity.

KINETIC ENERGY

Suppose that the object shown by the scenario in Fig. 8-4 is lifted a certain distance, imparting to it a potential energy E_p. What will happen if you let go of the rope, and the object falls back to the floor? First, the object might do damage, either to the floor or to itself, when it hits. Second, it will be moving, and in fact accelerating, when it hits. Third, all the potential energy that was imparted to the object in lifting it will be converted into other forms: vibration, sound, heat, and possibly the outward motion of flying chunks of concrete or linoleum.

Now think of the situation an infinitesimal moment—an instant—before the object strikes the floor. At this moment the kinetic energy possessed by the object will be just equal to the potential energy imparted to it by lifting it. All this kinetic energy is about to be dissipated or converted in the violence of impact. The kinetic energy is

$$E_k = Fq = ma_g q = 9.8mq$$

where F is the force applied, q is the distance the object was raised (and thus the distance it falls), m is the mass of the object, and a_g is the acceleration of gravity. Here we are taking a_g to be 9.8 m/s^2, accurate to only two significant figures.

There is another way to express E_k for a moving object having a mass m. This is

$$E_k = mv^2/2$$

where v is the speed of the object just before impact. We could use the formulas for displacement, speed, and acceleration from Chapter 7 to calculate the instantaneous speed of the object when it hits the floor, but there's no need. We already have a formula for kinetic energy in the example of Fig. 8-4. The mass-velocity formula is far more general and applies to any moving object, even if work is not done on it.

Another note should be made here. You will notice that we use the notation m (lowercase italic m) for mass and m (lowercase nonitalic m) for meter(s). It is easy to get careless and confuse these. Don't.

PROBLEM 8-6
Refer again to Fig. 8-4. Suppose that the object has mass m = 1.0 kg and is raised 4.0 m. Determine the kinetic energy it attains the instant before it strikes the floor, according to the force/displacement method. Use 9.8 m/s^2 as the value of a_g, the acceleration of gravity.

SOLUTION 8-6
This is done according to the formula $F = ma_g q$. Here, m = 1.0 kg, a_g = 9.8 m/s^2, and q = 4.0 m. Therefore:

$$E_k = 1.0 \text{ kg} \times 9.8 \text{ m/s}^2 \times 4.0 \text{ m}$$

$$= 39.2 \text{ kg} \cdot \text{m}^2/\text{s}^2 = 39.2 \text{ J}$$

We are given each input value to only two significant figures, so we should round this off to E_k = 39 J.

PROBLEM 8-7
In the example of Problem 8-6, determine the kinetic energy of the object the instant before it strikes the floor by using the mass/speed method.

Demonstrate that it yields the same final answer in the same units as the method used in Problem 8-6.

SOLUTION 8-7

The trick here is to calculate how fast the object is moving just before it strikes the floor. This involves doing the very exercise we avoided a few paragraphs ago. This will take some figuring. It's not complicated, only a little tedious.

First, let's figure out how long it takes for the object to fall. Recall from Chapter 7 that

$$q = a_{avg}\, t^2/2$$

where q is the displacement, a_{avg} is the average acceleration, and t is the time elapsed. Here, a_{avg} can be replaced with a_g because the Earth's gravitational acceleration never changes. (It is always at its average value.) We can then manipulate the preceding formula to solve for time:

$$t = (2q/a_g)^{1/2}$$

$$= [(2 \times 4.0 \text{ m})/(9.8 \text{ m/s}^2)]^{1/2}$$

$$= (8.0 \text{ m} \times 0.102 \text{ s}^2/\text{m})^{1/2}$$

"Wait!" you might say. "What have we done with a_g? Where does this s^2/m unit come from?" We are multiplying by the reciprocal of this quantity a_g, which is the same thing as dividing by the quantity a_g itself. When we take the reciprocal of a quantity that is expressed in terms of a unit, we also must take the reciprocal of the unit. This is where the s^2/m "unit" comes from. Continuing along now:

$$t = (8.0 \text{ m} \times 0.102 \text{ s}^2/\text{m})^{1/2}$$

$$= (0.816 \text{ m} \cdot \text{s}^2/\text{m})^{1/2}$$

$$= (0.816 \text{ s}^2)^{1/2} = 0.9033 \text{ s}$$

Meters cancel out in the preceding process. Units, just like numbers and variables, can cancel and become unity (the number 1) when they are divided by themselves. Let's not round this value, 0.9033 s, off just yet; we have more calculations to make.

Recall now from Chapter 7 the formula for the relationship among instantaneous velocity v_{inst}, acceleration a, and time t for an object that accelerates at a constant rate:

$$v_{inst} = at$$

Here, we can replace a with a_g, as before. We know both a_g and t already:

$$v_{inst} = (9.8 \text{ m/s}^2)\,(0.9033 \text{ s})$$

$$= 8.85234 \text{ m/s}$$

Don't worry about the fact that we keep getting more and more digits in our numbers. We'll round it off in the end.

There is only one more calculation to make, and this is using the formula for E_k in terms of v_{inst} and m. We now know that $v_{inst} = 8.85234$ m/s and $m = 1.0$ kg. Therefore:

$$E_k = mv_{inst}^2/2$$

$$= (1.0 \text{ kg}) (8.85234 \text{ m/s})^2/2$$

$$= (78.3639 \text{ kg} \cdot \text{m}^2/\text{s}^2)/2$$

$$= 39.18195 \text{ kg} \cdot \text{m}^2/\text{s}^2$$

The unit, $\text{kg} \cdot \text{m}^2/\text{s}^2$, is the same as the joule (J). And because we are entitled only to two significant figures, the numerical value must be rounded to 39. This gives the same final answer in the same units as the force/displacement method used in Problem 8-6.

Of course, given a choice, we would use the method of Problem 8-6 to determine kinetic energy in a scenario such as that illustrated by Fig. 8-4. We dragged ourselves through Problem 8-7 as an exercise to show that either method works. It's never a bad thing to verify the validity of a formula or concept!

Power

In the context of physics, *power* is the rate at which energy is expended or converted to another form. Mechanically, it is the rate at which work is done. The standard unit of power is the joule per second (J/s), more commonly known as the *watt* (W). Power is almost always associated with kinetic energy. Sometimes the rate at which potential energy is stored is referred to as power.

MECHANICAL POWER

In the examples shown by Fig. 8-4, the object acquires potential energy when it is lifted, and this potential energy is converted to kinetic energy as the object falls (if it is allowed to fall). The final burst of sound, shock waves, and perhaps outflying shrapnel is the last of the kinetic energy imparted to the object by lifting. Where does power fit into this scenario?

A slight variation on this theme can be used to talk about power. This is shown by Fig. 8-5. Suppose that instead of a free end of rope, you have a winch that you can turn to raise the object at the other end of the rope. The object starts out sitting on the floor, and you crank the winch (or use a motor to crank it) for the purpose of lifting the object. This, possibly in conjunction with a complex pulley system, will be necessary if you have a heavy object to lift. Then the pulley had better be strong! The same holds true for the rope. And let's not forget about the manner in which the pulley is anchored to the ceiling.

LET'S DO IT!

It will take energy to lift this object. You can crank the winch, imparting potential energy to the object. If the pulley system is complex, you might

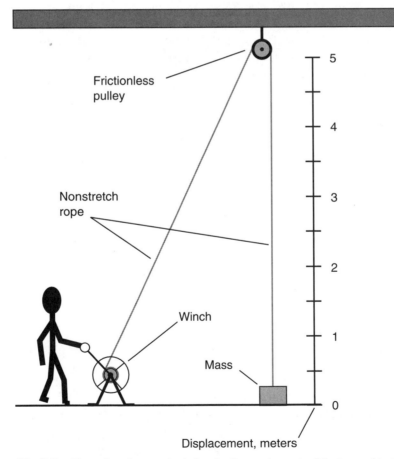

Fig. 8-5.   Illustration of power. A winch and pulley can be used to lift a heavy object.

reduce the force with which you have to bear down on the winch, but this will increase the number of times you must turn the winch to lift the object a given distance. The rate at which you expend energy cranking the winch can be expressed in watts and constitutes power. The faster you crank the winch for a given object mass, the higher is the expended power. The more massive the object for a given winch crank speed, the higher is the expended power. However, the power does not depend on how high the object is lifted. In theory, you could expend a little power for a long time and lift the mass 100 m, 1 km, or 100 km.

Assume that the winch and pulley system is frictionless and that the rope does not stretch. Suppose that you crank the winch at a constant rotational rate. The power you expend in terms of strain and sweat multiplied by the time spent applying it will equal the potential energy imparted to the object. If P is the power in watts and t is the time in seconds for which the constant power P is applied, then the potential energy imparted to the object E_p can be found according to this formula:

$$E_p = Pt$$

This can be rearranged to

$$P = E_p/t$$

We know that the potential energy is equal to the mass times the acceleration of gravity times the displacement q. Thus the power can be calculated directly by the following formula:

$$P = 9.8067mq/t$$

PROBLEM 8-8
Suppose that an object having mass 200 kg is to be lifted 2.50 m in a time of 7.00 s. What is the power required to perform this task? Take the acceleration of gravity to be 9.8 m/s^2.

SOLUTION 8-8
Simply use the preceding formula with the acceleration rounded to two significant figures:

$$P = 9.8 \times 200 \times 2.50/7.00$$

$$= 700 \text{ W}$$

When we say 700 W, we are technically justified in going to only two significant figures. How can we express this here? One way is to call it 7.0 × 10^2 W. Another way is to call it 0.70 kW, where kW stands for *kilowatt,* the equivalent of exactly 1,000 W. Yet another way is to say it is 700 W ± 5 W, meaning "700 W plus or minus 5 W." (This is the extent of the accuracy we can claim with two significant figures.) However, most physicists probably would accept our saying simply 700 W. There is a limit to how fussy we can get about these things without driving ourselves crazy.

You will notice that in Problem 8-8 we did not carry the units through the entire expression, multiplying and dividing them out along with the numbers. We don't have to do this once we know that a formula works, and we are certain to use units that are all consistent with each other in the context of the formula. In this case we are using all SI base units (meter, kilogram, second), so we know we'll come out all right in the end.

ELECTRICAL POWER

You might decide to spare yourself the tiring labor of cranking a winch to lift heavy objects over and over just to perform experiments to demonstrate the nature of power. Anyhow, it's hard to measure mechanical power directly, although it can be calculated theoretically as in Problem 8-8.

You might connect an electric motor to the winch, as shown in Fig. 8-6. If you then connect a wattmeter between the power source and the motor, you can measure the power directly. Of course, this assumes that the motor is 100 percent efficient, along with the other assumptions that the rope does not stretch and the pulley has no friction. All these assumptions are, of course, not representative of the real world. A real pulley does have friction, a real rope will stretch, and a real motor is less than perfectly efficient. As a result, the reading on a wattmeter connected as shown in Fig. 8-6 would be greater than the figure we would get if we used the scheme in Problem 8-8 to calculate the power.

SYSTEM EFFICIENCY

Suppose that we connect the apparatus of Fig. 8-6 and do the experiment described in Problem 8-8. The wattmeter might show something like 800 W. In this case we can calculate the efficiency of the whole system by dividing the actual mechanical power (700 W) by the measured input power (800 W). If we call the input power P_{in} and the actual mechanical power P_{out}, then the efficiency *Eff* is given by

$$Eff = P_{out}/P_{in}$$

If you want to calculate the efficiency in percent, $Eff_\%$, use this formula

$$Eff_\% = 100P_{out}/P_{in}$$

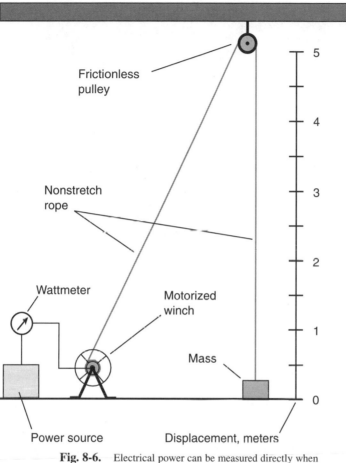

Fig. 8-6.   Electrical power can be measured directly when a motor is used to drive a winch to lift a heavy object.

PROBLEM 8-9

Consider the scenario of Problem 8-8 and Fig. 8-6. If the meter shows 800 W, what is the efficiency in percent?

SOLUTION 8-9

Use the second of the two efficiency formulas presented earlier:

$$Eff_\% = 100 P_{out}/P_{in}$$

$$Eff_\% = 100 \times 700/800$$

$$= 87.5 \text{ percent}$$

If you want to get formal and claim only two significant figures for the 700-W result in Problem 8-8, then you must round this efficiency figure off to 88 percent.

Quiz

Refer to the text in this chapter if necessary. A good score is eight correct. Answers are in the back of the book.

1. Consider a minor parking-lot accident. Car *A* backs out at 30 cm/s toward the west, and car *B* looks for a place to park, driving north at 40 cm/s. Both cars mass 1,000 kg. What is the total system momentum before the collision? Remember that momentum is a vector quantity. Also, be careful with your units.
   (a) 700 kg · m/s, generally northwest
   (b) 100 kg · m/s, generally northwest
   (c) 500 kg · m/s, generally northwest
   (d) There is not enough information to answer this.

2. Impulse is the product of
   (a) time and distance.
   (b) time, mass, and acceleration.
   (c) time, mass, and velocity.
   (d) time and velocity.

3. Consider a hockey player skating down the ice at 10.0 m/s. His mass is 82.0 kg. What is his kinetic energy?
   (a) 820 J
   (b) 410 J
   (c) 8.20×10^4 J
   (d) 4.10×10^3 J

4. An object whose mass is 10.0 kg is lifted through a distance of 4.000 m on a planet where the gravitational acceleration is 6.000 m/s^2. How much work is required to do this?
   (a) 60.0 J
   (b) 24.0 J
   (c) 40.0 J
   (d) 240 J

5. Suppose that an object is pushed with steady force along a frictionless surface. When you multiply the object's mass by the length of time for which it is pushed and then multiply the result by the object's acceleration over that period of time, you get
   (a) momentum.
   (b) velocity.
   (c) impulse.
   (d) a meaningless quantity.

6. According to the law of conservation of momentum, in an ideal closed system,
   (a) when two objects collide, the system neither loses nor gains any total momentum.
   (b) when two objects collide, neither object loses or gains any momentum.
   (c) when two objects collide, the magnitudes of their momentum vectors add.
   (d) when two objects collide, the magnitudes of their momentum vectors multiply.

7. When making calculations in which all the quantities have their units indicated throughout the entire process,
   (a) the units multiply and divide just like the numbers.
   (b) the units cannot cancel out.
   (c) the units can be multiplied but not divided.
   (d) the units can be added and subtracted, but not multiplied or divided.

8. One joule, reduced to base units, is equivalent to
   (a) one kilogram-meter per second squared.
   (b) one kilogram-meter.
   (c) one kilogram-meter squared per second squared.
   (d) one meter per second squared.

9. A 5,000-kg motorboat sits still on a frictionless lake. There is no wind to push against the boat. The captain starts the motor and runs it steadily for 10.00 seconds in a direction straight forward and then shuts the motor down. The boat has attained a speed of 5.000 meters per second straight forward. What is the impulse supplied by the motor?
   (a) $2.500 \times 10^5 \ \text{kg} \cdot \text{m/s}$
   (b) $2.500 \times 10^4 \ \text{kg} \cdot \text{m/s}$
   (c) $2,500 \ \text{kg} \cdot \text{m/s}$
   (d) $6.250 \times 10^4 \ \text{kg} \cdot \text{m/s}$

10. Which of the following does *not* have an effect on the momentum of a moving spherical particle?
    (a) The speed of the particle
    (b) The diameter of the particle
    (c) The direction in which the particle travels
    (d) The mass of the particle

CHAPTER 9

Particles of Matter

The idea that matter exists in the form of particles, rather than as a continuous mass, is many centuries old. How do we explain the fact that some substances are more dense than others, that some retain their shape while others flow freely, and that some are visible while others are not? The many ways in which matter can exist are difficult to explain in any other way than by means of a *particle theory*. This was the reasoning ancient alchemists used when they hypothesized that matter is comprised of tiny, invisible particles, or *atoms*.

Early Theories

All atoms are made up of countless smaller particles whizzing around. These *subatomic particles* are dense, but matter is mostly empty space. Matter seems solid and continuous because the particles are so small that we can't see them, and they move so fast that their individual motions would appear as a blur even if the particles themselves could be seen. However, the spaces within atoms are vastly greater than the particles that comprise them. If we could shrink ourselves down to the subatomic scale and also slow down time in proportion, a piece of metal would look something like a huge, hysterical swarm of gnats. Did the first chemical and physical scientists realize that the atoms they had dreamed up actually consisted of smaller particles, that these particles in turn consisted of tinier ones still, and that some people in future generations would come to believe that the particle sequence extends down to smaller and smaller scales *ad infinitum?*

THE SMALLEST PIECE

Millennia ago, scientists deduced the particle nature of matter from observing such things as water, rocks, and metals. These substances are much different from each other. However, any given material—copper, for example—is the same wherever it is found. Even without doing any complicated experiments, early physicists believed that substances could only have these consistent behaviors if they were made of unique types, or arrangements, of particles.

It was a long time before people began to realize how complicated this business really is. Even today, there are plenty of things that scientists don't know. For example, is there a smallest possible material particle? Or do particles keep on getting tinier as we deploy more and more powerful instruments to probe the depths of inner space? Either notion is difficult to comprehended intuitively. If there is something that represents the smallest possible particle, why can't it be cut in half? However, if particles can be cut into pieces forever and ever, then what is the ultimate elementary particle? A geometric point of zero volume? What would be the density of such a thing? Some mass divided by zero? This doesn't make sense! A literal and conclusive answer to this puzzle remains to be found. We may never know all there is to know about matter. It may not even be possible to know everything about matter.

THE ELEMENTS

Until about the year 1900, there were respected people who refused to believe the atomic theory of matter. Today, however, practically everyone accepts it. The atomic theory explains the behavior of matter better than any other scheme. Some people still think that matter is continuous; a few folks still believe that our planet Earth is flat, too.

Eventually, scientists identified 92 different kinds of fundamental substances in nature and called them *elements*. Later, more elements were made artificially. This process of discovery is still going on. Using machines known as *particle accelerators,* sometimes called *atom smashers,* nuclear physicists have fabricated human-made elements that can't exist in nature, at least not under conditions resembling anything we would imagine as normal.

EVERY ELEMENT IS UNIQUE

Atoms of different elements are always different. The slightest change in an atom can make a tremendous difference in its behavior. You can live by breathing pure oxygen but you cannot live off of pure nitrogen. Oxygen will cause metal to corrode, but nitrogen will not. Wood will burn furiously in an atmosphere of pure oxygen but will not even ignite in pure nitrogen. However, both look, smell, and feel exactly the same at normal temperature and pressure. Both are invisible gases, both are colorless, both are odorless, and both are just about equally heavy. These substances are so different because oxygen and nitrogen consist of different numbers of otherwise identical particles.

There are many other examples in nature where a tiny change in atomic structure makes a major difference in the way a substance behaves.

The Nucleus

The part of an atom that gives an element its identity is the *nucleus*. It is made up of two kinds of particles, the *proton* and the *neutron*. Both are extremely dense. Protons and neutrons have just about the same mass, but the proton has an electric charge, whereas the neutron does not.

THE PROTON

Protons are too small to be observed directly, even with the most powerful microscopes. All protons carry a positive electric charge, and the charge on every proton is the same as the charge on every other. Every proton at rest has the same mass as every other proton at rest. Most scientists accept the proposition that all protons are identical, at least in our part of the universe, although they, like all other particles, gain mass if accelerated to extreme speeds. This increase in mass takes place because of relativistic effects; you'll learn about this later.

While an individual proton is invisible and not massive enough to make much of an impact all by itself, a high-speed barrage of them can have considerable effects on matter. Protons are incredibly dense. If you could scoop up a level teaspoon of protons the way you scoop up a teaspoon of sugar—with the protons packed tightly together like the sugar crystals—the resulting sample would weigh tons in the Earth's gravitational field. A stone made of solid protons would fall into the Earth and cut through the crustal rocks like a lead shot falls through the air.

THE NEUTRON

A neutron has a mass slightly greater than that of a proton. Neutrons have no electrical charge, and they are roughly as dense as protons. However, while protons last for a long time all by themselves in free space, neutrons do not. The *mean life* of a neutron is only about 15 minutes. This means that if you gathered up a batch of, say, 1 million neutrons and let them float around in space, you would have only about 500,000 neutrons left after 15 minutes. After 30 minutes, you would have approximately 250,000 neutrons remaining; after 45 minutes, there would be only about 125,000 neutrons left.

Neutrons can last a long time when they are in the nuclei of atoms. This is a fortunate thing because if it weren't true, matter as we know it could not exist. Neutrons also can survive for a long time when a huge number of them are tightly squeezed together. This happens when large stars explode and then the remaining matter collapses under its own gravitation. The end product of this chain of events is a *neutron star.*

THE SIMPLEST ELEMENTS

The simplest element, hydrogen, has a nucleus made up of only one proton; there are usually no neutrons. This is the most common element in the universe. Sometimes a nucleus of hydrogen has a neutron or two along with the proton, but this does not occur very often. These mutant forms of hydrogen do, nonetheless, play significant roles in atomic physics.

The second most abundant element in the universe is helium. Usually, this atom has a nucleus with two protons and two neutrons. Hydrogen is changed into helium inside the Sun, and in the process, energy is given off. This makes the Sun shine. The process, called *atomic fusion* or *nuclear fusion,* is also responsible for the terrific explosive force of a hydrogen bomb.

ATOMIC NUMBER

According to modern atomic theory, every proton in the universe is exactly like every other. Neutrons are all alike too. The number of protons in an element's nucleus, the *atomic number,* gives that element its identity.

The element with three protons is *lithium,* a light metal that reacts easily with gases such as oxygen or chlorine. Lithium always has three protons; conversely, any element with three protons in its nucleus must be lithium. The element with four protons is *beryllium,* also a metal. Carbon has six protons in its nucleus, nitrogen has seven, and oxygen has eight. In general, as the number of protons in an element's nucleus increases, the number of neutrons also increases. Elements with high atomic numbers, such as lead, are therefore much more dense than elements with low atomic numbers, such as carbon. Perhaps you've compared a lead shot with a piece of coal of similar size and noticed this difference.

If you could somehow add two protons to the nucleus of every atom in a sample of carbon, you would end up with an equal number of atoms of oxygen. However, this is much easier said than done, even with a single atom. It is possible to change one element into another; the Sun does it all the time, fusing hydrogen into helium. The process is far from trivial, though. In ancient times, alchemists tried to do this; the most well-known example of their pursuits was the quest to turn lead (atomic number 82) into gold (atomic number 79). As far as anyone knows, they never succeeded. It was not until the 1940s, when the first atomic bombs were tested, that elements actually were "morphed" by human beings. The results were quite different from anything the alchemists ever strove for.

Table 9-1 lists all the known elements in alphabetical order, with the names of the elements in the first column, the standard chemical symbols in the second column, and the atomic numbers in the third column.

ISOTOPES

In the individual atoms of a given element, such as oxygen, the number of neutrons can vary. Regardless of the number of neutrons, however, the element keeps its identity based on the atomic number. Differing numbers of neutrons result in various *isotopes* for a specific material element.

Each element has one particular isotope that is found most often in nature. However, all elements have more than one isotope, and some have

Table 9-1   The Chemical Elements in Alphabetical Order by Name, Including Chemical Symbols and Atomic Numbers 1 through 118 (As of the time of writing, there were no known elements with atomic numbers 113, 115, or 117.)

Element name	Chemical symbol	Atomic number
Actinium	Ac	89
Aluminum	Al	13
Americium	Am	95
Antimony	Sb	51
Argon	Ar	18
Arsenic	As	33
Astatine	At	85
Barium	Ba	56
Berkelium	Bk	97
Beryllium	Be	4
Bismuth	Bi	83
Bohrium	Bh	107
Boron	B	5
Bromine	Br	35
Cadmium	Cd	48
Calcium	Ca	20
Californium	Cf	98
Carbon	C	6
Cerium	Ce	58
Cesium	Cs	55
Chlorine	Cl	17
Chromium	Cr	24
Cobalt	Co	27
Copper	Cu	29
Curium	Cm	96
Dubnium	Db	105
Dysprosium	Dy	66
Einsteinium	Es	99
Erbium	Er	68
Europium	Eu	63
Fermium	Fm	100
Fluorine	F	9
Francium	Fr	87
Gadolinium	Gd	64
Galliom	Ga	31

Table 9-1   (*Continued*) The Chemical Elements in Alphabetical Order by Name, Including Chemical Symbols and Atomic Numbers 1 through 118 (As of the time of writing, there were no known elements with atomic numbers 113, 115, or 117.)

Element name	Chemical symbol	Atomic number
Germanium	Ge	32
Gold	Au	79
Hafnium	Hf	72
Hassium	Hs	108
Helium	He	2
Holmium	Ho	67
Hydrogen	H	1
Indium	In	49
Iodine	I	53
Iridium	Ir	77
Iron	Fe	26
Krypton	Kr	36
Lanthanum	La	57
Lawrencium	Lr or Lw	103
Lead	Pb	82
Lithium	Li	3
Lutetium	Lu	71
Magnesium	Mg	12
Manganese	Mn	25
Meitnerium	Mt	109
Mendelevium	Md	101
Mercury	Hg	80
Molybdenum	Mo	42
Neodymium	Nd	60
Neon	Ne	10
Neptunium	Np	93
Nickel	Ni	28
Niobium	Nb	41
Nitrogen	N	7
Nobelium	No	102
Osmium	Os	76
Oxygen	O	8
Palladium	Pd	46
Phosphorus	P	15

Table 9-1   (*Continued*) The Chemical Elements in Alphabetical Order by Name, Including Chemical Symbols and Atomic Numbers 1 through 118 (As of the time of writing, there were no known elements with atomic numbers 113, 115, or 117.)

Element name	Chemical symbol	Atomic number
Platinum	Pt	78
Plutonium	Pu	94
Polonium	Po	84
Potassium	K	19
Praseodymium	Pr	59
Promethium	Pm	61
Protactinium	Pa	91
Radium	Ra	88
Radon	Rn	86
Rhenium	Re	75
Rhodium	Rh	45
Rubidium	Rb	37
Ruthenium	Ru	44
Rutherfordium	Rf	104
Samarium	Sm	62
Scandium	Sc	21
Seaborgium	Sg	106
Selenium	Se	34
Silicon	Si	14
Silver	Ag	47
Sodium	Na	11
Strontium	Sr	38
Sulfur	S	16
Tantalum	Ta	73
Technetium	Tc	43
Tellurium	Te	52
Terbium	Tb	65
Thallium	Tl	81
Thorium	Th	90
Thulium	Tm	69
Tin	Sn	50
Titanium	Ti	22
Tungsten	W	74

Table 9-1   (*Continued*) The Chemical Elements in Alphabetical Order by Name, Including Chemical Symbols and Atomic Numbers 1 through 118 (As of the time of writing, there were no known elements with atomic numbers 113, 115, or 117.)

Element name	Chemical symbol	Atomic number
Ununbium	Uub	112
Ununhexium	Uuh	116
Ununnilium	Uun	110
Ununoctium	Uuo	118
Ununquadium	Quq	114
Unununium	Uuu	111
Uranium	U	92
Vanadium	V	23
Xenon	Xe	54
Ytterbium	Yb	70
Yttrium	Y	39
Zinc	Zn	30
Zirconium	Zr	40

many. Changing the number of neutrons in an element's nucleus results in a difference in the mass, as well as a difference in the density, of the element. Thus, for example, hydrogen containing a neutron or two in the nucleus, along with the proton, is called *heavy hydrogen.* The naturally occurring form of uranium has three more neutrons in its nucleus than the type that is notorious for use in atomic weapons.

Adding or taking away neutrons from the nucleus of an element is not quite as farfetched a business as adding or taking away protons, but it is still a task generally relegated to high-energy physics. You can't simply take a balloon filled with air, which is approximately 78 percent nitrogen, and make it more massive by injecting neutrons into the nitrogen nuclei.

ATOMIC MASS

The *atomic mass,* sometimes called the *atomic weight,* of an element is approximately equal to the sum of the number of protons and the number of neutrons in the nucleus. This quantity is formally measured in atomic mass units (amu), where 1 amu is equal to exactly $1/12$ the mass of the nucleus of the carbon isotope having six neutrons. This is the most common isotope of carbon and is symbolized ^{12}C or carbon-12. Any proton or any neutron has a mass of approximately $^1/_{12}$ amu, but neutrons are a little more massive than protons.

Elements are uniquely defined by their atomic numbers, but the atomic mass of an element depends on the particular isotope of that element. A well-known isotope of carbon, ^{14}C, is found in trace amounts in virtually all carbon-containing substances. This fact has proven quite useful to geologists and archaeologists. The isotope ^{14}C is radioactive, whereas ^{12}C is not. The radioactivity of ^{14}C diminishes with time according to a well-known, predictable mathematical function. This makes it possible for researchers to determine when carbon-containing compounds were created and thus to find out how old various rocks, fossils, and artifacts are.

In nuclear reactions capable of producing energy, such as the reactions that take place inside stars, atomic bombs, and nuclear power plants, a certain amount of mass is always given up—and converted into energy—in the transactions between the atoms. This amount of mass can be exceedingly small yet produce an enormous burst of energy. The first person to formalize this relation was Albert Einstein, using his famous equation

$$E = mc^2$$

where E is the energy produced in joules, m is the total mass in kilograms lost during the reaction, and c is the speed of light in meters per second. The value of c^2 is gigantic: approximately 90 quadrillion meters squared per second squared (9×10^{16} m^2/s^2). This is why so much energy can be produced by an atomic reaction between two elemental samples of modest mass.

An excellent source of information concerning all the known elements, including atomic number, atomic mass, and various other characteristics, can be found at the following Web site:

http://www.chemicalelements.com/

If you have a computer with Internet access, it would be a good idea spend a while exploring this Web site right now.

PROBLEM 9-1
Suppose that the nucleus of an oxygen atom, which has eight protons and usually has eight neutrons, were split exactly in two. What element would be the result? How many atoms of this element would there be? Neglect, for simplicity, any energy that might be involved in the reaction.

SOLUTION 9-1
This reaction would produce two atoms of beryllium, each with four protons and four neutrons. This would not be the most common isotope, however; it turns out that beryllium usually has five neutrons in its nucleus.

Outside the Nucleus

Surrounding the nucleus of an atom are particles having electric charge opposite from the charge of the protons. These are *electrons*. Physicists arbitrarily call the electron charge negative and the proton charge positive.

The electron

An electron has exactly the same charge quantity as a proton but with opposite polarity. Electrons are far less massive than protons, however. It would take about 2,000 electrons to have the same mass as a single proton.

One of the earliest theories concerning the structure of the atom pictured the electrons embedded in the nucleus like raisins in a cake. Later, the electrons were imagined as orbiting the nucleus, making every atom like a miniature star system with the electrons as the planets (Fig. 9-1). Still later, this view was modified further. In today's model of the atom, the electrons are fast-moving, and they describe patterns so complex that it is impossible to pinpoint any individual particle at any given instant of time. All that can be done is to say that an electron just as likely will be inside a certain sphere as outside. These spheres are known as *electron shells*. The centers of the shells correspond to the position of the atomic nucleus. The greater a shell's radius, the more energy the electron has. Figure 9-2 is a greatly simplified drawing of what happens when an electron gains just enough energy to "jump" from one shell to another shell representing more energy.

Electrons can move rather easily from one atom to another in some materials. These substances are *electrical conductors*. In other substances, it is difficult to get electrons to move. These are called *electrical insulators*. In any case, however, it is far easier to move electrons than it is to move protons. Electricity almost always results, in some way, from the motions of electrons in a material.

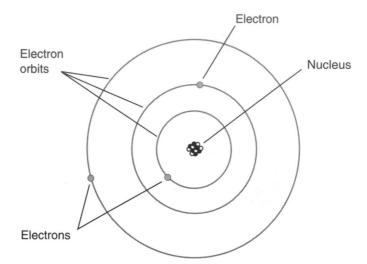

Fig. 9-1.   An early model of the atom, developed around the year 1900.

Generally, the number of electrons in an atom is the same as the number of protons. The negative charges therefore exactly cancel out the positive ones, and the atom is electrically neutral. Under some conditions, however, there can be an excess or shortage of electrons. High levels of radiant energy, extreme heat, or the presence of an electrical field (to be discussed later) can "knock" electrons loose from atoms, upsetting the balance.

IONS

If an atom has more or less electrons than protons, that atom acquires an electric charge. A shortage of electrons results in positive charge; an excess of electrons gives a negative charge. An element's identity remains the same, no matter how great the excess or shortage of electrons. In the extreme case, all the electrons may be removed from an atom, leaving only the nucleus. This will still represent the same element, however, as it would if it had all its electrons. A electrically charged atom is called an *ion.* When a substance contains many ions, the material is said to be *ionized.* If an atom has more electrons than protons, it is a *negative ion.* If it has fewer electrons than protons, it is a *positive ion.* If the number of electrons and protons is the same, then the atom is electrically neutral.

Ionization can take place when substances are heated to high temperatures or when they are placed in intense electrical fields. Ionization also can occur in

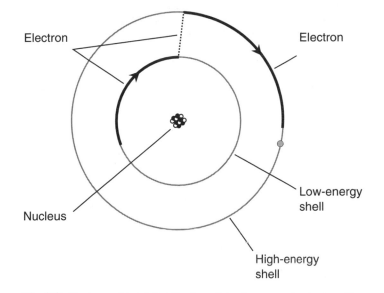

Fig. 9-2. Electrons exist at defined levels, each level corresponding to a specific, fixed energy state.

a substance as a result of exposure to ultraviolet light, x-rays, gamma rays, or high-speed subatomic particles such as neutrons, protons, helium nuclei, or electrons. So-called *ionizing radiation,* more often called *radioactivity,* ionizes the atoms in living tissue and can cause illness, death, and genetic mutations.

Lightning is the result of ionization of the air. An electric spark is caused by a large buildup of charges, resulting in forces on the electrons in the intervening medium. These forces pull the electrons away from individual atoms. Ionized atoms generally conduct electric currents with greater ease than electrically neutral atoms. The ionization, caused by a powerful electrical field, occurs along a jagged, narrow *channel,* as you have surely seen. After the lightning flash, the nuclei of the atoms quickly attract stray electrons back, and the air becomes electrically neutral again.

An element may be both an ion and an isotope different from the usual isotope. For example, an atom of carbon may have eight neutrons rather than the usual six, thus being the isotope ^{14}C, and it may have been stripped of an electron, giving it a positive unit electric charge and making it an ion.

The atmosphere of our planet becomes less dense with increasing altitude. Because of this, the amount of ultraviolet and x-ray energy received from the Sun gets greater as we go higher. At certain altitudes, the gases in the atmosphere become ionized by solar radiation. These regions comprise the *ionosphere* of the Earth. The ionosphere has a significant effect on the propagation of radio waves at certain frequencies. The ionized layers absorb or refract the waves. This makes long-distance communication possible on the so-called shortwave radio bands.

PROBLEM 9-2
Suppose that the nucleus of an oxygen atom is split exactly in two. As in Problem 9-1, neglect any energy that might be involved in the reaction. Suppose that the original oxygen atom is electrically neutral and that no electrons are gained or lost during the reaction. Is it possible for both the resulting atoms to be electrically neutral?

SOLUTION 9-2
Yes. The original oxygen atom must have eight electrons in order to be electrically neutral. If these eight electrons are equally divided between the two beryllium atoms, each of which has four protons in its nucleus, then both beryllium atoms will have four electrons, and both will be electrically neutral.

PROBLEM 9-3
Consider the preceding scenario in which the oxygen atom has been stripped of two of its electrons so that it is a positive ion. Can the resulting two beryllium atoms be electrically neutral?

SOLUTION 9-3
In this case, no. There must be eight electrons, in total, for both the beryllium atoms to end up neutral. It is possible for one of the beryllium atoms to be neutral, but at least one of them must be an ion.

Energy from Matter

The splitting up of an atomic nucleus is known as *nuclear fission*. This is, in a sense, the opposite of nuclear fusion, which occurs inside the Sun and other stars. The very first atomic bombs, developed in the 1940s, made use of fission reactions to produce energy. More powerful weapons, created in the 1950s, used atomic fission bombs to produce the temperatures necessary to generate hydrogen fusion.

HUMAN-CAUSED AND NATURAL FISSION

The preceding problems involving oxygen and beryllium are given for illustrative purposes, but the actual breaking up of atomic nuclei is not such a simple business. A physicist can't snap an atomic nucleus apart as if it were a toy. Nuclear reactions must take place under special conditions, and the results are not as straightforward as the foregoing problems suggest.

To split atomic nuclei in the laboratory, a *particle accelerator* is employed. This machine uses electric charges, magnetic fields, and other effects to hurl subatomic particles at extreme speeds at the nuclei of atoms to split them apart. The result is a fission reaction, often attended by the liberation of energy in various forms.

Some fission reactions occur spontaneously. Such a reaction can take place atom-by-atom over a long period of time, as is the case with the decay of radioactive minerals in the environment. The reaction can occur rapidly but under controlled conditions, as in a nuclear power plant. It can take place almost instantaneously and out of control, as in an atomic bomb when two sufficiently massive samples of certain radioactive materials are pressed together.

MATTER AND ANTIMATTER

The proton, the neutron, and the electron each has its own nemesis particle that occurs in the form of *antimatter*. These particles are called *antiparticles.* The antiparticle for the proton is the *antiproton*; for the neutron it is the *antineutron*; for the electron it is the *positron*. The antiproton has the same mass as the proton, but in a negative sort of way, and it has a negative electric charge that is equal but opposite to the positive electric charge of the proton. The antineutron has the same mass as the neutron, but again in a negative sense. Neither the neutron nor the antineutron have any electric charge. The positron has same mass as the electron, but in a negative sense, and it is positively charged to an extent equal to the negative charge on an electron.

You might have read or seen in science-fiction novels and movies that when a particle of matter collides with its nemesis, they annihilate each other. This is true. What, exactly, does this mean? Actually, the particles don't just vanish from the cosmos, but they change from matter into energy. The combined mass of the particle and the antiparticle is liberated completely according to the same Einstein formula that applies in nuclear reactions:

$$E = (m_+ + m_-) \, c^2$$

where E is the energy in joules, m_+ is the mass of the particle in kilograms, m_- is the mass of the antiparticle in kilograms, and c is the speed of light squared, which, as you recall, is approximately equal to 9×10^{16} m^2/s^2.

UNIMAGINABLE POWER

If equal masses of matter and antimatter are brought together, then in theory all the mass will be converted to energy. If there happens to be more matter than antimatter, there will be some matter left over after the encounter. Conversely, if there is more antimatter than matter, there will be some anti-matter remaining.

In a nuclear reaction, only a tiny fraction of the mass of the constituents is liberated as energy; plenty of matter is always left over, although its form has changed. You might push together two chunks of ^{235}U, the isotope of uranium whose atomic mass is 235 amu, and if their combined mass is great enough, an atomic explosion will take place. However, there will still be a considerable amount of matter remaining. We might say that the matter-to-energy conversion efficiency of an atomic explosion is low.

In a matter–antimatter reaction, if the masses of the samples are equal, the conversion efficiency is 100 percent. As you can imagine, a matter–antimatter bomb would make a conventional nuclear weapon of the same total mass look like a firecracker by comparison. A single matter–antimatter weapon of modest size could easily wipe out all life on Earth.

WHERE IS ALL THE ANTIMATTER?

Why don't we see antimatter floating around in the Universe? Why, for example, are the Earth, Moon, Venus, and Mars all made of matter and not antimatter? (If any celestial object were made of antimatter, then as soon as a spacecraft landed on it, the ship would vanish in a fantastic burst of energy.) This is an interesting question. We are not absolutely certain that all the distant stars and galaxies we see out there really are matter. However, we know that if there were any antimatter in our immediate vicinity, it would have long ago combined with matter and been annihilated. If there were both matter and antimatter in the primordial solar system, the mass of the matter was greater, for it prevailed after the contest.

Most astronomers are skeptical of the idea that our galaxy contains roughly equal amounts of matter and antimatter. If this were the case, we should expect to see periodic explosions of unimaginable brilliance or else a continuous flow of energy that could not be explained in any way other than matter–antimatter encounters. However, no one really knows the answers to questions about what comprises the distant galaxies and, in particular, the processes that drive some of the more esoteric objects such as quasars.

PROBLEM 9-4

Suppose that a 1.00-kg block of matter and a 1.00-kg block of antimatter are brought together. How much energy will be liberated? Will there be any matter or antimatter left over?

SOLUTION 9-4

We can answer the second question first: There will be no matter or antimatter left over because the masses of the two blocks are equal (and, in a sense, opposite). As for the first part of the question, the total mass involved in this encounter is 2.00 kg, so we can use the famous Einstein formula. For simplicity, let's round off the speed of light to $c = 3.00 \times 10^8$ m/s. Then the energy E, in joules, is

$$
\begin{aligned}
E &= mc^2 \\
&= 2.00 \times (3.00 \times 10^8)^2 \\
&= 2.00 \times 9.00 \times 10^{16} \\
&= 1.80 \times 10^{17} \text{ J}
\end{aligned}
$$

This is a lot of joules. It is not easy to conceive how great a burst of energy this represents because the number 1.80×10^{17}, or 180 quadrillion, is too large to envision. However, the quantity of energy represented by 1.80×10^{17} J can be thought of in terms of another problem.

PROBLEM 9-5

We know that 1 W = 1 J/s. How long would the energy produced in the preceding matter–antimatter reaction, if it could be controlled and harnessed, illuminate a 100-W light bulb?

SOLUTION 9-5

Divide the amount of energy in joules by the wattage of the bulb in joules per second. We know this will work because, in terms of units,

$$
\text{J/W} = \text{J/(J/s)} = \text{J} \cdot \text{(s/J)} = \text{s}
$$

The joules cancel out. Note also that the small dot ($\cdot$) is used to represent multiplication when dealing with units, as opposed to the slanted cross ($\times$) that is customarily used with numerals. Getting down to the actual numbers, let P be the power consumed by the bulb (100 W), let t_s be the number of seconds the 100-W light bulb will burn, and let E be the total energy produced by the matter–antimatter reaction, 1.80×10^{17} J. Thus

$$
\begin{aligned}
t_s &= E/P \\
&= 1.80 \times 10^{17}/100 \\
&= 1.80 \times 10^{17}/10^2 \\
&= 1.80 \times 10^{15} \text{ s}
\end{aligned}
$$

This is a long time, but how long in terms of, say, years? There are 60.0 seconds in a minute, 60.0 minutes in an hour, 24.0 hours in a day, and, on average, 365.25 days in a year. This makes 31,557,600, or 3.15576×10^7, seconds in a year. Let t_{yr} be the time in years that the light bulb burns. Then

$$t_{yr} = t_s/(3.15576 \times 10^7)$$
$$= (1.80 \times 10^{15})/(3.15576 \times 10^7)$$
$$= 0.570 \times 10^8$$
$$= 5.70 \times 10^7 \text{ yr}$$

This is 57.0 million years, rounded to three significant figures (the nearest 100,000 years), which is all the accuracy to which we are entitled based on the input data.

PROBLEM 9-6
Suppose that the amount of matter in the preceding two problems is doubled to 2.00 kg but the amount of antimatter remains 1.00 kg. How much energy will be liberated? Will there be any matter or antimatter left over?

SOLUTION 9-6
The amount of liberated energy will be the same as in the examples shown by the preceding two problems: 1.80×10^{17} J. There will be 1.00 kg of matter left over (the difference between the masses). However, assuming that the encounter produces an explosion, the matter won't remain in the form of a brick. It will be scattered throughout millions of cubic kilometers of space.

Compounds

Different elements can join together, sharing electrons. When this happens, the result is a chemical *compound*. One of the most common compounds on Earth is water, the result of two hydrogen atoms joining with an atom of oxygen. There are thousands of different chemical compounds that occur in nature.

NOT JUST A MIXTURE!

A compound is not the same thing as a mixture of elements. Sometimes, however, when elements are mixed (and, if necessary, given a jolt of energy), compounds result because the elements undergo *chemical reactions* with each other. If hydrogen and oxygen are mixed, the result is a colorless, odorless gas. A spark will cause the molecules to join together to form water vapor. This reaction will liberate energy in the form of light and heat. Under the right conditions, there will be an explosion because the two elements join eagerly. When atoms of elements join together to form a compound, the resulting particles are *molecules*. Figure 9-3 is a simplified diagram of a water molecule.

Compounds often, but not always, appear different from any of the elements that make them up. At room temperature and pressure, both hydrogen and oxygen are gases. But water under the same conditions is a liquid. The heat of the reaction just described, if done in the real world, would result in water vapor initially, and water vapor is a colorless, odorless gas. However, some of this vapor would condense into liquid water if the temperature got

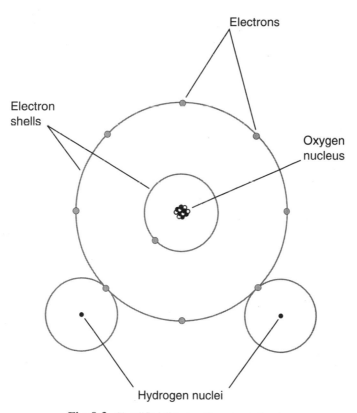

Fig. 9-3. Simplified diagram of a water molecule.

low enough for dew to form. Some of it would become solid, forming frost, snow, or ice if the temperature dropped below the freezing point of water.

A note of caution: *Do not try an experiment like this!* You could be severely burned. In the extreme, if enough of the hydrogen-oxygen air is inhaled, your lungs will be injured to the point where you may die of asphyxiation. We sometimes read or hear news reports about home experimenters who blew themselves up with chemistry sets. Don't become the subject matter for one of these stories!

Another common example of a compound is rust. This forms when iron joins with oxygen. Iron is a dull gray solid, and oxygen is a gas; however, iron rust is a maroon-red or brownish powder, completely unlike either of the elements from which it is formed. The reaction between iron and oxygen takes place slowly, unlike the rapid combination of hydrogen and oxygen when ignited. The rate of the iron-oxygen reaction can be sped up by the presence of water, as anyone who lives in a humid climate knows.

COMPOUNDS CAN BE SPLIT APART

The opposite of the element-combination process can occur with many compounds. Water is an excellent example. When water is *electrolyzed,* it separates into hydrogen and oxygen gases.

You can conduct the following *electrolysis* experiment at home. Make two electrodes out of large nails. Wrap some bell wire around each nail near the head. Add a cupful (a half-pint) of ordinary table salt to a bucket full of water, and dissolve the salt thoroughly to make the water into a reasonably good electrical conductor. Connect the two electrodes to opposite poles of a 12-volt (12-V) battery made from two 6-V lantern batteries or eight ordinary dry cells connected in series. (Do not use an automotive battery for this experiment.) Insert the electrodes into the water a few centimeters apart. You will see bubbles rising up from both electrodes. The bubbles coming from the negative electrode are hydrogen gas; the bubbles coming from the positive electrode are oxygen gas (Fig. 9-4). You probably will see a lot more hydrogen bubbles than oxygen bubbles.

Be careful when doing this experiment. Don't reach into the bucket and grab the electrodes. In fact, you shouldn't grab the electrodes or the battery terminals at all. The 12 V supplied by two lantern batteries is enough to give you a nasty shock when your hands are wet, and it can even be dangerous.

If you leave the apparatus shown in Fig. 9-4 running for a while, you will begin to notice corrosion on the exposed wire and the electrodes. This

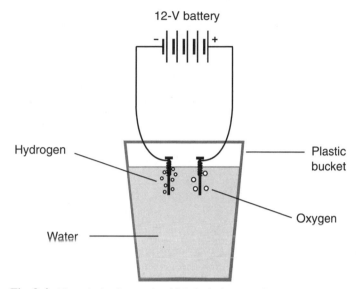

Fig. 9-4. Electrolysis of water, in which the hydrogen and oxygen atoms are split apart from the compound.

will especially take place on the positive electrode, where oxygen is attracted. Remember that you have added table salt to the water; this will attract chlorine ions as well. Both oxygen and chlorine combine readily with the copper in the wire and the iron in the nail. The resulting compounds are solids that will tend to coat the wire and the nail after a period of time. Ultimately, this coating will act as an electrical insulator and reduce the current flowing through the saltwater solution.

ALWAYS IN MOTION

Figure 9-3 shows an example of a molecule of water, consisting of three atoms put together. However, molecules also can form from two or more atoms of a single element. Oxygen tends to occur in pairs most of the time in Earth's atmosphere. Thus an oxygen molecule is sometimes denoted by the symbol O_2, where the O represents oxygen, and the subscript 2 indicates that there are two atoms per molecule. The water molecule is symbolized H_2O because there are two atoms of hydrogen and one atom of oxygen in each molecule. Sometimes oxygen atoms are by themselves; then we denote the molecule simply as O. Sometimes there are three atoms of oxygen grouped together. This is the gas called *ozone* that has received attention in environmental news. It is written O_3.

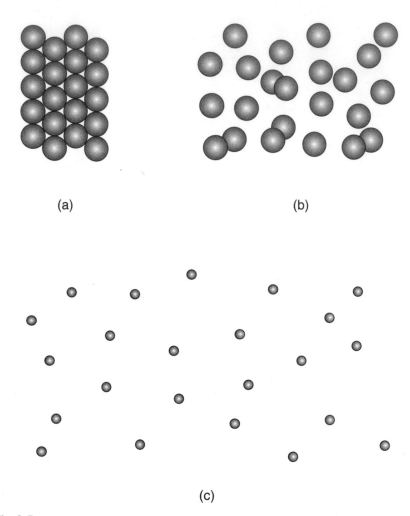

(a)   (b)

(c)

Fig. 9-5. Simplified rendition of molecules in a solid (*a*), in a liquid (*b*), in a gas (*c*). The gas molecules are shown smaller for illustrative purposes only.

Molecules are always moving. The speed with which they move depends on the temperature. The higher the temperature, the more rapidly the molecules move around. In a solid, the molecules are interlocked in a sort of rigid pattern, although they vibrate continuously (Fig. 9-5*a*). In a liquid, they slither and slide around (see Fig. 9-5*b*). In a gas, they are literally whizzing all over the place, bumping into each other and into solids and liquids adjacent to the gas (see Fig. 9-5*c*). We'll look at solids, liquids, and gases more closely in the next chapter.

 **Quiz**

Refer to the text in this chapter if necessary. A good score is eight correct. Answers are in the back of the book.

1. Suppose that an isotope of nitrogen contains seven electrons and seven neutrons. What, approximately, is the atomic mass of this element?
   (a) 7 amu
   (b) 14 amu
   (c) 49 amu
   (d) It cannot be determined from this information.

2. A mysterious particle X collides with a proton, and the two completely annihilate each other in a burst of energy. We can conclude that particle X was
   (a) a positron.
   (b) a neutron.
   (c) an electron.
   (d) an antiproton.

3. Neutrons are
   (a) stable all by themselves but unstable when in the nuclei of atoms.
   (b) unstable all by themselves but stable when in the nuclei of atoms.
   (c) stable under all conditions.
   (d) unstable under all conditions.

4. The atoms in a compound
   (a) share a single nucleus.
   (b) share protons.
   (c) share electrons.
   (d) share neutrons.

5. Examine Fig. 9-3. How many electrons are in the outer shell of the oxygen atom when the two atoms of hydrogen each share an electron with it?
   (a) 2
   (b) 6
   (c) 8
   (d) 10

6. Two different elements can never have the same number of
   (a) protons.
   (b) neutrons.
   (c) electrons.
   (d) nuclei.

7. The number of neutrons in an element's nucleus determines the
    (a) isotope of the element.
    (b) ion of the element.
    (c) atomic number of the element.
    (d) No! Neutrons never exist in atomic nuclei.

8. The mass of a neutron
    (a) is slightly greater than the mass of an electron.
    (b) is much greater than the mass of an electron.
    (c) is slightly less than the mass of a proton.
    (d) is much less than the mass of a proton.

9. Suppose that an atom of argon, whose atomic number is 18, has 16 electrons. This atom is
    (a) a positive ion.
    (b) a negative ion.
    (c) a positive isotope.
    (b) a negative isotope.

10. There were 92 different kinds of atoms discovered when scientists began to refine the atomic theory. These 92 unique entities are known as
    (a) molecules.
    (b) compounds.
    (c) isotopes.
    (d) elements.

CHAPTER 10

Basic States of Matter

Thousands of years ago, in the time of the ancient Greek and Roman civilizations, scientists believed that all things in the material universe consisted of combinations of four "elements": *earth, water, air,* and *fire*. According to this theory, different proportions of these four "elements" give materials their unique properties. This was used to explain why gold is different from salt, which in turn is different from oil. This seems primitive to us, but the ancients had keen minds. They were especially good at observing things and at seeing the "big picture."

It is interesting to speculate on what might have happened if those scientists had been allowed to expand on their knowledge for all the time between, say, 100 A.D. and today. However, such unimpeded progress did not take place. After the Roman civilization declined, the entire Western world came under a sort of collective trance in which superstition and religious dogma prevailed. At its worst, this regime was so strict that a philosopher, mathematician, or scientist who voiced an opinion different from the conventional wisdom was punished severely. Some were even put to death.

During and after the Renaissance, when scientific reasoning became a respected mode of thought once again, physical scientists discovered that there are many more than four elements and that even these elements are not the fundamental constituents of matter. However, there are three basic *states of matter* recognized by scientists today, and these are analogous, in a crude sort of way, to three of the original "elements." These states, also called *phases,* are known as *solid* (the analog of earth), *liquid* (the analog of water), and *gaseous.*

The Solid Phase

A sample of matter in the solid phase will retain its shape unless subjected to violent impact, placed under stress, or subjected to high temperatures. Examples of solids are rock, steel at room temperature, water ice, salt, wood, and plastic at room temperature.

THE ELECTRICAL FORCE

What makes a solid behave as it does? Why, if you place a concrete block on a concrete floor, does the block not gradually sink into the floor or meld with the floor so that you can't pick it up again later? Why, if you strike a brick wall with your fist, are you likely to hurt yourself rather than having your fist go into the bricks? Internally, atoms are mostly empty space; this is true even in the most dense solids we see on Earth. Why can't solid objects pass through one another the way galaxies sometimes do in outer space or the way dust clouds do in the atmosphere? They're mostly empty space too, and they can pass through each other easily.

The answer to this question lies in the nature of the electrical forces within and around atoms. Every atomic nucleus is surrounded by "shells" of electrons, all of which are negatively charged. Objects with electrical charges of the same polarity (negative-negative or positive-positive) always repel. The closer together two objects with like charge come to each other, the more forcefully they repel. Thus, even when an atom has an equal number of electrons and protons so that it is electrically neutral as a whole, the charges are concentrated in different places. The positive charge is contained in the nucleus, and the negative charge surrounds the nucleus in one or more concentric spheres.

Suppose that you could shrink down to submicroscopic size and stand on the surface of a sheet of, say, elemental aluminum. What would you see? Below you, the surface would appear something like a huge field full of basketballs (Fig. 10-1). You would find it difficult to walk on this surface because it would be irregular. However, you would find the balls quite resistant to penetration by other balls. All the balls would be negatively charged, so they would all repel each other. This would keep them from passing through each other and also would keep the surface in a stable, fixed state. The balls would be mostly empty space inside, but there wouldn't be much space in between them. They would be just about as tightly packed as spheres can be.

Outer
electron
shells

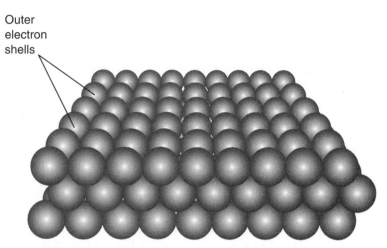

Fig. 10-1.   In a solid, the outer electron shells of the atoms are
packed tightly. (This drawing is greatly oversimplified.)

The foregoing is an oversimplification, but it should give you an idea of the reason why solids normally don't pass through each other and in fact why many solids resist penetration even by liquids such as water or gases such as air.

BRITTLENESS, MALLEABILITY, AND DUCTILITY OF SOLIDS

The atoms of elemental solids can "stack up" in various ways. This is evident in the shapes of the crystals we observe in many different solid substances. Salt, for example, has a characteristic cubical crystalline shape. The same is true of sugar. Ice crystals, however, can appear in a fantastic variety of shapes, but they always have six sides, axes, or facets. Some substances, such as iron, don't seem to form crystals under normal circumstances. Some materials, such as glass, break away along smooth but curved boundaries. Some solids can be ground up into a fine powder, whereas others defy all attempts to pulverize them.

Crystalline solids are brittle. If a sample of such a material is subjected to a blow with enough force, it will crack or shatter. These types of solids cannot be stretched or squashed or bent out of shape very much without breaking. Glass is an example, although you may have noticed that glass has a little bit of "give." You can observe the flexibility of glass if you watch the reflections from large window panes on a windy day. However, you cannot bend a straight glass rod into a donut shape.

Soft copper wire, in contrast to glass, is malleable (it can be pounded flat) and ductile (it can be stretched and bent). The same is true to some extent of iron. Gold is one of the most malleable known metals. It is expensive but can be pounded into sheets so thin that towers of buildings can be gold-plated without breaking the government budget. Aluminum is more ductile and malleable than glass, but not to the extent of soft copper or gold. Wood can be bent to a variable extent, depending on its water content, but can't be pounded into thin sheets or stretched into wire.

The brittleness, ductility, and malleability of some solids depend on the temperature. Glass, copper, and gold can be made more malleable and ductile by heating. The professional glass blower takes advantage of this phenomenon, as does the coin minter and the wire manufacturer. A person who works with wood has no such luck. If you heat wood, it gets drier and less flexible. Ultimately, if you heat glass, copper, or gold enough, it will turn into a liquid. As wood is heated, it will remain solid; then at a certain temperature it will undergo *combustion,* a rapid form of *oxidation.* That is, it will catch on fire.

HARDNESS OF SOLIDS

Some solids are literally "more solid" than others. A quantitative means of expressing hardness, known as the *Mohs scale,* classifies solids from 1 to 10. The lower numbers represent softer solids, and the higher numbers represent harder ones. The standard substances used in the Mohs scale, along with their hardness numbers, are shown in Table 10-1. The test of hardness is simple and twofold: (1) a substance always scratches something less hard than itself, and (2) a substance never scratches anything harder than itself.

An example of a soft solid is talc, which can be crumbled in the hand. Chalk is another soft solid. Wood is somewhat harder than either of these. Limestone is harder still. Then, in increasing order of hardness, there are glass, quartz, and diamond. The hardness of a solid always can be determined according to which samples scratch other samples.

Many substances have hardness numbers that change with temperature. In general, colder temperatures harden these materials. Ice is a good example. It is a fairly soft solid on a skating rink, but on the surface of Charon, the bitterly cold moon of the planet Pluto, water ice is as hard as granite.

Table 10-1   The Mohs Scale of Hardness
(Higher numbers represent harder substances.
Relative hardness is determined by attempting
to scratch one substance with another.)

Hardness Number	Standard Substance
1	Talc
2	Gypsum
3	Calcite
4	Fluorite
5	Apatite
6	Orthoclase
7	Quartz
8	Topaz
9	Corundum
10	Diamond

Hardness is measured by maintaining laboratory samples of each of the 10
substances noted in Table 10-1. A *scratch* must be a permanent mark, not just
a set of particles transferred from one substance to the other. Substances commonly have hardness values that fall between two whole numbers on the
scale. The Mohs hardness scale is not especially precise, and many scientists
prefer more elaborate methods of defining and measuring hardness.

DENSITY OF SOLIDS

The *density* of a solid is measured in terms of the number of kilograms contained in a cubic meter. That is, density is equal to mass divided by volume.
The kilogram per meter cubed (kg/m^3 or $kg \cdot m^{-3}$) is the measure of density in the International System (SI). It's a rather awkward unit in most real-
life situations. Imagine trying to determine the density of sandstone by
taking a cubical chunk of the stuff measuring 1 m on an edge and placing
it on a laboratory scale! You'd need a construction crane to lift the boulder,
and it would smash the scale.

Because of the impracticality of measuring density directly in standard
international units, the centimeter-gram-second (cgs) unit is sometimes
used instead. This is the number of grams contained in 1 cubic centimeter
(cm^3) of the material in question. Technically, it is called the *gram per centimeter cubed* (g/cm^3 or $g \cdot cm^{-3}$). To convert from grams per centimeter

cubed to kilograms per meter cubed, multiply by 1,000. Conversely, multiply by 0.001.

You certainly can think of solids that are extremely dense, such as lead. Iron is quite dense too. Aluminum is not so dense. Rocks are less dense than most common metals. Glass is about the same density as silicate rock, from which it is made. Wood and most plastics are not very dense.

PROBLEM 10-1
A sample of solid matter has a volume of 45.3 cm^3 and a mass of 0.543 kg. What is the density in grams per centimeter cubed?

SOLUTION 10-1
This problem is a little tricky because two different systems of units are used, SI for the volume and cgs for the mass. To get a meaningful answer, we must be consistent with our units. The problem requires that we express the answer in the cgs system, so we convert kilograms to grams. This means that we have to multiply the mass figure by 1,000, which tells us that the sample mass is 543 g. Determining the density in grams per centimeter cubed is now a simple arithmetic problem: Divide the mass by the volume. If d is density, m is mass, and v is volume,

$$d = m/v$$

In this case,

$$d = 543/45.3 = 12.0 \text{ g/cm}^3$$

This answer is rounded to three significant figures.

PROBLEM 10-2
Calculate the density of the sample from Problem 10-1 in kilograms per meter cubed. Do not use the conversion factor on the result of Problem 10-1. Start from scratch.

SOLUTION 10-2
This requires that we convert the volume to units in SI, that is, to meters cubed. There are 1 million, or 10^6, centimeters cubed in a meter cubed. Therefore, in order to convert this cgs volume to volume in SI, we must divide by 10^6 or multiply by 10^{-6}. This gives us 45.3 $\times$ 10^{-6} m^3, or 4.53 $\times$ 10^{-5} m^3 in standard scientific notation, as the volume of the object. Now we can divide the mass by the volume directly:

$$d = m/v$$
$$= 0.543/(4.53 \times 10^{-5})$$
$$= 0.120 \times 10^5$$
$$= 1.20 \times 10^4 \text{ kg/m}^3$$

This was rounded to three significant figures when the numerical division was performed.

MEASURING SOLID VOLUME

Suppose, in the preceding problem, that the object in question is irregular. How can we know that its volume is 45.3 cm^3? It would be easy to figure out the volume if the object were a perfect sphere or a perfect cube or a rectangular prism. Suppose, however, that it's a knobby little thing?

Scientists have come up with a clever way of measuring the volumes of irregular solids: Immerse them in a liquid. First, measure the amount of liquid in a container (Fig. 10-2a). Then measure the amount of liquid that is

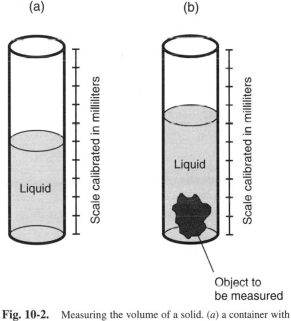

Fig. 10-2.   Measuring the volume of a solid. (*a*) a container with liquid but without the sample; (*b*) a container with the sample totally submerged in the liquid.

displaced when the object is completely submerged. This will show up as an increase in the apparent amount of liquid in the container (see Fig. 10-2b). One milliliter (1 ml) of water happens to be exactly equal to 1 cm^3, and any good chemist is bound to have a few containers marked off in milliliters. This is the way to do it, then, provided the solid does not dissolve in the liquid and that none of the liquid is absorbed into the solid.

SPECIFIC GRAVITY OF SOLIDS

Another important characteristic of a solid is its density relative to that of pure liquid water at 4°C (about 39°F). Water is at its most dense at this temperature and is assigned a relative density of 1. Substances with relative density greater than 1 will sink in pure water at 4°C, and substances with relative density less than 1 will float in pure water at 4°C. The relative density of a solid, defined in this way, is called the *specific gravity*. You often will see this abbreviated as sp gr. It is also known as relative density.

You certainly can think of substances whose specific gravity numbers are greater than 1. Examples include most rocks and virtually all metals. However, pumice, a volcanic rock that is filled with air pockets, floats on water. Most of the planets, their moons, and the asteroids and meteorites in our solar system have specific gravities greater than 1, with the exception of Saturn, which would float if a lake big enough could be found in which to test it!

Interestingly, water ice has specific gravity of less than 1, so it floats on liquid water. This property of ice is more significant than you might at first suppose. It allows fish to live underneath the frozen surfaces of lakes in the winter in the temperate and polar regions of the Earth because the layer of ice acts as an insulator against the cold atmosphere. If ice had specific gravity of greater than 1, it would sink to the bottoms of lakes during the winter months. This would leave the surfaces constantly exposed to temperatures below freezing, causing more and more of the water to freeze, until shallow lakes would become ice from the surface all the way to the bottom. In such an environment, all the fish would die during the winter because they wouldn't be able to extract the oxygen they need from the solid ice, nor would they be able to swim around in order to feed themselves. It is difficult to say how life on Earth would have evolved if water ice had a specific gravity of greater than 1.

ELASTICITY OF SOLIDS

Some solids can be stretched or compressed more easily than others. A piece of copper wire, for example, can be stretched, although a similar length of rubber band can be stretched much more. However, there is a difference in the stretchiness of these two substances that goes beyond mere extent. If you let go of a rubber band after stretching it, it will spring back to its original length, but if you let go of a copper wire, it will stay stretched.

The *elasticity* of a substance is the extent of its ability to return to its original dimensions after a sample of it has been stretched or compressed. According to this definition, rubber has high elasticity, and copper has low elasticity. Note that elasticity, defined in this way, is qualitative (it says something about how a substance behaves) but is not truly quantitative (we aren't assigning specific numbers to it). Scientists can and sometimes do define elasticity according to a numerical scheme, but we won't worry about that here. It is worth mentioning that there is no such thing as a perfectly elastic or perfectly inelastic material in the real world. Both these extremes are theoretical ideals.

This being said, suppose that there does exist a perfectly elastic substance. Such a material will obey a law concerning the extent to which it can be stretched or compressed when an external force is applied. This is called *Hooke's law:* The extent of stretching or compression of a sample of any substance is directly proportional to the applied force. Mathematically, if F is the magnitude of the applied force in newtons and s is the amount of stretching or compression in meters, then

$$s = kF$$

where k is a constant that depends on the substance. This can be written in vector form as

$$\mathbf{s} = k\mathbf{F}$$

to indicate that the stretching or compression takes place in the same direction as the applied force.

Perfectly elastic stuff can't be found in the real world, but there are plenty of materials that come close enough so that Hooke's law can be considered valid in a practical sense, provided that the applied force is not so great that a test sample of the material breaks or is crushed.

PROBLEM 10-3
Suppose that an elastic bungee cord has near perfect elasticity as long as the applied stretching force does not exceed 5.00 N. When no force is applied to the cord, it is 1.00 m long. When the applied force is 5.00 N, the band stretches to a length of 2.00 m. How long will the cord be if a stretching force of 2.00 N is applied?

SOLUTION 10-3
Applying 5.00 N of force causes the cord to become 1.00 m longer than its length when there is no force. We are assured that the cord is "perfectly elastic" as long as the force does not exceed 5.00 N. Therefore, we can calculate the value of the constant k, called the *spring constant*, in meters per newton (m/N) by rearranging the preceding formula:

$$s = kF$$

$$k = s/F$$

$$k = (1.00\ \text{m})/(5.00\ \text{N}) = 0.200\ \text{m/N}$$

provided that $F \leq 5.00$ N. Therefore, the formula for displacement as a function of force becomes

$$s = 0.200F$$

If $F = 2.00$ N, then

$$s = 0.200\ \text{m/N} \times 2.00\ \text{N} = 0.400\ \text{m}$$

This is the additional length by which the cord will "grow" when the force of 2.00 N is applied. Because the original length, with no applied force, is 1.00 m, the length with the force applied is 1.00 m + 0.400 m = 1.400 m. Theoretically, we ought to round this off to 1.40 m.

   The behavior of this bungee cord, for stretching forces between 0 and 5.00 N, can be illustrated graphically as shown in Fig. 10-3. This is a linear function; it appears as a straight line when graphed in standard rectangular coordinates. If the force exceeds 5.00 N, according to the specifications for this particular bungee cord, we have no assurance that the function of displacement versus force will remain linear. In the extreme, if the magnitude of the stretching force F is great enough, the cord will snap, and the displacement s will skyrocket to indeterminate values.

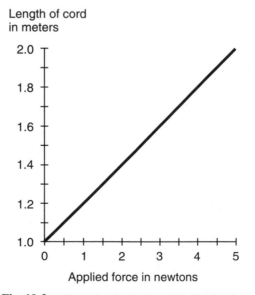

Fig. 10-3.   Illustration for Problem 10-3. The function
is linear within the range of forces shown here.

The Liquid Phase

In the liquid state or phase, a substance has two properties that distinguish it from the solid phase. First, a liquid changes shape so that it conforms to the inside boundaries of any container in which it is placed. Second, a liquid placed in an open container (such as a jar or bucket) flows to the bottom of the container and develops a defined, flat surface. At least this is the way a sample of liquid behaves in an environment where there is gravity.

DIFFUSION OF LIQUIDS

Imagine a jar on board a space ship in which the environment is weightless (there is no acceleration force). Suppose that the jar is filled with liquid, and then another liquid that does not react chemically with the first liquid is introduced into the jar. Gradually, the two liquids blend together until the mixture is uniform throughout the jar. This blending process is called *diffusion.*

In a liquid, diffusion takes place rather slowly; some liquids undergo the process faster than others. Alcohol diffuses into water at room temperature much more quickly than heavy motor oil into light motor oil. Eventually, however, when any two liquids are mixed (as long as they don't react chemically, as do an acid and a base), the mixture will become uniform throughout any container of finite size. This happens without the need for shaking the container because the molecules of a liquid are always in motion, and this motion literally causes them to push and jostle each other until they become uniformly mixed.

If the same experiment is conducted in a bucket on Earth where there is acceleration force produced by gravity, diffusion will occur, but "heavier" liquids will sink toward the bottom and "lighter" liquids will rise toward the surface. Alcohol, for example, will float on water. However, the "surface" between the alcohol and water will not be sharply defined, as is the surface between the water and the air. The motion of the molecules constantly tries to mix the two liquids. However, gravitation prevents the mixture from becoming uniform throughout the bucket unless the two liquids are of exactly the same density. We'll talk about the meaning of density for liquids shortly.

VISCOSITY OF LIQUIDS

Some liquids are "runnier" than others. You know there is a difference at room temperature between, say, water and thick molasses. If you fill a glass with water and another glass with an equal amount of molasses and then pour the contents of both glasses into the sink, the glass containing the water will empty much faster. The molasses is said to have higher *viscosity* than the water at room temperature. On an extremely hot day, the difference is less obvious than it is on a cold day, unless, of course, you have air conditioning that keeps the air in your house at the same temperature all the time.

Some liquids are far more viscous even than thick molasses. An example of a liquid with extremely high viscosity is hot tar as it is poured to make the surface of a new highway. Another example is warm petroleum jelly. These substances meet the criteria as defined above to qualify as liquids, but they are thick indeed. As the temperature goes down, these substances become less and less liquid-like and more solid-like. In fact, it's impossible to draw an exact line between the liquid and the solid phases for either of these two substances. They aren't like water; they don't freeze into ice and change state in an obvious way. As hot tar cools, where do we draw the line? How can we say, "Now, this stuff is liquid," and then 1 second later say, "Now, this stuff is solid," and be sure of the exact point of transition?

LIQUID OR SOLID?

There is not always a defined answer to the question, "Is this substance a solid or a liquid?" It can depend on the observer's point of reference. Some substances can be considered solid in the short-term time sense but liquid in the long-term sense. An example is the mantle of the Earth, the layer of rock between the crust and the core. In a long-term time sense, pieces of the crust, known as *tectonic plates,* float around on top of the mantle like scum on the surface of a hot vat of liquid. This is manifested as *continental drift* and is apparent when the Earth is evaluated over periods of millions of years. From one moment (as we perceive it) to the next, however, and even from hour to hour or from day to day, the crust seems rigidly fixed on the mantle. The mantle behaves like a solid in the short-term sense but like a liquid in the long-term sense.

Imagine that we could turn ourselves into creatures whose life spans were measured in trillions (units of 10^{12}) of years so that 1 million years seemed to pass like a moment. Then, from our point of view, Earth's mantle would behave like a liquid with low viscosity, just as water seems to us in our actual state of time awareness. If we could become creatures whose entire lives lasted only a tiny fraction of a second, then liquid water would seem to take eons to get out of a glass tipped on its side, and we would conclude that this substance was solid, or a else a liquid with extremely high viscosity.

The way we define the state of a substance can depend on the temperature, and it also can depend on the time frame over which the substance is observed.

DENSITY OF LIQUIDS

The density of a liquid is defined in three ways: *mass density, weight density,* and *particle density.* The difference between these quantities might seem theoretically subtle, but in practical situations, the difference becomes apparent.

Mass density is defined in terms of the number of kilograms per meter cubed (kg/m^3) in a sample of liquid. Weight density is defined in newtons per meter cubed (N/m^3) and is equal to the mass density multiplied by the acceleration in meters per second squared (m/s^2) to which the sample is subjected. Particle density is defined as the number of moles of atoms per meter cubed (mol/m^3), where 1 mol $\approx 6.02 \times 10^{23}$.

Let d_m be the mass density of a liquid sample (in kilograms per meter cubed), let d_w be the weight density (in newtons per meter cubed), and let d_p be the particle density (in moles per meter cubed). Let m represent the mass of the sample (in kilograms), let V represent the volume of the sample (in meters cubed), and let N represent the number of moles of atoms in the sample. Let a be the acceleration (in meters per second squared) to which the sample is subjected. Then the following equations hold:

$$d_m = m/V$$

$$d_w = ma/V$$

$$d_p = N/V$$

Alternative definitions for mass density, weight density, and particle density use the *liter,* which is equal to a thousand centimeters cubed ($1000\ cm^3$) or one-thousandth of a meter cubed ($0.001\ m^3$), as the standard unit of volume. Once in awhile you'll see the centimeter cubed (cm^3), also known as the *milliliter* because it is equal to 0.001 liter, used as the standard unit of volume.

These are simplified definitions because they assume that the density of the liquid is uniform throughout the sample.

PROBLEM 10-4

A sample of liquid measures 0.275 m^3. Its mass is 300 kg. What is its mass density in kilograms per meter cubed?

SOLUTION 10-4

This is straightforward because the input quantities are already given in SI. There is no need for us to convert from grams to kilograms, from milliliters to meters cubed, or anything like that. We can simply divide the mass by the volume:

$$d_m = m/V$$

$$= 300 \text{ kg}/0.275 \text{ m}^3$$

$$= 1090 \text{ kg/m}^3$$

We're entitled to go to three significant figures here because our input numbers are both given to three significant figures.

PROBLEM 10-5

Given that the acceleration of gravity at the Earth's surface is 9.81 m/s^2, what is the weight density of the sample of liquid described in Problem 10-4?

SOLUTION 10-5

All we need to do in this case is multiply our mass density answer by 9.81 m/s^2. This gives us

$$d_w = 1090 \text{ kg/m}^3 \times 9.81 \text{ m/s}^2$$

$$= 10,700 \text{ N/m}^3 = 1.07 \times 10^4 \text{ N/m}^3$$

Note the difference here between the nonitalicized uppercase N, which represents newtons, and the italicized uppercase *N*, which represents the number of moles of atoms in a sample.

MEASURING LIQUID VOLUME

The volume of a liquid sample is usually measured by means of a test tube or flask marked off in milliliters or liters. However, there's another way to measure the volume of a liquid sample, provided we know its chemical composition and the weight density of the substance in question. This is to weigh the sample of liquid and then divide the weight by the weight density. We must, of course, pay careful attention to the units. In particular, the weight must be expressed in newtons, which is equal to the mass in kilograms times the acceleration of gravity (9.81 m/s^2).

Let's do a mathematical exercise to show why we can measure volume in this way. Let d_w be the known weight density of a huge sample of liquid too large for its volume to be measured using a flask or test tube. Suppose that this substance has a weight of w, in newtons. If V is the volume in meters cubed, we know from the preceding formula that

$$d_w = w/V$$

because $w = ma$, where a is the acceleration of gravity. If we divide both sides of this equation by w, we get

$$d_w/w = 1/V$$

Then we can invert both sides of this equation and exchange the left-hand and the right-hand sides to obtain

$$V = w/d_w$$

All this is based on the assumption that V, w, and d_w are all nonzero quantities. This is always true in the real world; all materials occupy at least some volume, have at least some weight because of gravitation, and have some density because there is some "stuff" in a finite amount of physical space.

PRESSURE IN LIQUIDS

Have you read or been told that liquid water can't be compressed? In a simplistic sense, this is true, but it doesn't mean liquid water never exerts pressure. Liquids can and do exert pressure, as anyone who has been in a flood or a hurricane or a submarine will tell you. You can experience "water pressure" for yourself by diving down several feet in a swimming pool and noting the sensation the water produces as it presses against your eardrums.

In a fluid, the pressure, which is defined in terms of force per unit area, is directly proportional to the depth. Pressure is also directly proportional to the weight density of the liquid. Let d_w be the weight density of a liquid (in newtons per meter cubed), and let s be the depth below the surface (in meters). Then the pressure P (in newtons per meter squared) is given by

$$P = d_w s$$

If we are given the mass density d_m (in kilograms per meter cubed) rather than the weight density, the formula becomes

$$P = 9.81 d_m s$$

PROBLEM 10-6

Liquid water generally has a mass density of 1000 kg/m³. How much force is exerted on the outer surface of a cube measuring 10.000 cm on an edge that is submerged 1.00 m below the surface of a body of water?

SOLUTION 10-6

First, figure out the total surface area of the cube. It measures 10.000 cm, or 0.10000 m, on an edge, so the surface area of one face is 0.10000 m × 0.10000 m = 0.010000 m². There are six faces on a cube, so the total surface area of the object is 0.010000 m² × 6.0000 = 0.060000 m². (Don't be irritated by the "extra" zeroes here. They are important. They indicate that the length of the edge of the cube has been specified to five significant figures.)

Next, figure out the weight density of water (in newtons per meter cubed). This is 9.81 times the mass density, or 9,810 N/m³. This is best stated as 9.81 × 10³ N/m³ because we are given the acceleration of gravity to only three significant figures, and scientific notation makes this fact clear. From this point on let's revert to power-of-10 notation so that we don't fall into the trap of accidentally claiming more accuracy than that to which we're entitled.

The cube is at a depth of 1.00 m, so the water pressure at that depth is 9.81 × 10³ N/m³ × 1.00 m = 9.81 × 10³ N/m². The force F (in newtons) on the cube is therefore equal to this number multiplied by the surface area of the cube:

$$F = 9.81 \times 10^3 \text{ N/m}^2 \times 6.00000 \times 10^{-2} \text{ m}^2$$

$$= 58.9 \times 10^1 \text{ N} = 589 \text{ N}$$

PASCAL'S LAW FOR INCOMPRESSIBLE LIQUIDS

Imagine a watertight, rigid container. Suppose that there are two pipes of unequal diameters running upward out of this container. Imagine that you fill the container with an incompressible liquid such as water so that the container is completely full and the water rises partway up into the pipes. Suppose that you place pistons in the pipes so that they make perfect water seals, and then you leave the pistons to rest on the water surface (Fig. 10-4).

Because the pipes have unequal diameters, the surface areas of the pistons are different. One of the pistons has area A_1 (in meters squared), and the other has area A_2. Suppose that you push downward on piston number 1 (the one whose area is A_1) with a force F_1 (in newtons). How much upward force F_2 is produced at piston number 2 (the one whose area is A_2)? *Pascal's law* provides the answer: The forces are directly proportional to the areas of the piston faces in terms of their contact with the liquid. In the example shown by Fig. 10-4, piston number 2 is smaller than piston number 1, so the force F_2 is proportionately less than the force F_1. Mathematically, the following equations both hold:

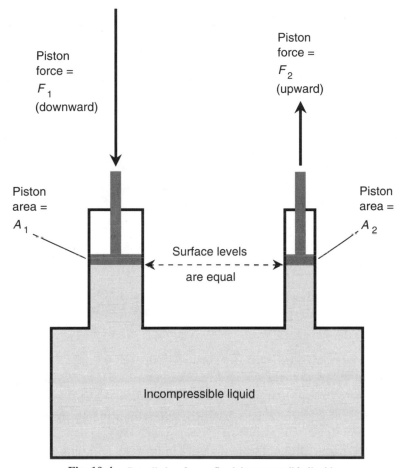

Fig. 10-4. Pascal's law for confined, incompressible liquids. The forces are directly proportional to the areas of the pistons.

$$F_1/F_2 = A_1/A_2$$

$$A_1F_2 = A_2F_1$$

When using either of these equations, we must be consistent with units throughout the calculations. In addition, the top equation is meaningful only as long as the force exerted is nonzero.

PROBLEM 10-7
Suppose that the areas of the pistons shown in Fig. 10-4 are $A_1 = 12.00$ cm^2 and $A_2 = 15.00$ cm^2. (This does not seem to agree with the illustration, where

piston number 2 looks smaller than piston number 1, but forget about that while we solve this problem.) If you press down on piston number 1 with a force of 10.00 N, how much upward force will result at piston number 2?

SOLUTION 10-7
At first, you might think that we have to convert the areas of the pistons to meters squared in order to solve this problem. In this case, however, it is sufficient to find the ratio of the areas of the pistons because both areas are given to us in the same units:

$$A_1/A_2 = 12.00 \text{ cm}^2/15.00 \text{ cm}^2$$

$$= 0.8000$$

Thus we know that $F_1/F_2 = 0.8000$. We are given $F_1 = 10.00$ N, so it is easy to solve for F_2:

$$10.00/F_2 = 0.8000$$

$$1/F_2 = 0.08000$$

$$F_2 = 1/0.08000 = 12.50 \text{ N}$$

We are entitled to four significant figures throughout this calculation because all the input data were provided to this degree of precision.

The Gaseous Phase

The gaseous phase of matter is similar to the liquid phase insofar as a gas will conform to the boundaries of a container or enclosure. However, a gas is much less affected by gravity than a liquid. If you fill up a bottle with a gas, there is no discernible surface to the gas. Another difference between liquids and gases is the fact that gases generally are compressible.

GAS DENSITY

The density of a gas can be defined in three ways, exactly after the fashion of liquids. Mass density is defined in terms of the number of kilograms per meter cubed (kg/m^3) that a sample of gas has. The weight density is defined in newtons per meter cubed (N/m^3) and is equal to the mass density multiplied by the acceleration in meters per second squared (m/s^2) to which the sample is subjected. The particle density is defined as the number of moles of atoms per meter cubed (mol/m^3) in a parcel or sample of gas, where 1 mol $\approx 6.02 \times 10^{23}$.

DIFFUSION IN SMALL CONTAINERS

Imagine a rigid enclosure, such as a glass jar, from which all the air has been pumped. Suppose that this jar is placed somewhere out in space, far away from the gravitational effects of stars and planets and where space itself is a near vacuum (compared with conditions on Earth anyhow). Suppose that the temperature is the same as that in a typical household. Now suppose that a certain amount of elemental gas is pumped into the jar. The gas distributes itself quickly throughout the interior of the jar.

Now suppose that another gas that does not react chemically with the first gas is introduced into the chamber to mix with the first gas. The diffusion process occurs rapidly, so the mixture is uniform throughout the enclosure after a short time. It happens so fast because the atoms in a gas move around furiously, often colliding with each other, and their motion is so energetic that they spread out inside any container of reasonable size (Fig. 10-5a).

What would happen if the same experiment were performed in the presence of a gravitational field? As you can guess, the gases would still mix uniformly inside the jar. This happens with all gases in containers of reasonable size.

Planetary atmospheres, such as that of our own Earth, consist of mixtures of various gases. In the case of our planet, approximately 78 percent of the gas in the atmosphere at the surface is nitrogen, 21 percent is oxygen, and 1 percent is made up of many other gases, including argon, carbon dioxide, carbon monoxide, hydrogen, helium, ozone (oxygen molecules with three atoms rather than the usual two), and tiny quantities of some gases that would be poisonous in high concentrations, such as chlorine and methane. These gases blend uniformly in containers of reasonable size, even though some of them have atoms that are far more massive than others. Diffusion, again, is responsible.

GASES NEAR A PLANET

Now imagine the gaseous shroud that surrounds a reasonably large planet, such as our own Earth. Gravitation attracts some gas from the surrounding space. Other gases are ejected from the planet's interior during volcanic activity. Still other gases are produced by the biologic activities of plants and animals, if the planet harbors life. In the case of Earth, some gases are produced by industrial activity and by the combustion of fossil fuels.

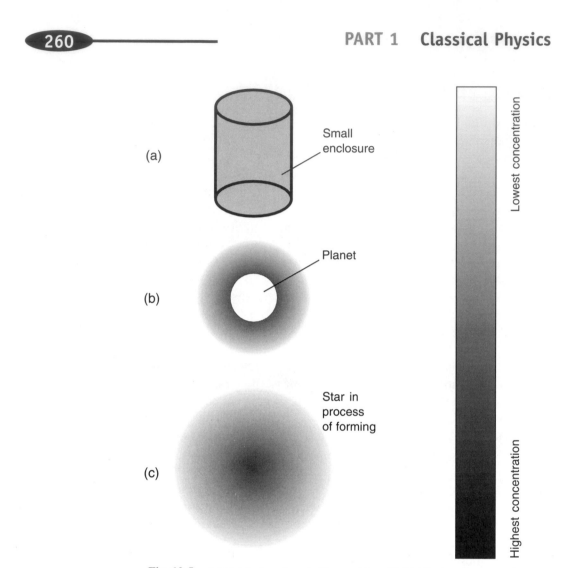

Fig. 10-5. (*a*) Distribution of gas inside a container. (*b*) Distribution of gas around
a planet with an atmosphere. (*c*) Distribution of gas in a star as it is forming.
Darkest shading indicates highest concentration.

All the gases in the Earth's atmosphere tend to diffuse, but because there
is an essentially unlimited amount of "outer space" and only a finite amount
of gas, and because the gravitational pull of the Earth is greater near the surface than far out in space, the diffusion takes place in a different way than
inside a small container. The greatest concentration of gas molecules (particle density) occurs near the surface, and it decreases with increasing altitude
(see Fig. 10-5*b*). The same is true of the number of kilograms per meter
cubed of the atmosphere, that is, the mass density of the gas.

On the large scale of the Earth's atmosphere, yet another effect takes place. For a given number of atoms or molecules per meter cubed, some gases are more massive than others. Hydrogen is the least massive; helium is light too. Oxygen is more massive, and carbon dioxide is more massive still. The most massive gases tend to sink toward the surface, whereas the least massive gases rise up high, and some of their atoms escape into outer space or are not captured permanently by the Earth's gravitation.

There are no distinct boundaries, or layers, from one type of gas to another in the atmosphere. Instead, the transitions are gradual and vague. This is good, because if the gases of the atmosphere were stratified in a defined way, we would have no oxygen down here on the surface. Instead, we'd be smothered in some noxious gas such as carbon dioxide or sulfur dioxide.

GASES IN OUTER SPACE

Outer space was once believed to be a perfect vacuum. However, this is not the case. There is plenty of stuff out there, and much of it is hydrogen and helium gas. (There are also trace amounts of heavier gases and some solid rocks and ice chunks as well.) All the atoms in outer space interact gravitationally with all the others. This might be hard to imagine at first, but if you think about it, there's no escaping it. Even a single atom of hydrogen exerts a gravitational pull on another atom 1 million km away.

The motion of atoms in outer space is almost random but not quite. The slightest perturbation in the randomness of the motion gives gravitation a chance to cause the gas to clump into huge clouds. Once this process begins, it can continue until a globe of gas forms in which the central particle density is significant (see Fig. 10-5c). As gravitation continues to pull the atoms in toward the center, the mutual attraction among the atoms there becomes greater and greater. If the gas cloud has some spin, it flattens into an oblate spherical shape and eventually into a disk with a bulge at the center. A vicious circle ensues, and the density in the central region skyrockets. The gas pressure in the center rises, and this causes it to heat up. Ultimately, it gets so hot that *nuclear fusion* begins, and a star is born. Similar events among the atoms of the gas on a smaller scale can result in the formation of asteroids, planets, and planetary moons.

GAS PRESSURE

Unlike most liquids, gases can be compressed. This is why it is possible to fill up hundreds of balloons with a single, small tank of helium gas and why it is possible for a scuba diver to breathe for a long time from a single small tank of air.

Imagine a container whose volume (in meters cubed) is equal to V. Suppose that there are N moles of atoms of a particular gas inside this container, which is surrounded by a perfect vacuum. We can say certain things about the pressure P, in newtons per meter squared, that the gas exerts outward on the walls of the container. First, P is proportional to N, provided that V is held constant. Second, if V increases while N remains constant, P will decrease. These things are apparent intuitively.

There is another important factor—temperature—involved when it comes to gases under pressure when they expand and contract. The involvement of temperature T, generally measured in degrees above absolute zero (the absence of all heat), is significant and inevitable in gases. When a parcel of gas is compressed, it heats up; when it is decompressed, it cools off. Heating up a parcel of gas will increase the pressure, if all other factors are held constant, and cooling it off will reduce the pressure. The behavior of matter, especially liquids and gases, under conditions of varying temperature and pressure is a little complicated, so the entire next chapter is devoted to this subject.

 **Quiz**

Refer to the text in this chapter if necessary. A good score is eight correct. Answers are in the back of the book.

1. Suppose that a sample of gas has 5.55×10^{18} atoms in 1 cubic centimeter. What is the particle density?
   (a) 922 mol/m^3
   (b) 9.22 mol/m^3
   (c) 1.08 mol/m^3
   (d) 33.4 mol/m^3

2. Suppose that a rubber band has a spring constant of 0.150 m/N for stretching forces ranging from 0 to 10 N. If the band measures 1.00 m when 3.00 N of stretching force is applied, how long with the band be when 5.00 N of stretching force is applied?

(a) 1.30 m

(b) 1.67 m

(c) 0.66 m

(d) It cannot be determined from this information.

3. Refer to Fig. 10-4. Suppose that the areas of the pistons are $A_1 = 0.0600$ m^2 and $A_2 = 0.0300$ m^2. If you press down on piston number 1 with a force of 5.00 N, how much upward force will result at piston number 2?

(a) 30.0 N

(b) 10.0 N

(c) 3.00 N

(d) 2.50 N

4. The Mohs scale is based on a solid's ability or tendency to

(a) boil when heated.

(b) fracture under stress.

(c) be stretched or compressed.

(d) scratch or be scratched.

5. A solid object with a specific gravity of less than 1 will

(a) float on liquid water.

(b) mix evenly and stay mixed with liquid water.

(c) sink in liquid water.

(d) dissolve in liquid water.

6. For a perfectly elastic substance,

(a) the extent of stretching is inversely proportional to the applied force.

(b) the extent of stretching is independent of the applied force.

(c) the extent of stretching is directly proportional to the applied force.

(d) the amount of force necessary to break the object in half is inversely proportional to the length of the object.

7. A substance with high malleability

(a) can be pounded into a thin, flat layer.

(b) is extremely brittle.

(c) readily fills any container into which it is poured.

(d) diffuses easily into other liquids.

8. Diffusion of gases at room temperature occurs because

(a) there are not many atoms per unit volume.

(b) the atoms or molecules move rapidly.

(c) gases always have high specific gravity.

(d) gases dissolve easily in one another.

9. Suppose that a sample of substance has a mass density of 8.6×10^3 kg/m^3 on Earth. If this sample is taken to Mars, where gravity is only about 37 percent as strong as it is on Earth, the mass density will be

(a) 3.2 kg/m^3.

(b) 8.6 kg/m^3.

(c) 23 kg/m^3.

(d) Impossible to calculate based on the information given.

10. A vat contains 100.00 m^3 of liquid, and the liquid masses of 2.788×10^5 kg. What is the mass density of the liquid?

(a) 2.788×10^7 kg/m^3

(b) 2,788 g/cm^3

(c) 2,788 kg/m^3

(d) It is impossible to answer this based on the data given.

CHAPTER 11

Temperature, Pressure, and Changes of State

When a confined sample of gas gets hotter, its pressure increases. The converse of this is also true: When a gas is put under increasing pressure, it gets hotter. However, what do we mean when we talk about *heat* and *temperature?* What effects do heat and temperature have on matter? In this chapter we will find out. We'll also see how matter can change state with changes in temperature or pressure.

What Is Heat?

Heat is a special kind of energy transfer that can take place from one material object, place, or region to another. For example, if you place a kettle of water on a hot stove, heat is transferred from the burner to the water. This is *conductive heat,* also called *conduction* (Fig. 11-1*a*). When an infrared lamp, sometimes called a *heat lamp,* shines on your sore shoulder, energy is transferred to your skin surface from the filament of the lamp; this is *radiative heat,* also called *radiation* (see Fig. 11-1*b*). When a blower-type electric heater warms up a room, air passes through the heating elements and is blown by a fan into the room, where the heated air rises and mixes with the rest of the air in the room. This is *convective heat,* also called *convection* (see Fig. 11-1*c*).

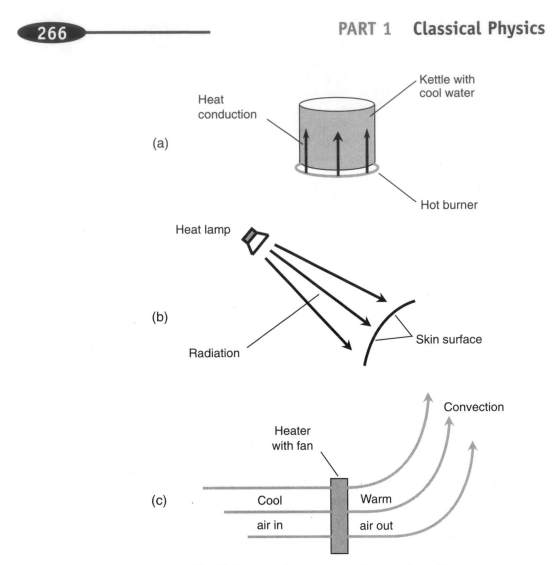

Fig. 11-1. Examples of energy transfer, in the form of heat, by conduction (*a*), radiation (*b*), and convection (*c*).

Heat is not quite the same thing as energy, although the units of heat and energy are defined in the same physical dimensions. Heat is the transfer of energy that occurs when conduction, radiation, and/or convection take place. Sometimes the energy transfer takes place in only one of these three modes, but sometimes it occurs in two or all three.

THE CALORIE

The unit of heat used by physicists is the *calorie*. You've heard and read this word many times (probably too often, but that's a subject for another book).

The calorie that scientists use is a much smaller unit than the *calorie* used by nutritionists—only 1/1,000 as large—and the scientific use of the term usually refers to inanimate things, whereas the nutritional term involves biologic processes.

The calorie (cal) in which we, as physicists, are interested is the amount of energy transfer that raises or lowers the temperature of exactly one gram (1 g) of pure liquid water by exactly one degree Celsius (1°C). The *kilocalorie* (kcal), equivalent to the nutritionist's *calorie,* is the amount of energy transfer that will raise or lower the temperature of 1 kg, or 1,000 g, of pure liquid water by 1°C. This holds true only as long as the water is liquid during the entire process. If any of the water freezes, thaws, boils, or condenses, this definition falls apart. At standard atmospheric pressure at Earth's surface, in general, this definition holds for temperatures between approximately 0°C (the freezing point of water) and 100°C (the boiling point).

SPECIFIC HEAT

Pure liquid water requires 1 calorie per gram (1 cal/g) to warm it up or cool it down by 1°C (provided it is not at the melting/freezing temperature or the vaporization/condensation temperature, as we shall shortly see.) However, what about oil, alcohol, or salt water? What about solids such as steel or wood? What about gases such as air? It is not so simple then. A certain, fixed amount of heat energy will raise or lower the temperatures of fixed masses of some substances more than others. Some matter takes more than 1 cal/g to get hotter or cooler by 1°C; some matter takes less. Pure liquid water takes exactly 1 cal/g to warm up or cool down by 1°C simply because this is the substance on which the definition of the calorie is based. It is one of those things scientists call a *convention.*

Suppose that we have a sample of some mysterious liquid. Call it *substance X.* We measure out 1 gram (1.00 g), accurate to three significant figures, of this liquid by pouring some of it into a test tube placed on a laboratory balance. Then we transfer 1 calorie (1.00 cal) of energy to substance X. Suppose that, as a result of this energy transfer, substance X increases in temperature by 1.20°C? Obviously, substance X is not water because it behaves differently from water when it receives a transfer of energy. In order to raise the temperature of 1.00 g of this stuff by 1.00°C, it takes somewhat less than 1.00 cal of heat. To be exact, at least insofar as we are allowed by the rules of significant figures, it will take 1.00/1.20 = 0.833 cal to raise the temperature of this material by 1.00°C.

Now suppose that we have a sample of another material, this time a solid. Let's call it *substance Y.* We carve a chunk of it down until we have a piece that masses 1.0000 g, accurate to five significant figures. Again, we can use our trusty laboratory balance for this purpose. We transfer 1.0000 cal of energy to substance Y. Suppose that the temperature of this solid goes up by 0.80000°C? This material accepts heat energy in a manner different from either liquid water or substance X. It takes a little more than 1.0000 cal of heat to raise the temperature of 1.0000 g of this material by 1.0000°C. Calculating to the allowed number of significant figures, we can determine that it takes 1.0000/0.80000 = 1.2500 cal to raise the temperature of this material by 1.0000°C.

We're onto something here: a special property of matter called the *specific heat,* defined in units of calories per gram per degree Celsius (cal/g/°C). Let's say that it takes c calories of heat to raise the temperature of exactly 1 gram of a substance by exactly 1°C. For water, we already know that $c = 1$ cal/g/°C, to however many significant figures we want. For substance X, $c = 0.833$ cal/g/°C (to three significant figures), and for substance Y, $c = 1.2500$ cal/g/°C (to five significant figures).

Alternatively, c can be expressed in kilocalories per kilogram per degree Celsius (kcal/kg/°C), and the value for any given substance will be the same. Thus, for water, $c = 1$ kcal/kg/°C, to however many significant figures we want. For substance X, $c = 0.833$ kcal/kg/°C (to three significant figures), and for substance Y, $c = 1.2500$ kcal/kg/°C (to five significant figures).

THE BRITISH THERMAL UNIT (BTU)

In some applications, a completely different unit of heat is used: the *British thermal unit* (Btu). You've heard this unit mentioned in advertisements for furnaces and air conditioners. If someone talks about Btus alone in regard to the heating or cooling capacity of a furnace or air conditioner, this is an improper use of the term. They really mean to quote the rate of energy transfer in Btus per hour, not the total amount of energy transfer in Btus.

The Btu is defined as the amount of heat that will raise or lower the temperature of exactly one pound (1 lb) of pure liquid water by one degree Fahrenheit (1°F). Does something seem flawed about this definition? If you're uneasy about it, you have a good reason. What is a *pound?* It depends where you are. How much water weighs 1 lb? On the Earth's surface, it's approximately 0.454 kg or 454 g. On Mars, however, it takes about 1.23 kg of liquid water to weigh 1 lb. In a weightless environment, such as on board a space vessel orbiting the Earth or coasting through deep space, the definition of Btu is meaningless because there is no such thing as a pound at all.

Despite these flaws, the Btu is still used once in awhile, so you should be acquainted with it. Specific heat is occasionally specified in Btus per pound per degree Fahrenheit (Btu/lb/°F). In general, this is not the same number, for any given substance, as the specific heat in cal/g/°C.

PROBLEM 11-1
Suppose that you have 3.00 g of a certain substance. You transfer 5.0000 cal of energy to it, and the temperature goes up uniformly throughout the sample by 1.1234°C. It does not boil, condense, freeze, or thaw during this process. What is the specific heat of this stuff?

SOLUTION 11-1
Let's find out how much energy is accepted by 1.00 g of the matter in question. We have 3.00 g of the material, and it gets 5.0000 cal, so we can conclude that each gram gets ⅓ of this 5.0000 cal, or 1.6667 cal.

We're told that the temperature rises uniformly throughout the sample. This is to say, it doesn't heat up more in some places than in other places. It gets hotter to exactly the same extent everywhere. Therefore, 1.00 g of this stuff goes up in temperature by 1.1234°C when 1.6667 cal of energy is transferred to it. How much heat is required to raise the temperature by 1.0000°C? This is the number c we seek, the specific heat. To get c, we must divide 1.6667 cal/g by 1.1234°C. This gives us $c = 1.4836$ cal/g/°C. Because we are given the mass of the sample to only three significant figures, we must round this off to 1.48 cal/g/°C.

Temperature

Now that we've defined heat, what do we mean by the term *temperature?* You have an intuitive idea of this; the temperature is generally higher in the summer than in the winter, for example. Temperature is a quantitative expression of the average kinetic energy contained in matter. This is the most familiar definition. In general, for any given substance, the higher the temperature, the faster the atoms and molecules dance around.

Temperature can be expressed in another way. For example, to measure the temperatures of distant stars, planets, and nebulae in outer space, astronomers look at the way they emit electromagnetic (EM) energy in the form of visible light, infrared, ultraviolet, and even radio waves and x-rays. By examining the intensity of this radiation as a function of the wavelength, astronomers come up with a value for the *spectral temperature* of the distant matter or object.

When energy is allowed to flow from one substance into another in the form of heat, the temperatures try to equalize. Ultimately, if the energy-transfer process is allowed to continue for a long enough time, the temperatures of the two objects will become the same, unless one of the substances is driven away (for example, steam boiling off of a kettle of water). The kinetic energy of everything in the entire universe is trying to level off to a state of equilibrium. It won't succeed in your lifetime or mine or even during the lifetime of the Sun and solar system, but it will keep trying anyway, and gradually it is succeeding. This process is known as *heat entropy.*

THE CELSIUS (OR CENTIGRADE) SCALE

Up to now, we've been talking rather loosely about temperature and usually have expressed it in terms of the Celsius or centigrade scale (°C). This is based on the behavior of water at the surface of the Earth under normal atmospheric pressure and at sea level.

If you have a sample of ice that is extremely cold and you begin to warm it up, it will eventually start to melt as it accepts heat from the environment. The ice, and the liquid water produced as it melts, is assigned a temperature value of 0°C by convention (Fig. 11-2a). As you continue to pump energy into the chunk of ice, more and more of it will melt, and its temperature will stay at 0°C. It won't get any hotter because it is not yet all liquid and doesn't yet obey the rules for pure liquid water.

Once all the water has become liquid and as you keep pumping energy into it, its temperature will start to increase (see Fig. 11-2b). For awhile, the water will remain liquid and will get warmer and warmer, obeying the 1 cal/g/°C rule. Eventually, however, a point will be reached where the water starts to boil, and some of it changes to the gaseous state. The liquid water temperature, and the water vapor that comes immediately off of it, is then assigned a value of 100°C by convention (see Fig. 11-2c).

Now there are two definitive points—the *freezing point* of water and the *boiling point*—at which there exist two specific numbers for temperature. We can define a scheme to express temperature based on these two points. This is the *Celsius temperature scale,* named after the scientist who supposedly first came up with the idea. Sometimes it is called the *centigrade temperature scale* because one degree of temperature in this scale is equal to 1/100 of the difference between the melting temperature of pure water at sea level and the boiling temperature of pure water at sea level. The prefix multiplier *centi-* means "1/100," so *centigrade* literally means "graduations of 1/100."

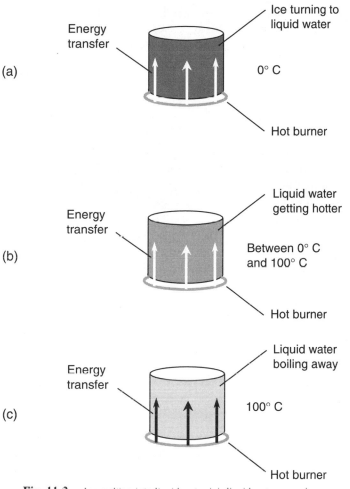

Fig. 11-2.   Ice melting into liquid water (*a*), liquid water warming up without boiling (*b*), and liquid water starting to boil (*c*).

THE KELVIN SCALE

Of course, it is possible to freeze water and keep cooling it down or boil it all away into vapor and then keep heating it up. Temperatures can plunge far below 0°C and can rise far above 100°C. Are there limits to how low or how high the temperature can get?

Interestingly, there is an absolute limit to how low the temperature in degrees Celsius can become, but there is no limit on the upper end of the scale. We might take extraordinary efforts to cool a chunk of ice down to see how cold we can make it, but we can never chill it down to a temperature any lower than approximately 273 degrees Celsius below zero (−273°C).

This is known as *absolute zero*. An object at absolute zero can't transfer energy to anything else because it possesses no energy to transfer. There is believed to be no such object in our universe, although some atoms in the vast reaches of intergalactic space come close.

Absolute zero is the basis for the Kelvin temperature scale (K). A temperature of $-273.15°C$ is equal to 0 K. The size of the Kelvin degree is the same as the size of the Celsius degree, so $0°C = 273.15$ K, and $+ 100°C = 373.15$ K. Note that the degree symbol is not used with K.

On the high end, it is possible to keep heating matter up indefinitely. Temperatures in the cores of stars rise into the millions of degrees Kelvin. No matter what the actual temperature, the difference between the Kelvin temperature and the Celsius temperature is always 273.15 degrees.

Sometimes, Celsius and Kelvin figures can be considered equivalent. When you hear someone say that a particular star's core has a temperature of 30 million K, it means the same thing as 30 million °C for the purposes of most discussions because ±273.15 is a negligible difference value relative to 30 million.

THE RANKINE SCALE

The Kelvin scale isn't the only one that exists for defining absolute temperature, although it is by far the most commonly used. Another scale, called the *Rankine scale* (°R), also assigns the value zero to the coldest possible temperature. The difference is that the Rankine degree is exactly $5/9$ as large as the Kelvin degree. Conversely, the Kelvin degree is exactly $9/5$, or 1.8 times, the size of the Rankine degree.

A temperature of 50 K is the equivalent of 90°R; a temperature of 360°R is the equivalent of 200 K. To convert any reading in °R to its equivalent in K, multiply by $5/9$. Conversely, to convert any reading in K to its equivalent in °R, multiply by $9/5$, or exactly 1.8.

The difference between the Kelvin and the Rankine scales is significant at extreme readings. If you hear someone say that a star's core has a temperature of 30 million °R, they are talking about the equivalent of approximately 16.7 million K. However, you are not likely to hear anyone use Rankine numbers.

THE FAHRENHEIT SCALE

In much of the English-speaking world, and especially in the United States, the Fahrenheit temperature scale (°F) is used by laypeople. A Fahrenheit degree is the same size as a Rankine degree. However, the scale is situated differently. The melting temperature of pure water ice at sea level is +32°F, and the boiling point of pure liquid water is +212°F. Thus, +32°F corresponds to 0°C, and +212°F corresponds to +100°C. Absolute zero is approximately −459.67°F.

The most common temperature conversions you are likely to perform involve changing a Fahrenheit reading to Celsius, or vice versa. Formulas have been developed for this purpose. Let F be the temperature in °F, and let C be the temperature in °C. Then, if you need to convert from °F to °C, use this formula:

$$F = 1.8C + 32$$

If you need to convert a reading from °C to °F, use this formula:

$$C = \frac{5}{9}(F - 32)$$

While the constants in these equations are expressed only to one or two significant figures (1.8, $\frac{5}{9}$, and 32), they can be considered mathematically exact for calculation purposes.

Figure 11-3 is a nomograph you can use for approximate temperature conversions in the range from −50°C to +150°C.

When you hear someone say that the temperature at the core of a star is 30 million °F, the Rankine reading is about the same, but the Celsius and Kelvin readings are only about $\frac{5}{9}$ as great.

PROBLEM 11-2
What is the Celsius equivalent of a temperature of 72°F?

SOLUTION 11-2
To solve this, simply use the preceding formula for converting Fahrenheit temperatures to Celsius temperatures:

$$C = \frac{5}{9}(F - 32)$$

Thus, in this case:

$$C = \frac{5}{9}(72 - 32)$$

$$= \frac{5}{9} \times 40 = 22.22°C$$

We are justified in carrying this out to only two significant figures because this is the extent of the accuracy of our input data. Thus we can conclude that the Celsius equivalent is 22°C.

PROBLEM 11-3
What is the Kelvin equivalent of a temperature of 80.0°F?

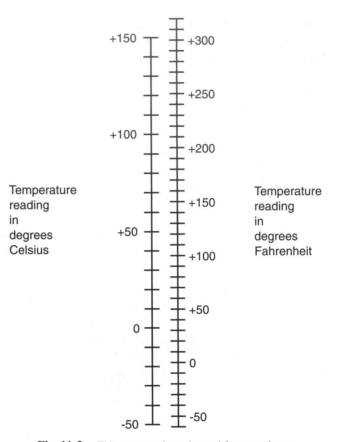

+150 ∓   ∓ +300

                          ∓ +250

+100 ∓
                          ∓ +200

Temperature   Temperature
reading   ∓ +150   reading
in   in
degrees   +50 ∓   degrees
Celsius   ∓ +100   Fahrenheit

                          ∓ +50

   0 ∓
                          ∓ 0

                          ∓ -50
 -50 ∓

Fig. 11-3.   This nomograph can be used for approximate
conversions between temperatures in °F and °C.

SOLUTION 11-3

There are two ways to approach this problem. The first is to convert the
Fahrenheit reading to Rankine and then convert this figure to Kelvin. The second is to convert the Fahrenheit reading to Celsius and then convert this figure to Kelvin. Let's use the second method because the Rankine scale is
hardly ever used for anything.

Using the preceding formula to convert from °F to °C, we get

$$C = \tfrac{5}{9}(80.0 - 32)$$

$$= \tfrac{5}{9} \times 48.0 = 26.67°C$$

Let's not round our answer off yet because we have another calculation to
perform. Remember that the difference between readings in °C and K is
always equal to 273.15 degrees. The Kelvin figure is the greater of the two.

Thus we must add 273.15 to our Celsius reading. If *K* represents the temperature in K, then

$$K = C + 273.15$$
$$= 26.67 + 273.15$$
$$= +299.82 \text{ K}$$

Now we should round our answer off. Because we are given our input data to three significant figures, we can say that the Kelvin temperature equivalent is +300 K.

Some Effects of Temperature

Temperature can affect the volume of or the pressure exerted by a sample of matter. You are familiar with the fact that most metals expand when they are heated; some expand more than others.

TEMPERATURE, VOLUME, AND PRESSURE

A sample of gas confined to a rigid container will exert more and more pressure on the walls of the container as the temperature goes up. If the container is flexible, such as a balloon, the volume of the gas will increase. Similarly, if you take a container with a certain amount of gas in it and suddenly make the container bigger without adding any more gas, the drop in pressure will produce a decrease in temperature. If you have a rigid container with gas in it and then some of the gas is allowed to escape (or is pumped out), the drop in pressure will chill the container. This is why, for example, a compressed-air canister gets cold when you use it to blow dust out of your computer keyboard.

Liquids behave a little more strangely. The volume of the liquid water in a kettle and the pressure it exerts on the kettle walls don't change when the temperature goes up and down unless the water freezes or boils. Some liquids, however, unlike water, expand when they heat up. Mercury is an example. This is how an old-fashioned thermometer works.

Solids, in general, expand when the temperature rises and contract when the temperature falls. In many cases you don't notice this expansion and contraction. Does your desk look bigger when the room is 30°C than it does when the room is only 20°C? Of course not. But it is! You don't see the difference because it is microscopic. However, the bimetallic strip in the thermostat, which controls the furnace or air conditioner, bends considerably when one of its metals expands or contracts just a tiny bit more than the other. If you hold such a strip near a hot flame, you actually can watch it curl up or straighten out.

STANDARD TEMPERATURE AND PRESSURE (STP)

To set a reference for temperature and pressure against which measurements can be made and experiments conducted, scientists have defined *standard temperature and pressure* (STP). This is a more or less typical state of affairs at sea level on the Earth's surface when the air is dry.

The standard temperature is 0°C (32°F), which is the freezing point or melting point of pure liquid water. Standard pressure is the air pressure that will support a column of mercury 0.760 m (just a little less than 30 in) high. This is the proverbial 14.7 pounds per inch squared (lb/in^2), which translates to approximately 1.01×10^5 newtons per meter squared (N/m^2).

Air is surprisingly massive. We don't think of air as having significant mass, but this is because we're immersed in it. When you dive only a couple of meters down in a swimming pool, you don't feel a lot of pressure and the water does not feel massive, but if you calculate the huge amount of mass above you, it might scare you out of the water! The density of dry air at STP is approximately 1.29 kg/m^3. A parcel of air measuring 4.00 m high by 4.00 m deep by 4.00 m wide, the size of a large bedroom, masses 82.6 kg. In Earth's gravitational field, that translates to 182 pounds, the weight of a good-sized, full-grown man.

THERMAL EXPANSION AND CONTRACTION

Suppose that we have a sample of solid material that expands when the temperature rises. This is the usual case, but some solids expand more per degree Celsius than others. The extent to which the height, width, or depth of a solid (its *linear dimension*) changes per degree Celsius is known as the *thermal coefficient of linear expansion.*

For most materials, within a reasonable range of temperatures, the coefficient of linear expansion is constant. This means that if the temperature changes by 2°C, the linear dimension will change twice as much as it would if the temperature variation were only 1°C. However, there are limits to this, of course. If you heat a metal up to a high enough temperature, it will become soft and ultimately will melt or even burn or vaporize. If you cool the mercury in a thermometer down enough, it will freeze. Then the simple length-versus-temperature rule no longer applies.

In general, if s is the difference in linear dimension (in meters) produced by a temperature change of T (in degrees Celsius) for an object whose linear dimension (in meters) is d, then the thermal coefficient of linear expansion, symbolized by the lowercase Greek letter alpha (α), is given by this equation:

$$\alpha = s/(dT)$$

When the linear dimension increases, consider s to be positive; when it decreases, consider s to be negative. Rising temperatures produce positive values of T; falling temperatures produce negative values of T.

The coefficient of linear expansion is defined in meters per meter per degree Celsius. The meters cancel out in this expression of units, so the technical quantity is per degree Celsius, symbolized /°C.

PROBLEM 11-4
Imagine a metal rod 10.000 m long at 20.00°C. Suppose that this rod expands to a length of 10.025 m at 25.00°C. What is the thermal coefficient of linear expansion?

SOLUTION 11-4
This rod increases in length by 0.025 m for a temperature increase of 5.00°C. Therefore, $s = 0.025$, $d = 10$, and $T = 5.00$. Plugging these numbers into the preceding formula, we get

$$\alpha = 0.025/(10 \times 5.00)$$
$$- 0.00050/°C - 5.0 \times 10^{-4}/°C$$

We are justified in going to only two significant figures here because that is as accurate as our data are for the value of s.

PROBLEM 11-5
Suppose that $\alpha = 2.50 \times 10^{-4}/°C$ for a certain substance. Imagine a cube of this substance whose volume V_1 is 8.000 m^3 at a temperature of 30.0°C. What will be the volume V_2 of the cube if the temperature falls to 20.0°C?

SOLUTION 11-5
It is important to note the word *linear* in the definition of α. This means that the length of each edge of the cube of this substance will change according to the thermal coefficient of linear expansion.

We can rearrange the preceding general formula for α so that it solves for the change in linear dimension s as follows:

$$s = \alpha dT$$

where *T* is the temperature change (in degrees Celsius) and *d* is the initial linear dimension (in meters). Because our object is a cube, the initial length *d* of each edge is 2.000 m (the cube root of 8.000, or 8.000$^{1/3}$). Because the temperature falls, $T = -10.0$. Therefore,

$$s = 2.50 \times 10^{-4} \times (-10.0) \times 2.000$$

$$= -2.50 \times 10^{-3} \times 2.000$$

$$= -5.00 \times 10^{-3} \text{m} = -0.00500 \text{ m}$$

This means that the length of each side of the cube at 20°C will be 2.000 − 0.00500 = 1.995 m. The volume of the cube at 20.0°C is therefore 1.995^3 = 7.940149875 m^3. Because our input data are given to only three significant figures, we must round this off to 7.94 m^3.

Temperature and States of Matter

When matter is heated or cooled, it often does things other than simply expanding or contracting, or exerting increased or decreased pressure. Sometimes it undergoes a *change of state*. This happens when solid ice melts into liquid water or when water boils into vapor, for example.

THAWING AND FREEZING

Consider our old friend, water. Imagine that it is late winter in a place such as northern Wisconsin and that the temperature of the water ice on the lake is exactly 0°C. The ice is not safe to skate on, as it was in the middle of the winter, because the ice has become "soft." It is more like slush than ice. It is partly solid and partly liquid. Nevertheless, the temperature of this soft ice is 0°C.

As the temperature continues to rise, the slush gets softer. It becomes proportionately more liquid water and less solid ice. However, its temperature remains at 0°C. Eventually, all the ice melts into liquid. This can take place with astonishing rapidity. You might leave for school one morning and see the lake nearly "socked in" with slush and return in the evening to find it almost entirely thawed. Now you can get the canoe out! But you won't want to go swimming. The liquid water will stay at 0°C until all the ice is gone. Only then will the temperature begin to rise slowly.

Consider now what happens in the late autumn. The weather, and the water, is growing colder. The water finally drops to 0°C. The surface begins to freeze. The temperature of this new ice is 0°C. Freezing takes place until the whole lake surface is solid ice. The weather keeps growing colder (a lot colder if you live in northern Wisconsin). Once the surface is entirely solid ice, the temperature of the ice begins to fall below 0°C, although it remains at 0°C at the boundary just beneath the surface where solid ice meets liquid water. The layer of ice gets thicker. The ice near the surface can get much colder than 0°C. How much colder depends on various factors, such as the severity of the winter and the amount of snow that happens to fall on top of the ice and insulate it against the bitter chill of the air.

The temperature of water does not follow exactly along with the air temperature when heating or cooling takes place in the vicinity of 0°C. Instead, the water temperature follows a curve something like that shown in Fig. 11-4. In part *a*, the air temperature is getting warmer; in part *b*, it is getting colder. The water "stalls" as it thaws or freezes. Other substances exhibit this same property when they thaw or freeze.

HEAT OF FUSION

It takes a certain amount of energy to change a sample of solid matter to its liquid state, assuming that the matter is of the sort that can exist in either of these two states. (Water, glass, most rocks, and most metals fill this bill, but

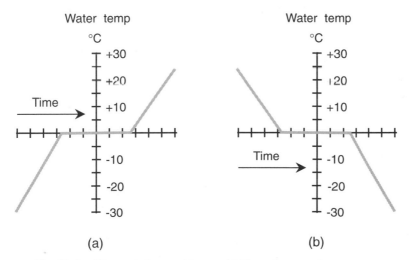

Fig. 11-4. Water as it thaws and freezes. (*a*) The environmental temperature is getting warmer, and the ice is thawing. (*b*) The environmental temperature is getting colder, and the liquid water is freezing.

wood does not.) In the case of ice formed from pure water, it takes 80 cal to convert 1 g of ice at 0°C to 1 g of pure liquid water at 0°C. This quantity varies for different substances and is called the *heat of fusion* for the substance.

In the reverse scenario, if 1 g of pure liquid water at 0°C freezes completely solid and becomes ice at 0°C, it gives up 80 cal of heat. The heat of fusion is thus expressed in calories per gram (cal/g). It also can be expressed in kilocalories per kilogram (kcal/kg) and will yield exactly the same numbers as the cal/g figures for all substances. When the substance is something other than water, then the freezing/melting point of that substance must be substituted for 0°C in the discussion.

Heat of fusion is sometimes expressed in calories per mole (cal/mol) rather than in calories per gram. However, unless it is specifically stated that the units are intended to be expressed in calories per mole, you should assume that they are expressed in calories per gram.

If the heat of fusion (in calories per gram) is symbolized h_f, the heat added or given up by a sample of matter (in calories) is h, and the mass of the sample (in grams) is m, then the following formula holds:

$$h_f = h/m$$

PROBLEM 11-6
Suppose that a certain substance melts and freezes at +400°C. Imagine a block of this material whose mass is 1.535 kg, and it is entirely solid at +400°C. It is subjected to heating, and it melts. Suppose that it takes 142,761 cal of energy to melt the substance entirely into liquid at +400°C. What is the heat of fusion for this material?

SOLUTION 11-6
First, we must be sure we have our units in agreement. We are given the mass in kilograms; to convert it to grams, multiply by 1,000. Thus m = 1,535 g. We are given that h = 142,761 cal. Therefore, we can use the preceding formula directly:

$$h_f = 142,761/1535 = 93.00 \text{ cal/g}$$

This is rounded off to four significant figures because this is the extent of the accuracy of our input data.

BOILING AND CONDENSING

Let's return to the stove, where a kettle of water is heating up. The temperature of the water is exactly +100°C, but it has not yet begun to boil. As heat is continually applied, boiling begins. The water becomes proportionately more vapor and less liquid. However, the temperature remains at +100°C. Eventually, all the liquid has boiled away, and only vapor is left. Imagine that we have captured all this vapor in an enclosure, and in the process of the water's boiling away, all the air has been driven out of the enclosure and replaced by water vapor. The stove burner, an electric type, keeps on heating the water even after all of it has boiled into vapor.

At the moment when the last of the liquid vanishes, the temperature of the vapor is $+100°C$. Once all the liquid is gone, the vapor can become hotter than $+100°C$. The ultimate extent to which the vapor can be heated depends on how powerful the burner is and on how well insulated the enclosure is.

Consider now what happens if we take the enclosure, along with the kettle, off the stove and put it into a refrigerator. The environment, and the water vapor, begins to grow colder. The vapor temperature eventually drops to $+100°C$. It begins to condense. The temperature of this liquid water is $+100°C$. Condensation takes place until all the vapor has condensed. (But hardly any of it will condense back in the kettle. What a mess!) We allow a bit of air into the chamber near the end of this experiment to maintain a reasonable pressure inside. The chamber keeps growing colder; once all the vapor has condensed, the temperature of the liquid begins to fall below $+100°C$.

As is the case with melting and freezing, the temperature of water does not follow exactly along with the air temperature when heating or cooling takes place in the vicinity of $+100°C$. Instead, the water temperature follows a curve something like that shown in Fig. 11-5. In part *a,* the air temperature

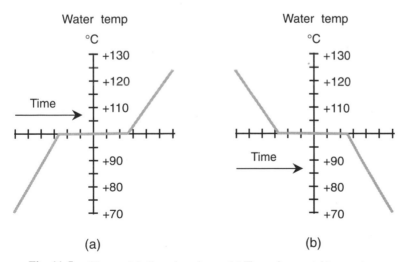

Fig. 11-5.   Water as it boils and condenses. (*a*) The environmental temperature is getting warmer, and the liquid water is boiling. (*b*) The environmental temperature is getting colder, and the water vapor is condensing.

is getting warmer; in part *b,* it is getting colder. The water temperature "stalls" as it boils or condenses. Other substances exhibit this same property when they boil or condense.

HEAT OF VAPORIZATION

It takes a certain amount of energy to change a sample of liquid to its gaseous state, assuming that the matter is of the sort that can exist in either of these two states. In the case of water, it takes 540 cal to convert 1 g of liquid at $+100°C$ to 1 g of pure water vapor at $+100°C$. This quantity varies for different substances and is called the *heat of vaporization* for the substance.

In the reverse scenario, if 1 g of pure water vapor at $+100°C$ condenses completely and becomes liquid water at $+100°C$, it gives up 540 cal of heat. The heat of vaporization is expressed in the same units as heat of fusion, that is, in calories per gram (cal/g). It also can be expressed in kilocalories per kilogram (kcal/kg) and will yield exactly the same numbers as the cal/g figures for all substances. When the substance is something other than water, then the boiling/condensation point of that substance must be substituted for $+100°C$.

Heat of vaporization, like heat of fusion, is sometimes expressed in calories per mole (cal/mol) rather than in cal/g. However, this is not the usual case.

If the heat of vaporization (in calories per gram) is symbolized h_v, the heat added or given up by a sample of matter (in calories) is h, and the mass of the sample (in grams) is m, then the following formula holds:

$$h_v = h/m$$

This is the same as the formula for heat of fusion, except that h_v has been substituted for h_f.

PROBLEM 11-7
Suppose that a certain substance boils and condenses at $+500°C$. Imagine a beaker of this material whose mass is 67.5 g, and it is entirely liquid at $+500°C$. It heat of vaporization is specified as 845 cal/g. How much heat, in calories and in kilocalories, is required to completely boil away this liquid?

SOLUTION 11-7
Our units are already in agreement: grams for m and calories per gram for h_v. We must manipulate the preceding formula so that it expresses the heat h (in calories) in terms of the other given quantities. This can be done by multiplying both sides by m, giving us this formula:

$$h = h_v m$$

Now it is simply a matter of plugging in the numbers:

$$h = 845 \times 67.5$$
$$= 5.70 \times 10^4 \text{ cal} = 57.0 \text{ kcal}$$

This has been rounded off to three significant figures, the extent of the accuracy of our input data.

Quiz

Refer to the text in this chapter if necessary. A good score is eight correct. Answers are in the back of the book.

1. A decrease in temperature can cause a gas to
   (a) boil away into vapor.
   (b) turn into a liquid.
   (c) exert increased pressure in a rigid container.
   (d) do nothing; it will remain a gas no matter what.

2. Suppose that there is a vessel containing 1.000 kg of liquid. It has a specific heat of 1.355 cal/g/°C. Suppose that it is exactly at its vaporization temperature of +235.0°C, and 5,420 cal of energy is transferred to the liquid in the form of heat. The temperature of the liquid in the vessel after the application of this heat will be
   (a) +235.0°C.
   (b) +239.0°C.
   (c) +231.0°C.
   (d) impossible to calculate from this information.

3. The British thermal unit
   (a) expresses rate of energy transfer, not total energy transfer.
   (b) is the unit of temperature preferred by scientists in England.
   (c) is equal to 1,000 cal.
   (d) is based on weight and therefore varies in size depending on gravitation.

4. A rod of metal is 4.5653100 m long at a temperature of 36.000°C. The temperature is lowered until the rod shrinks to 4.5643000 m. The temperature is measured as 35.552°C. What, approximately, is the thermal coefficient of linear expansion for this metal?
   (a) 0.00225/°C
   (b) 4.94 × 10^4/°C
   (c) 2.21 × 10^{-4}/°C
   (d) It cannot be determined from the information given here.

5. Suppose that a substance boils and condenses at $+217°C$. Imagine a beaker of this material whose mass is 135 g, and it is entirely liquid at $+217°C$. Its heat of vaporization is 451 cal/g. How much heat, in kilocalories, is required to completely boil away this liquid?

  (a) 6.089×10^4
  (b) 3.341
  (c) 60.89
  (d) 0.2993

6. The heat of fusion of a substance refers to

  (a) the temperature necessary to produce a nuclear fusion reaction.
  (b) the heat required to liquefy a vapor at its condensation temperature.
  (c) the heat required to liquefy a solid at its melting temperature.
  (d) the temperature at which a liquid becomes a gas.

7. The coldest possible temperature is

  (a) $0°R$.
  (b) $0°C$.
  (c) $0°F$.
  (d) meaningless; there is no coldest possible temperature.

8. You develop a severe cough and feel weak, dizzy, and exhausted. It is mid-winter, and the temperature is below $0°F$ outside. You take your temperature with a thermometer that registers $40.2°C$. You don't recall the formulas for converting Celsius to Fahrenheit, but you do remember that normal body temperature is about $98.6°F$. You call your doctor and tell him the reading of $40.2°C$. What is he likely to say?

  (a) "Don't worry, your temperature is normal. Drink some water and take a nap."
  (b) "You have a high fever. Have someone drive you to my office or to urgent care right now. Don't try to drive yourself."
  (c) "Your temperature is a little bit below normal. Have some hot soup."
  (d) "What did you do? Spend all day out in the cold without a coat on? You have hypothermia (dangerously low body temperature). Have someone drive you to the emergency room. Don't try to drive yourself."

9. The hottest possible temperature is

  (a) $+30,000,000°F$.
  (b) $+30,000,000°C$.
  (c) $+30,000,000$ K.
  (d) meaningless; there is no known hottest possible temperature.

10. The kilocalorie is a unit of

  (a) temperature.
  (b) power.
  (c) heat.
  (d) pressure.

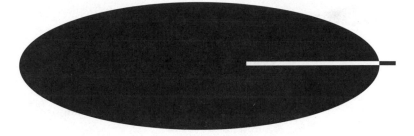

Test: Part One

Do not refer to the text when taking this test. A good score is at least 37 correct. Answers are in the back of the book. It is best to have a friend check your score the first time so that you won't memorize the answers if you want to take the test again.

1. The joule is equivalent to
   (a) a newton-meter.
   (b) a kilogram-meter.
   (c) a watt.
   (d) a candela.
   (e) an erg.

2. The acceleration vector of the Earth in its orbit around the Sun points
   (a) straight out away from the Sun.
   (b) in the same direction as the instantaneous motion of the Earth.
   (c) straight inward toward the Sun.
   (d) at a right angle to the plane of the Earth's orbit around the Sun.
   (e) nowhere; it is the zero vector.

3. Which of the following is not expressible as a vector quantity?
   (a) Displacement
   (b) Velocity
   (c) Acceleration
   (d) Mass
   (e) Force

4. A car travels 200 km in 3 hours (3.00 hr). What is its average speed?
   (a) 18.5 m/s
   (b) 0.0540 m/s
   (c) 54.0 m/s
   (d) 66.7 m/s
   (e) It can't be calculated based on this information.

5. What is the difference between a chemical reaction and an atomic reaction?
   (a) A chemical reaction involves fission or fusion of nuclei, but an atomic reaction does not.
   (b) An atomic reaction involves fission or fusion of nuclei, but a chemical reaction does not.
   (c) An atomic reaction requires antimatter, but a chemical reaction does not.
   (d) A chemical reaction requires an atomic reaction to set it off.
   (e) There is no difference; chemical and atomic reactions are exactly the same thing.

6. What is the distinction between mass and weight?
   (a) Nothing. They are different names for the same thing.
   (b) Weight is the force produced by gravity on an object having mass.
   (c) Mass is the force produced by gravity on an object having weight.
   (d) Mass depends on the speed of an object, but weight does not.
   (e) Mass is an expression of the resistance of an object to movement, but weight is an expression of the number of atoms in an object.

7. A highly malleable substance
   (a) can be pounded into thin sheets.
   (b) evaporates at a low temperature.
   (c) changes state directly from solid to gaseous.
   (d) does not melt when heated but burns instead.
   (e) is extremely brittle.

8. Suppose that an object has a mass of 540 g and is lifted 25.5 m. How much potential energy will it attain? Take the value of the magnitude of Earth's gravitational acceleration to be 9.81 m/s^2.
   (a) 0.208 J
   (b) 135 J
   (c) 208 J
   (d) 463 J
   (e) 1.35×10^5 J

9. Which of the following two types of particles have roughly the same mass?
   (a) A proton and an electron
   (b) A neutron and an electron
   (c) A proton and a neutron
   (d) A proton and a helium nucleus
   (e) A neutron and a helium nucleus

10. The newton is a unit of
    (a) mass.
    (b) frequency.
    (c) gravitational acceleration.
    (d) temperature.
    (e) None of the above.

11. There are 1.806×10^{24} atoms in a sample of liquid measuring 100.0 ml in volume. What is the mass density of this sample?
    (a) 1.806×10^{28} mol/cm^3
    (b) 1.806×10^{28} g/cm^3
    (c) 0.03000 mol/cm^3
    (d) 0.003000 mol/m^3
    (e) It cannot be calculated from the information given.

12. How long does it take a ray of light to travel 3.00×10^6 km through free space?
    (a) 100 s
    (b) 10.0 s
    (c) 1.00 s
    (d) 0.100 s
    (e) 0.0100 s

13. Pascal's law involves the behavior of
    (a) confined incompressible liquids.
    (b) objects in gravitational fields.
    (c) substances that are cooled to extremely low temperatures.
    (d) substances when they change from one phase of matter to another.
    (e) molecules in a vacuum.

14. An example of diffusion is illustrated by
    (a) the way molasses is less "runny" than water.
    (b) the way liquid dye gradually disperses in a glass of water without stirring or shaking.
    (c) the way water in a lake freezes at the surface but not underneath.
    (d) the way a liquid develops a flat surface in an environment where there is gravity.
    (e) any of the above.

15. Approximately how many kilometers are there between the Earth's north geographic pole and the equator, as measured in a great circle over the surface?
    (a) 10 million km
    (b) 1 million km
    (c) 100,000 km
    (d) 10,000 km
    (e) 1,000 km

16. A spherical ball bearing has a radius of 0.765 cm. The mass of this ball bearing is 25.5 g. What is the density?
    (a) 7.12 g/cm^3
    (b) 33.3 g/cm^3
    (c) 57.0 g/cm^3
    (d) 13.6 g/cm^3
    (e) It cannot be calculated from this information.

17. The acceleration of a moving object has a constant magnitude of $a = 3.00$ m/s^2. The object starts out from a dead stop at $t = 0.00$ s and moves in a straight-line path. How far will it have traveled from its starting point at $t = 5.00$ s?
    (a) 0.120 m
    (b) 7.50 m
    (c) 15.0 m
    (d) 37.5 m
    (e) It can't be calculated from this information.

18. In an ideal system,
    (a) there is no heat.
    (b) there is no mass.
    (c) there is no friction.
    (d) all the objects move at the same speed.
    (e) all the objects move in the same direction.

19. The rate at which energy is expended can be defined in terms of
    (a) joules.
    (b) newton-meters.
    (c) newtons per meter.
    (d) kilogram-meters.
    (e) joules per second.

20. An object whose mass is 2.00 kg is lifted upward a distance of 3.55 m against the pull of gravity on a planet where the gravitational acceleration is 5.70 m/s^2. How much work is done?
    (a) 40.5 kg $\cdot$ m^2/s^2
    (b) 7.10 kg $\cdot$ m^2/s^2
    (c) 11.4 kg $\cdot$ m^2/s^2
    (d) 1.25 kg $\cdot$ m^2/s^2
    (e) It cannot be calculated from this information.

21. Specific heat can be expressed in
    (a) calories per second.
    (b) kilocalories per hour.
    (c) Btus per hour.
    (d) calories per gram.
    (e) calories per gram per degree Celsius.

22. The momentum vector of a moving object is directly affected by all the following except
    (a) the speed of the object.
    (b) the velocity of the object.
    (c) the mass of the object.
    (d) the direction in which the object moves.
    (e) the temperature of the object.

23. Suppose that a rod measuring 1.00 m long of a certain metal has a thermal coefficient of linear expansion of $3.32 \times 10^{-5}/°C$. If the rod is heated from 10 to 20°C, how much longer will the rod become?
    (a) 0.0000332 m
    (b) 0.000332 m
    (c) 0.00332 m
    (d) 0.032 m
    (e) No! The rod will not lengthen. It will shorten.

24. You are told that a sample of matter masses 365 μg. How much is this in kilograms?
    (a) 3.65×10^{5}
    (b) 36.5
    (c) 0.365
    (d) 3.65×10^{-7}
    (e) It depends on the intensity of the gravitational field in which the mass is measured.

25. Nuclear physicists commonly use particle accelerators to
    (a) weigh heavy objects such as boulders.
    (b) determine the masses of distant stars and galaxies.
    (c) fabricate elements that don't occur naturally.
    (d) evacuate all the air from an enclosure.
    (e) generate powerful beams of light.

26. The Einstein equation $E = mc^2$ might be applied directly to calculate
    (a) the energy produced by a matter-antimatter reaction.
    (b) the energy produced by the electrolysis of water.
    (c) the energy produced when oxygen and iron react to form rust.
    (d) the mass produced when two atoms of hydrogen and one atom of oxygen combine to form a molecule of water.
    (e) the mass of the chlorine liberated by the electrolysis of salt water.

27. The velocity of gravity at the surface of the Earth is
    (a) approximately 9.8 m.
    (b) approximately 9.8 m/s.
    (c) approximately 9.8 m/s^2.
    (d) approximately 9.8 m/s^3.
    (e) none of the above; the expression "velocity of gravity" is meaningless.

28. Suppose that a certain substance melts and freezes at +200°C. Imagine a block of this material whose mass is 500 g, and it is entirely solid at +200°C. It is subjected to heating, and it melts. Suppose that it takes 50,000 cal of energy to melt the substance entirely into liquid at +200°C. What is the heat of fusion for this material?
    (a) It cannot be determined from this information.
    (b) 0.100 cal/g

    (c) 1.00 cal/g

    (d) 10.0 cal/g

    (e) 100 cal/g

29. The term *heat of vaporization* refers to
    - (a) the amount of heat necessary to convert a certain amount of liquid matter to the gaseous state.
    - (b) the amount of heat necessary to convert a certain amount of solid matter to the liquid state.
    - (c) the heat produced when a substance vaporizes.
    - (d) the heat absorbed by a substance when it liquefies.
    - (e) a device used for vaporizing water.

30. The base International Unit of visible-light brightness is the
    - (a) lumen.
    - (b) lux.
    - (c) candela.
    - (d) joule.
    - (e) watt.

31. What is a fundamental difference between speed and velocity?
    - (a) Velocity depends on gravitation, but speed does not.
    - (b) Velocity depends on mass, but speed does not.
    - (c) Velocity depends on force, but speed does not.
    - (d) Velocity depends on direction, but speed does not.
    - (e) There is no difference; speed and velocity are exactly the same thing.

32. Potential energy can be defined in terms of
    - (a) newton-meters.
    - (b) meters per second squared.
    - (c) kilograms per second.
    - (d) kilograms per meter.
    - (e) kilogram-meters.

33. A car whose mass is 900 kg travels east along a highway at 50.0 km/h. What is the magnitude of the momentum vector of this car?
    - (a) 450 kg · m/s
    - (b) 1.25×10^4 kg · m/s
    - (c) 4.50×10^6 kg · m/s
    - (d) 2.25×10^6 kg · m/s
    - (e) 6.48×10^4 kg · m/s

34. Refer to test question 28. What is the heat of vaporization for this material?
    - (a) It cannot be determined from this information.
    - (b) 0.100 cal/g
    - (c) 1.00 cal/g
    - (d) 10.0 cal/g
    - (e) 100 cal/g

35. A marble massing 1.5 g and a large brick massing 5.5 kg are dropped from the same height on the moon. Which object will strike the surface of the moon with greater force?
    (a) The marble; it concentrates its mass in a smaller volume.
    (b) The brick; it has greater mass and is "pulled down" with greater force.
    (c) Neither; they will strike with the same amount of force.
    (d) This is a meaningless question because it involves units that don't agree.
    (e) We need more information to determine the answer.

36. The Rankine scale
    (a) is the same as the centigrade scale.
    (b) has degrees that are the same size as centigrade degrees, but the zero point is different.
    (c) has degrees that are the same size as Fahrenheit degrees, but the zero point is different.
    (d) is commonly used by laypeople in European countries.
    (e) is preferred when talking about extremely high temperatures.

37. Which of the following statements is *not always* true?
    (a) Acceleration is a quantitative representation of the change in velocity of a moving object.
    (b) The acceleration vector of a moving object always points in the same direction as the velocity vector.
    (c) The instantaneous velocity of a moving object can change even if the direction remains constant.
    (d) The instantaneous velocity of a moving object can change even if the speed remains constant.
    (e) Speed is a scalar quantity.

38. If an electron is stripped from an electrically neutral atom, the result is
    (a) a different isotope of the same element.
    (b) a different element altogether.
    (c) a nuclear reaction.
    (d) a change in the atomic number.
    (e) none of the above.

39. A room becomes warmer by 10 K. How much warmer has it become in degrees Fahrenheit?
    (a) 18°F
    (b) 5.6°F
    (c) 10°F
    (d) 273.15°F
    (e) It cannot be calculated from this information.

40. The atomic mass of an element is approximately equal to
    (a) the sum of the number of protons and neutrons in the nucleus.
    (b) the number of protons in the nucleus.

    (c) the number of neutrons in the nucleus.

    (d) the sum of the number of protons and electrons.

    (e) the sum of the number of neutrons and electrons.

41. A typical carbon atom has six neutrons and six protons in its nucleus. If one of the protons is taken out of the nucleus somehow but no other aspect of the atom is changed, which of the following best describes the new atom?

    (a) It will be a different isotope of carbon.

    (b) It will be a negative carbon ion.

    (c) It will be a positive carbon ion.

    (d) It will be an atom of a different element.

    (e) None of the above.

42. Suppose that there is an airtight chamber that can be enlarged and reduced in size. The chamber is located in a laboratory on the Earth's surface. The chamber contains N moles of oxygen molecules. The volume of the chamber is reduced rapidly without adding or removing any molecules. All the following things will happen *except*

    (a) the temperature of the oxygen will go down.

    (b) the mass density of the oxygen will increase.

    (c) the oxygen will exert increased pressure on the walls of the chamber.

    (d) the particle density of the oxygen will increase.

    (e) the weight density of the oxygen will increase.

43. Heat is an expression of

    (a) energy radiation.

    (b) energy convection.

    (c) energy conduction.

    (d) energy transfer.

    (e) kinetic energy.

44. Energy and mass are intimately and absolutely related, according to Albert Einstein's hypothesis, by

    (a) gravitation.

    (b) the rate of energy transfer.

    (c) the rate of mass transfer.

    (d) the speed of light squared.

    (e) the intensity of acceleration.

45. Suppose that a motor is used to drive a mechanical system. The motor draws 500 W from the power source that runs it, and the mechanical power produced by the system is 400 W. What is the efficiency of this system, expressed as a ratio?

    (a) 0.800

    (b) 1.25

    (c) 80.0

    (d) 125

    (e) It cannot be calculated from this information.

46. A sample of matter is placed on a table. It retains its shape. Based only on this information, we can be certain that this material is
    (a) a gas.
    (b) a liquid.
    (c) a solid.
    (d) frozen.
    (e) less dense than the table.

47. The unit of force in the International System is the
    (a) gram.
    (b) dyne.
    (c) pound.
    (d) kilogram.
    (e) newton.

48. A positron is the same thing as
    (a) a proton.
    (b) an antiproton.
    (c) an electron.
    (d) an antielectron.
    (e) nothing; there is no such thing as a positron.

49. Fill in the blanks so that the following sentence is true: "A substance that appears as a liquid with low viscosity in one _____ can appear to be a liquid with high viscosity, even at the same temperature and pressure, when observed in another _____."
    (a) gravitational field
    (b) container
    (c) quantity
    (d) time sense
    (e) state of matter

50. The megahertz (MHz) is a unit of
    (a) mass.
    (b) time.
    (c) speed.
    (d) quantity.
    (e) none of the above.

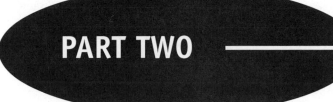

PART TWO

Electricity, Magnetism, and Electronics

CHAPTER 12

Direct Current

You now have a solid grasp of physics math, and you know the basics of classical physics. It is time to delve into the workings of things that can't be observed directly. These include particles, and forces among them, that make it possible for you to light your home, communicate instantly with people on the other side of the world, and in general do things that would have been considered magical a few generations ago.

What Does Electricity Do?

When I took physics in middle school, they used 16-millimeter celluloid film projectors. Our teacher showed us several films made by a well-known professor. I'll never forget the end of one of these lectures, in which the professor said, "We evaluate electricity not by knowing what it is, but by scrutinizing what it does." This was a great statement. It really expresses the whole philosophy of modern physics, not only for electricity but also for all phenomena that aren't directly tangible. Let's look at some of the things electricity does.

CONDUCTORS

In some materials, electrons move easily from atom to atom. In others, the electrons move with difficulty. And in some materials, it is almost impossible to get them to move. An *electrical conductor* is a substance in which the electrons are highly mobile.

The best conductor, at least among common materials, at room temperature is pure elemental silver. Copper and aluminum are also excellent electrical conductors. Iron, steel, and various other metals are fair to good conductors of electricity. Some liquids are good conductors. Mercury is one example. Salt water is a fair conductor. Gases are, in general, poor conductors because the atoms or molecules are too far apart to allow a free exchange of electrons. However, if a gas becomes ionized, it can be a fair conductor of electricity.

Electrons in a conductor do not move in a steady stream like molecules of water through a garden hose. They pass from atom to atom (Fig. 12-1). This happens to countless atoms all the time. As a result, trillions of electrons pass a given point each second in a typical electric circuit.

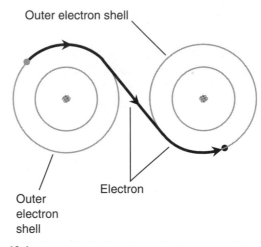

Fig. 12-1. In an electrical conductor, electrons pass easily from atom to atom. This drawing is greatly simplified.

Imagine a long line of people, each one constantly passing a ball to his or her neighbor on the right. If there are plenty of balls all along the line, and if everyone keeps passing balls along as they come, the result is a steady stream of balls moving along the line. This represents a good conductor. If the people become tired or lazy and do not feel much like passing the balls along, the rate of flow decreases. The conductor is no longer very good.

INSULATORS

If the people refuse to pass balls along the line in the preceding example, the line represents an *electrical insulator.* Such substances prevent electric currents from flowing, except in very small amounts under certain circumstances.

Most gases are good electrical insulators (because they are poor conductors). Glass, dry wood, paper, and plastics are other examples. Pure water is a good electrical insulator, although it conducts some current when minerals are dissolved in it. Metal oxides can be good insulators, even though the metal in pure form is a good conductor.

An insulating material is sometimes called a *dielectric*. This term arises from the fact that it keeps electric charges apart, preventing the flow of electrons that would equalize a charge difference between two places. Excellent insulating materials can be used to advantage in certain electrical components such as capacitors, where it is important that electrons not be able to flow steadily. When there are two separate regions of electric charge having opposite polarity (called *plus* and *minus, positive* and *negative,* or + and −) that are close to each other but kept apart by an insulating material, that pair of charges is called an *electric dipole.*

RESISTORS

Some substances, such as carbon, conduct electricity fairly well but not very well. The conductivity can be changed by adding impurities such as clay to a carbon paste. Electrical components made in this way are called *resistors.* They are important in electronic circuits because they allow for the control of current flow. The better a resistor conducts, the lower is its resistance; the worse it conducts, the higher is the resistance.

Electrical resistance is measured in *ohms,* sometimes symbolized by the uppercase Greek letter omega (Ω). In this book we'll sometimes use the symbol Ω and sometimes spell out the word *ohm* or *ohms,* so that you'll get used to both expressions. The higher the value in ohms, the greater is the resistance, and the more difficult it is for current to flow. In an electrical system, it is usually desirable to have as low a resistance, or *ohmic value,* as possible because resistance converts electrical energy into heat. This heat is called *resistance loss* and in most cases represents energy wasted. Thick wires and high voltages reduce the resistance loss in long-distance electrical lines. This is why gigantic towers, with dangerous voltages, are employed in large utility systems.

CURRENT

Whenever there is movement of charge carriers in a substance, there is an *electric current.* Current is measured in terms of the number of *charge carriers,* or particles containing a unit electric charge, passing a single point in 1 second.

Charge carriers come in two main forms: electrons, which have a unit negative charge, and *holes,* which are electron absences within atoms and which carry a unit positive charge. Ions can act as charge carriers, and in some cases, atomic nuclei can too. These types of particles carry whole-number multiples of a unit electric charge. Ions can be positive or negative in polarity, but atomic nuclei are always positive.

Usually, a great many charge carriers go past any given point in 1 second, even if the current is small. In a household electric circuit, a 100-W light bulb draws a current of about 6 quintillion (6×10^{18}) charge carriers per second. Even the smallest minibulb carries a huge number of charge carriers every second. It is ridiculous to speak of a current in terms of charge carriers per second, so usually it is measured in *coulombs per second* instead. A *coulomb* (symbolized C) is equal to approximately 6.24×10^{18} electrons or holes. A current of 1 coulomb per second (1 C/s) is called an *ampere* (symbolized A), and this is the standard unit of electric current. A 60-W bulb in a common table lamp draws about 0.5 A of current.

When a current flows through a resistance—and this is always the case, because even the best conductors have resistance—heat is generated. Sometimes visible light and other forms of energy are emitted as well. A light bulb is deliberately designed so that the resistance causes visible light to be generated. However, even the best incandescent lamp is inefficient, creating more heat than light energy. Fluorescent lamps are better; they produce more light for a given amount of current. To put this another way, they need less current to give off a certain amount of light.

In physics, electric current is theoretically considered to flow from the positive to the negative pole. This is known as *conventional current.* If you connect a light bulb to a battery, therefore, the conventional current flows out of the positive terminal and into the negative terminal. However, the electrons, which are the primary type of charge carrier in the wire and the bulb, flow in the opposite direction, from negative to positive. This is the way engineers usually think about current.

STATIC ELECTRICITY

Charge carriers, particularly electrons, can build up or become deficient on objects without flowing anywhere. You've experienced this when walking on a carpeted floor during the winter or in a place where the humidity is low. An excess or shortage of electrons is created on and in your body. You acquire a charge of *static electricity*. It's called *static* because it doesn't go anywhere. You don't feel this until you touch some metallic object that is connected to an electrical ground or to some large fixture, but then there is a discharge, accompanied by a spark and a small electric shock. It is the current, during this discharge, that causes the sensation.

If you were to become much more charged, your hair would stand on end because every hair would repel every other one. Objects that carry the same electric charge, caused by either an excess or a deficiency of electrons, repel each other. If you were massively charged, the spark might jump several centimeters. Such a charge is dangerous. Static electric (also called *electrostatic*) charge buildup of this magnitude does not happen with ordinary carpet and shoes, fortunately. However, a device called a *Van de Graaff generator,* found in some high-school physics labs, can cause a spark this large. You have to be careful when using this device for physics experiments.

On the grand scale of the Earth's atmosphere, lightning occurs between clouds and between clouds and the surface. This spark is a greatly magnified version of the little spark you get after shuffling around on a carpet. Until the spark occurs, there is an electrostatic charge in the clouds, between different clouds, or between parts of a cloud and the ground. In Fig. 12-2, four types of lightning are shown. The discharge can occur within a single cloud (*intracloud lightning,* part *a*), between two different clouds (*intercloud lightning,* part *b*), or from a cloud to the surface (*cloud-to-ground lightning,* part *c*), or from the surface to a cloud (*ground-to-cloud lightning,* part *d*). The direction of the current flow in these cases is considered to be the same as the direction in which the electrons move. In cloud-to-ground or ground-to-cloud lightning, the charge on the Earth's surface follows along beneath the thunderstorm cloud like a shadow as the storm is blown along by the prevailing winds.

The current in a lightning stroke can approach 1 million A. However, it takes place only for a fraction of a second. Still, many coulombs of charge are displaced in a single bolt of lightning.

ELECTROMOTIVE FORCE

Current can flow only if it gets a "push." This push can be provided by a buildup of electrostatic charges, as in the case of a lightning stroke. When the charge builds up, with positive polarity (shortage of electrons) in one place and negative polarity (excess of electrons) in another place, a powerful

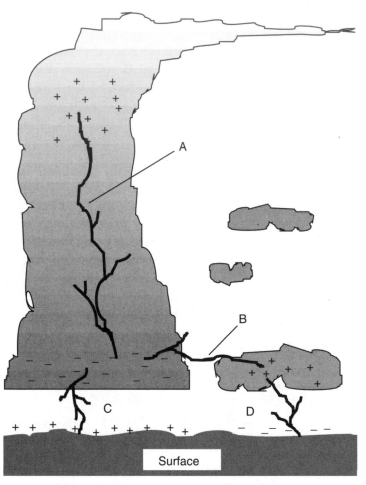

Fig. 12-2. (*a*) Lightning can occur within a single cloud (intracloud),
(*b*) between clouds (intercloud), or between a cloud and the surface
(*c*) cloud to ground or (*d*) ground to cloud.

electromotive force (emf) exists. This effect, also known as *voltage* or *electrical potential,* is measured in *volts* (symbolized V).

Ordinary household electricity has an effective voltage of between 110 and 130 V; usually it is about 117 V. A car battery has an emf of 12 V (6 V in some older systems). The static charge that you acquire when walking on a carpet with hard-soled shoes can be several thousand volts. Before a discharge of lightning, millions of volts exist.

An emf of 1 V, across a resistance of 1 Ω, will cause a current of 1 A to flow. This is a classic relationship in electricity and is stated generally as

Ohm's law. If the emf is doubled, the current is doubled. If the resistance is doubled, the current is cut in half. This law of electricity will be covered in detail a little later.

It is possible to have an emf without having current flow. This is the case just before a lightning bolt occurs and before you touch a metallic object after walking on the carpet. It is also true between the two prongs of a lamp plug when the lamp switch is turned off. It is true of a dry cell when there is nothing connected to it. There is no current, but a current can flow if there is a conductive path between the two points.

Even a large emf might not drive much current through a conductor or resistance. A good example is your body after walking around on the carpet. Although the voltage seems deadly in terms of numbers (thousands), not many coulombs of charge normally can accumulate on an object the size of your body. Therefore, not many electrons flow through your finger, in relative terms, when you touch the metallic object. Thus you don't get a severe shock.

Conversely, if there are plenty of coulombs available, a moderate voltage, such as 117 V (or even less), can result in a lethal flow of current. This is why it is dangerous to repair an electrical device with the power on. The utility power source can pump an unlimited number of coulombs of charge through your body if you are foolish enough to get caught in this kind of situation.

Electrical Diagrams

To understand how electric circuits work, you should be able to read electrical wiring diagrams, called *schematic diagrams*. These diagrams use *schematic symbols*. Here are the basic symbols. Think of them as something like an alphabet in a language such as Chinese or Japanese, where things are represented by little pictures. However, before you get intimidated by this comparison, rest assured that it will be easier for you to learn schematic symbology than it would be to learn Chinese (unless you already know Chinese!).

BASIC SYMBOLS

The simplest schematic symbol is the one representing a wire or electrical conductor: a straight solid line. Sometimes dashed lines are used to represent conductors, but usually, broken lines are drawn to partition diagrams into constituent circuits or to indicate that certain components interact with each other or operate in step with each other. Conductor lines are almost always drawn either horizontally across or vertically up and down the page so that the imaginary charge carriers are forced to march in formation like soldiers. This keeps the diagram neat and easy to read.

When two conductor lines cross, they are not connected at the crossing point unless a heavy black dot is placed where the two lines meet. The dot always should be clearly visible wherever conductors are to be connected, no matter how many of them meet at the junction.

A *resistor* is indicated by a zigzaggy line. A variable resistor, such as a *rheostat* or *potentiometer,* is indicated by a zigzaggy line with an arrow through it or by a zigzaggy line with an arrow pointing at it. These symbols are shown in Fig. 12-3.

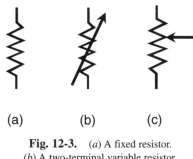

(a)   (b)   (c)

Fig. 12-3.   (*a*) A fixed resistor.
(*b*) A two-terminal variable resistor.
(*c*) A three-terminal potentiometer.

An *electrochemical cell* is shown by two parallel lines, one longer than the other. The longer line represents the positive terminal. A *battery,* or combination of cells in series, is indicated by an alternating sequence of parallel lines, long-short-long-short. The symbols for a cell and a battery are shown in Fig. 12-4.

SOME MORE SYMBOLS

Meters are indicated as circles. Sometimes the circle has an arrow inside it, and the meter type, such as mA (milliammeter) or V (voltmeter), is written alongside the circle, as shown in Fig. 12-5a. Sometimes the meter type is indicated inside the circle, and there is no arrow (see Fig. 12-5b). It doesn't

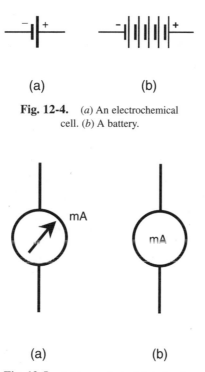

Fig. 12-4. (*a*) An electrochemical
cell. (*b*) A battery.

Fig. 12-5. Meter symbols: (*a*) designator
outside; (*b*) designator inside.

matter which way it's done as long as you are consistent everywhere in a
given diagram.

Some other common symbols include the *lamp,* the *capacitor,* the *air-core
coil,* the *iron-core coil,* the *chassis ground,* the *earth ground,* the *alternating-
current* (AC) *source,* the set of *terminals,* and the *black box* (which can
stand for almost anything), a rectangle with the designator written inside.
These are shown in Fig. 12-6.

Voltage/Current/Resistance Circuits

Most direct current (dc) circuits can be boiled down ultimately to three major
components: a voltage source, a set of conductors, and a resistance. This is
shown in the schematic diagram of Fig. 12-7. The voltage of the emf source

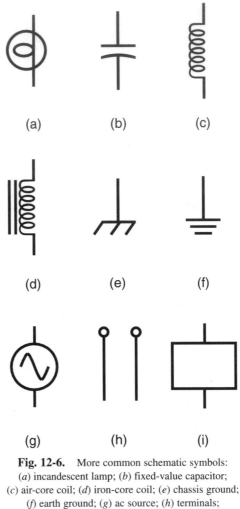

Fig. 12-6.   More common schematic symbols:
(*a*) incandescent lamp; (*b*) fixed-value capacitor;
(*c*) air-core coil; (*d*) iron-core coil; (*e*) chassis ground;
(*f*) earth ground; (*g*) ac source; (*h*) terminals;
and (*i*) and black box.

is called *E* (or sometimes *V*); the current in the conductor is called *I*; the resistance is called *R*. The standard units for these components are the volt (V), the ampere (A), and the ohm (Ω), respectively. Note which characters here are italicized and which are not. Italicized characters represent mathematical variables; nonitalicized characters represent symbols for units.

You already know that there is a relationship among these three quantities. If one of them changes, then one or both of the others also will change. If you make the resistance smaller, the current will get larger. If you make

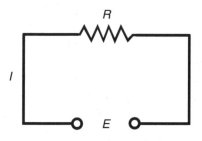

Fig. 12-7.   A simple dc circuit. The
voltage is E, the current is I,
and the resistance is R.

the emf source smaller, the current will decrease. If the current in the circuit increases, the voltage across the resistor will increase. There is a simple arithmetic relationship between these three quantities.

OHM'S LAW

The interdependence among current, voltage, and resistance in dc circuits is called *Ohm's law,* named after the scientist who supposedly first expressed it. Three formulas denote this law:

$$E = IR$$

$$I = E/R$$

$$R = E/I$$

You need only remember the first of these formulas to be able to derive the others. The easiest way to remember it is to learn the abbreviations E for emf, I for current, and R for resistance; then remember that they appear in alphabetical order with the equals sign after the E. Thus $E = IR$.

It is important to remember that you must use units of volts, amperes, and ohms in order for Ohm's law to work right. If you use volts, milliamperes (mA), and ohms or kilovolts (kV), microamperes (μA), and megohms (MΩ), you cannot expect to get the right answers. If the initial quantities are given in units other than volts, amperes, and ohms, you must convert to these units and then calculate. After that, you can convert the units back again to whatever you like. For example, if you get 13.5 million ohms as a calculated resistance, you might prefer to say that it is 13.5 megohms. However, in the calculation, you should use the number 13.5 million (or 1.35×10^7) and stick to ohms for the units.

CURRENT CALCULATIONS

The first way to use Ohm's law is to find current values in dc circuits. In order to find the current, you must know the voltage and the resistance or be able to deduce them.

Refer to the schematic diagram of Fig. 12-8. It consists of a variable dc generator, a voltmeter, some wire, an ammeter, and a calibrated wide-range potentiometer. Actual component values are not shown here, but they can be assigned for the purpose of creating sample Ohm's law problems. While calculating the current in the following problems, it is necessary to mentally "cover up" the meter.

PROBLEM 12-1
Suppose that the dc generator (see Fig. 12-8) produces 10 V and that the potentiometer is set to a value of 10 Ω. What is the current?

SOLUTION 12-1
This is solved easily by the formula $I = E/R$. Plug in the values for E and R; they are both 10, because the units are given in volts and ohms. Then $I = 10/10 = 1.0$ A.

PROBLEM 12-2
The dc generator (see Fig. 12-8) produces 100 V, and the potentiometer is set to 10.0 kΩ. What is the current?

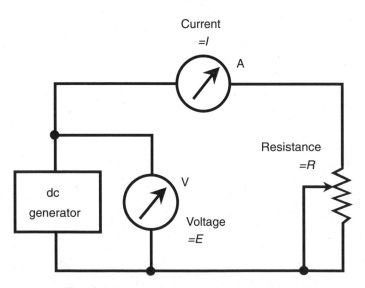

Fig. 12-8. Circuit for working Ohm's law problems.

SOLUTION 12-2
First, convert the resistance to ohms: 10.0 kΩ = 10,000 Ω. Then plug the values in: I = 100/10,000 = 0.0100 A.

VOLTAGE CALCULATIONS

The second use of Ohm's law is to find unknown voltages when the current and the resistance are known. For the following problems, uncover the ammeter and cover the voltmeter scale in your mind.

PROBLEM 12-3
Suppose that the potentiometer (see Fig. 12-8) is set to 100 Ω, and the measured current is 10.0 mA. What is the dc voltage?

SOLUTION 12-3
Use the formula $E = IR$. First, convert the current to amperes: 10.0 mA = 0.0100 A. Then multiply: E = 0.0100 $\times$ 100 = 1.00 V. This is a low, safe voltage, a little less than what is produced by a flashlight cell.

RESISTANCE CALCULATIONS

Ohm's law can be used to find a resistance between two points in a dc circuit when the voltage and the current are known. For the following problems, imagine that both the voltmeter and ammeter scales in Fig. 12-8 are visible but that the potentiometer is uncalibrated.

PROBLEM 12-4
If the voltmeter reads 24 V and the ammeter shows 3.0 A, what is the value of the potentiometer?

SOLUTION 12-4
Use the formula $R = E/I$, and plug in the values directly because they are expressed in volts and amperes: R = 24/3.0 = 8.0 Ω.

POWER CALCULATIONS

You can calculate the power P (in watts, symbolized W) in a dc circuit such as that shown in Fig. 12-8 using the following formula:

$$P = EI$$

where E is the voltage in volts and I is the current in amperes. You may not be given the voltage directly, but you can calculate it if you know the current and the resistance.

Remember the Ohm's law formula for obtaining voltage: $E = IR$. If you know I and R but don't know E, you can get the power P by means of this formula:

$$P = (IR)\, I = I^2 R$$

That is, take the current in amperes, multiply this figure by itself, and then multiply the result by the resistance in ohms.

You also can get the power if you aren't given the current directly. Suppose that you're given only the voltage and the resistance. Remember the Ohm's law formula for obtaining current: $I = E/R$. Therefore, you can calculate power using this formula:

$$P = E\,(E/R) = E^2/R$$

That is, take the voltage, multiply it by itself, and divide by the resistance.

Stated all together, these power formulas are

$$P = EI = I^2 R = E^2/R$$

Now we are all ready to do power calculations. Refer once again to Fig. 12-8.

PROBLEM 12-5
Suppose that the voltmeter reads 12 V and the ammeter shows 50 mA. What is the power dissipated by the potentiometer?

SOLUTION 12-5
Use the formula $P = EI$. First, convert the current to amperes, getting $I = 0.050$ A. Then $P = EI = 12 \times 0.050 = 0.60$ W.

How Resistances Combine

When electrical components or devices containing dc resistance are connected together, their resistances combine according to specific rules. Sometimes the combined resistance is more than that of any of the components or devices alone. In other cases the combined resistance is less than that of any of the components or devices all by itself.

RESISTANCES IN SERIES

When you place resistances in series, their ohmic values add up to get the total resistance. This is intuitively simple, and it's easy to remember.

PROBLEM 12-6
Suppose that the following resistances are hooked up in series with each other: 112 ohms, 470 ohms, and 680 ohms (Fig. 12-9). What is the total resistance of the series combination?

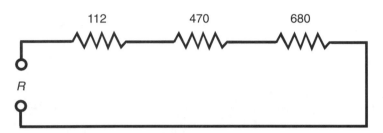

Fig. 12-9.   An example of three specific resistances in series.

SOLUTION 12-6
Just add the values, getting a total of 112 + 470 + 680 = 1,262 ohms. You can round this off to 1,260 ohms. It depends on the *tolerances* of the components—how much their actual values are allowed to vary, as a result of manufacturing processes, from the values specified by the vendor. Tolerance is more of an engineering concern than a physics concern, so we won't worry about that here.

RESISTANCES IN PARALLEL

When resistances are placed in parallel, they behave differently than they do in series. In general, if you have a resistor of a certain value and you place other resistors in parallel with it, the overall resistance decreases. Mathematically, the rule is straightforward, but it can get a little messy.

One way to evaluate resistances in parallel is to consider them as *conductances* instead. Conductance is measured in units called *siemens*, sometimes symbolized S. (The word *siemens* serves both in the singular and the plural sense). In older documents, the word *mho* (*ohm* spelled backwards) is used instead. In parallel, conductances add up in the same way as resistances add in series. If you change all the ohmic values to siemens, you can add these figures up and convert the final answer back to ohms.

The symbol for conductance is *G*. Conductance in siemens is the reciprocal of resistance in ohms. This can be expressed neatly in the following two formulas. It is assumed that neither *R* nor *G* is ever equal to zero:

$$G = 1/R$$

$$R = 1/G$$

PROBLEM 12-7
Consider five resistors in parallel. Call them R_1 through R_5, and call the total resistance R, as shown in the diagram of Fig. 12-10. Let R_1 = 100 ohms, R_2 = 200 ohms, R_3 = 300 ohms, R_4 = 400 ohms, and R_5 = 500 ohms, respectively. What is the total resistance R of this parallel combination?

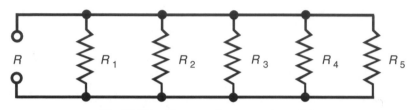

Fig. 12-10. Five general resistances in parallel.

SOLUTION 12-7
Converting the resistances to conductance values, you get G_1 = 1/100 = 0.0100 siemens, G_2 = 1/200 = 0.00500 siemens, G_3 = 1/300 = 0.00333 siemens, G_4 = 1/400 = 0.00250 siemens, and G_5 = 1/500 = 0.00200 siemens. Adding these gives G = 0.0100 + 0.00500 + 0.00333 + 0.00250 + 0.00200 = 0.02283 siemens. The total resistance is therefore R = 1/G = 1/0.02283 = 43.80 ohms. Because we're given the input numbers to only three significant figures, we should round this off to 43.8 ohms.

When you have resistances in parallel and their values are all equal, the total resistance is equal to the resistance of any one component divided by the number of components. In a more general sense, the resistances in Fig. 12-10 combine like this:

$$R = 1/(1/R_1 + 1/R_2 + 1/R_3 + 1/R_4 + 1/R_5)$$

If you prefer to use exponents, the formula looks like this:

$$R = (R_1^{-1} + R_2^{-1} + R_3^{-1} + R_4^{-1} + R_5^{-1})^{-1}$$

These resistance formulas are cumbersome for some people to work with, but mathematically they represent the same thing we just did in Problem 12-7.

CURRENT THROUGH SERIES RESISTANCES

Have you ever used those tiny holiday lights that come in strings? If one bulb burns out, the whole set of bulbs goes dark. Then you have to find out which bulb is bad and replace it to get the lights working again. Each bulb works with something like 10 V, and there are about a dozen bulbs in the string. You plug in the whole bunch, and the 120-V utility mains drive just the right amount of current through each bulb.

In a series circuit such as a string of light bulbs, the current at any given point is the same as the current at any other point. An ammeter can be connected in series at any point in the circuit, and it will always show the same reading. This is true in any series dc circuit, no matter what the components actually are and regardless of whether or not they all have the same resistance.

If the bulbs in a string are of different resistances, some of them will consume more power than others. In case one of the bulbs burns out and its socket is shorted out instead of filled with a replacement bulb, the current through the whole chain will increase because the overall resistance of the string will go down. This will force each of the remaining bulbs to carry too much current. Another bulb will burn out before long as a result of this excess current. If it, too, is replaced by a short circuit, the current will be increased still further. A third bulb will blow out almost right away. At this point it would be wise to buy some new bulbs!

VOLTAGES ACROSS SERIES RESISTANCES

In a series circuit, the voltage is divided up among the components. The sum total of the potential differences across each resistance is equal to the dc power-supply or battery voltage. This is always true, no matter how large or how small the resistances and whether or not they're all the same value.

If you think about this for a moment, it's easy to see why this is true. Look at the schematic diagram of Fig. 12-11. Each resistor carries the same current. Each resistor R_n has a potential difference E_n across it equal to the product of the current and the resistance of that particular resistor. These E_n values are in series, like cells in a battery, so they add together. What if the E_n values across all the resistors added up to something more or less than the supply voltage E? Then there would be a "phantom emf" someplace, adding or taking away voltage. However, there can be no such thing. An emf cannot come out of nowhere.

Look at this another way. The voltmeter V in Fig. 12-11 shows the voltage E of the battery because the meter is hooked up across the battery. The meter V also shows the sum of the E_n values across the set of resistors simply because the meter is connected across the set of resistors. The meter

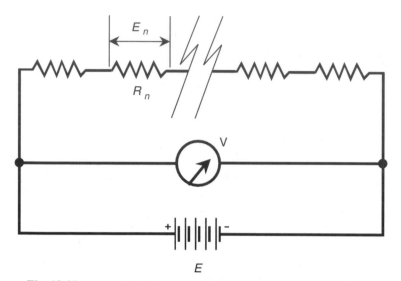

Fig. 12-11.   Analysis of voltage in a series dc circuit. See text for discussion.

says the same thing whether you think of it as measuring the battery voltage E or as measuring the sum of the E_n values across the series combination of resistors. Therefore, E is equal to the sum of the E_n values.

This is a fundamental rule in series dc circuits. It also holds for common utility ac circuits almost all the time.

How do you find the voltage across any particular resistor R_n in a circuit like the one in Fig. 12-11? Remember Ohm's law for finding voltage: $E = IR$. The voltage is equal to the product of the current and the resistance. Remember, too, that you must use volts, ohms, and amperes when making calculations. In order to find the current in the circuit I, you need to know the total resistance and the supply voltage. Then $I = E/R$. First find the current in the whole circuit; then find the voltage across any particular resistor.

PROBLEM 12-8
In Fig. 12-11, suppose that there are 10 resistors. Five of them have values of 10 ohms, and the other 5 have values of 20 ohms. The power source is 15 V dc. What is the voltage across one of the 10-ohm resistors? Across one of the 20-ohm resistors?

SOLUTION 12-8
First, find the total resistance: $R = (10 \times 5) + (20 \times 5) = 50 + 100 = 150$ ohms. Then find the current: $I = E/R = 15/150 = 0.10$ A. This is the current through each of the resistors in the circuit. If $R_n = 10$ ohms, then

$$E_n = I(R_n) = 0.10 \times 10 = 1.0 \text{ V}$$

If $R_n = 20$ ohms, then

$$E_n = I(R_n) = 0.10 \times 20 = 2.0 \text{ V}$$

You can check to see whether all these voltages add up to the supply voltage. There are 5 resistors with 1.0 V across each, for a total of 5.0 V; there are also 5 resistors with 2.0 V across each, for a total of 10 V. Thus the sum of the voltages across the 10 resistors is 5.0 V + 10 V = 15 V.

VOLTAGE ACROSS PARALLEL RESISTANCES

Imagine now a set of ornamental light bulbs connected in parallel. This is the method used for outdoor holiday lighting or for bright indoor lighting. You know that it's much easier to fix a parallel-wired string of holiday lights if one bulb should burn out than it is to fix a series-wired string. The failure of one bulb does not cause catastrophic system failure. In fact, it might be awhile before you notice that the bulb is dark because all the other ones will stay lit, and their brightness will not change.

In a parallel circuit, the voltage across each component is always the same and is always equal to the supply or battery voltage. The current drawn by each component depends only on the resistance of that particular device. In this sense, the components in a parallel-wired circuit work independently, as opposed to the series-wired circuit, in which they all interact.

If any branch of a parallel circuit is taken away, the conditions in the other branches remain the same. If new branches are added, assuming that the power supply can handle the load, conditions in previously existing branches are not affected.

CURRENTS THROUGH PARALLEL RESISTANCES

Refer to the schematic diagram of Fig. 12-12. The total parallel resistance in the circuit is R. The battery voltage is E. The current in branch n, containing resistance R_n, is measured by ammeter A and is called I_n.

The sum of all the I_n values in the circuit is equal to the total current I drawn from the source. That is, the current is divided up in the parallel circuit, similarly to the way that voltage is divided up in a series circuit.

PROBLEM 12-9
Suppose that the battery in Fig. 12-12 delivers 12 V. Further suppose that there are 12 resistors, each with a value of 120 ohms in the parallel circuit.

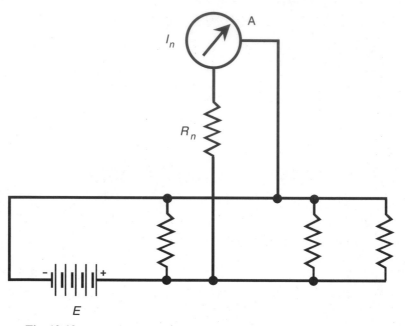

Fig. 12-12. Analysis of current in a parallel dc circuit. See text for discussion.

What is the total current *I* drawn from the battery?

SOLUTION 12-9
First, find the total resistance. This is easy because all the resistors have the same value. Divide R_n = 120 by 12 to get R = 10 ohms. Then the current *I* is found by Ohm's law:

$$I = E/R = 12/10 = 1.2 \text{ A}$$

PROBLEM 12-10
In the circuit of Fig. 12-12, what does the ammeter A say, given the same component values as exist in the scenario of the preceding problem?

SOLUTION 12-10
This involves finding the current in any given branch. The voltage is 12 V across every branch; R_n = 120. Therefore, I_n, the ammeter reading, is found by Ohm's law:

$$I_n = E/R_n = 12/120 = 0.10 \text{ A}$$

Let's check to be sure all the I_n values add to get the total current *I*. There are 12 identical branches, each carrying 0.10 A; therefore, the sum is 0.10 × 12 = 1.2 A. It checks out.

POWER DISTRIBUTION IN SERIES CIRCUITS

Let's switch back now to series circuits. When calculating the power in a circuit containing resistors in series, all you need to do is find out the current I, in amperes, that the circuit is carrying. Then it's easy to calculate the power P_n, in watts, dissipated by any particular resistor of value R_n, in ohms, based on the formula $P_n = I^2 R_n$.

The total power dissipated in a series circuit is equal to the sum of the wattages dissipated in each resistor. In this way, the distribution of power in a series circuit is like the distribution of the voltage.

PROBLEM 12-11
Suppose that we have a series circuit with a supply of 150 V and three resistors: $R_1 = 330$ ohms, $R_2 = 680$ ohms, and $R_0 = 910$ ohms. What is the power dissipated by R_2?

SOLUTION 12-11
Find the current in the circuit. To do this, calculate the total resistance first. Because the resistors are in series, the total is resistance is $R - 330 + 680 + 910 = 1920$ ohms. Therefore, the current is $I = 150/1920 = 0.07813$ A $= 78.1$ mA. The power dissipated by R_2 is

$$P_2 = I^2 R_2 = 0.07813 \times 0.07813 \times 680 = 4.151 \text{ W}$$

We must round this off to three significant figures, getting 4.15 W.

POWER DISTRIBUTION IN PARALLEL CIRCUITS

When resistances are wired in parallel, they each consume power according to the same formula, $P = I^2 R$. However, the current is not the same in each resistance. An easier method to find the power P_n dissipated by resistor of value R_n is by using the formula $P_n = E^2 / R_n$, where E is the voltage of the supply. This voltage is the same across every resistor.

In a parallel circuit, the total power consumed is equal to the sum of the wattages dissipated by the individual resistances. This is, in fact, true for any dc circuit containing resistances. Power cannot come out of nowhere, nor can it vanish.

PROBLEM 12-12
A circuit contains three resistances $R_1 = 22$ ohms, $R_2 = 47$ ohms, and $R_3 = 68$ ohms, all in parallel across a voltage $E = 3.0$ V. Find the power dissipated by each resistor.

SOLUTION 12-12
First find E^2, the square of the supply voltage: $E^2 = 3.0 \times 3.0 = 9.0$. Then $P_1 = 9.0/22 = 0.4091$ W, $P_2 = 9.0/47 = 0.1915$ W, and $P_3 = 9.0/68 = 0.1324$ W. These should be rounded off to $P_1 = 0.41$ W, $P_2 = 0.19$ W, and $P_3 = 0.13$ W, respectively.

Kirchhoff's Laws

The physicist Gustav Robert Kirchhoff (1824–1887) was a researcher and experimentalist in electricity, back in the time before radio, before electric lighting, and before much was understood about how electric currents flow.

KIRCHHOFF'S CURRENT LAW

Kirchhoff reasoned that current must work something like water in a network of pipes and that the current going into any point has to be the same as the current going out. This is true for any point in a circuit, no matter how many branches lead into or out of the point (Fig. 12-13).

In a network of water pipes that does not leak and into which no water is added along the way, the total number of cubic meters going in has to be the same as the total volume going out. Water cannot form from nothing, nor can it disappear, inside a closed system of pipes. Charge carriers, thought Kirchhoff, must act the same way in an electric circuit.

PROBLEM 12-13
In Fig. 12-13, suppose that each of the two resistors below point Z has a value of 100 ohms and that all three resistors above Z have values of 10.0 ohms. The current through each 100-ohm resistor is 500 mA (0.500 A). What is the current through any of the 10.0-ohm resistors, assuming that the current is equally distributed? What is the voltage, then, across any of the 10.0-ohm resistors?

SOLUTION 12-13
The total current into Z is 500 mA + 500 mA = 1.00 A. This must be divided three ways equally among the 10-ohm resistors. Therefore, the current through any one of them is 1.00/3 A = 0.333 A = 333 mA. The voltage across any one of the 10.0-ohm resistors is found by Ohm's law: $E = IR = 0.333 \times 10.0 = 3.33$ V.

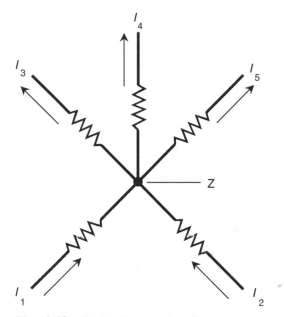

Fig. 12-13.   Kirchhoff's current law. The current enter-
ing point *Z* is equal to the current leaving point *Z*. In this
case, $I_1 + I_2 = I_3 + I_4 + I_5$.

KIRCHHOFF'S VOLTAGE LAW

The sum of all the voltages, as you go around a circuit from some fixed
point and return there from the opposite direction, and taking polarity into
account, is always zero. At first thought, some people find this strange.
Certainly there is voltage in your electric hair dryer, radio, or computer!
Yes, there is—between different points in the circuit. However, no single
point can have an electrical potential with respect to itself. This is so simple that it's trivial. A point in a circuit is always shorted out to itself.

What Kirchhoff was saying when he wrote his voltage law is that voltage cannot appear out of nowhere, nor can it vanish. All the potential differences must balance out in any circuit, no matter how complicated and no
matter how many branches there are.

Consider the rule you've already learned about series circuits: The voltages across all the resistors add up to the supply voltage. However, the
polarities of the emfs across the resistors are opposite to that of the battery.
This is shown in Fig. 12-14. It is a subtle thing, but it becomes clear when

a series circuit is drawn with all the components, including the battery or other emf source, in line with each other, as in Fig. 12-14.

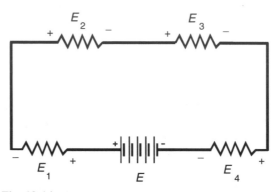

Fig. 12-14.   Kirchhoff's voltage law. The sum of the voltages $E + E_1 + E_2 + E_3 + E_4 = 0$, taking polarity into account.

PROBLEM 12-14
Refer to the diagram of Fig. 12-14. Suppose that the four resistors have values of 50, 60, 70, and 80 ohms and that the current through them is 500 mA (0.500 A). What is the supply voltage E?

SOLUTION 12-14
Find the voltages E_1, E_2, E_3, and E_4 across each of the resistors. This is done using Ohm's law. In the case of E_1, say, with the 50-ohm resistor, calculate $E_1 = 0.500 \times 50 = 25$ V. In the same way, you can calculate $E_2 = 30$ V, $E_3 = 35$ V, and $E_4 = 40$ V. The supply voltage is the sum $E_1 + E_2 + E_3 + E_4 = 25 + 30 + 35 + 40$ V $= 130$ V.

Quiz

Refer to the text in this chapter if necessary. A good score is eight correct. Answers are in the back of the book.

1. Suppose that 5.00×10^{17} electrical charge carriers flow past a point in 1.00 s. What is the electrical voltage?
   (a) 0.080 V
   (b) 12.5 V

(c) 5.00 V

(d) It cannot be calculated from this information.

2. An ampere also can be regarded as

(a) an ohm per volt.

(b) an ohm per watt.

(c) a volt per ohm.

(d) a volt-ohm.

3. Suppose that there are two resistances in a series circuit. One of the resistors has a value of 33 kΩ (that is, 33,000 or 3.3×10^4 ohms). The value of the other resistor is not known. The power dissipated by the 33-kΩ resistor is 3.3 W. What is the current through the unknown resistor?

(a) 0.11 A

(b) 10 mA

(c) 0.33 mA

(d) It cannot be calculated from this information.

4. If the voltage across a resistor is E (in volts) and the current through that resistor is I (in milliamperes), then the power P (in watts) is given by the following formula:

(a) $P = EI$.

(b) $P = EI \times 10^3$.

(c) $P = EI \times 10^{-3}$.

(d) $P = E/I$.

5. Suppose that you have a set of five 0.5-W flashlight bulbs connected in parallel across a dc source of 3.0 V. If one of the bulbs is removed or blows out, what will happen to the current through the other four bulbs?

(a) It will remain the same.

(b) It will increase.

(c) It will decrease.

(d) It will drop to zero.

6. A good dielectric is characterized by

(a) excellent conductivity.

(b) fair conductivity.

(c) poor conductivity.

(d) variable conductivity.

7. Suppose that there are two resistances in a parallel circuit. One of the resistors has a value of 100 ohms. The value of the other resistor is not known. The power dissipated by the 100-ohm resistor is 500 mW (that is, 0.500 W). What is the current through the unknown resistor?

(a) 71 mA

(b) 25 A

(c) 200 A

(d) It cannot be calculated from this information.

8. Conventional current flows
   (a) from the positive pole to the negative pole.
   (b) from the negative pole to the positive pole.
   (c) in either direction; it doesn't matter.
   (d) nowhere; current does not flow.

9. Suppose that a circuit contains 620 ohms of resistance and that the current in the circuit is 50.0 mA. What is the voltage across this resistance?
   (a) 12.4 kV
   (b) 31.0 V
   (c) 8.06×10^{-5} V
   (d) It cannot be calculated from this information.

10. Which of the following cannot be an electric charge carrier?
    (a) A neutron
    (b) An electron
    (c) A hole
    (d) An ion

CHAPTER 13

Alternating Current

Direct current (dc) can be expressed in terms of two variables: the *polarity* (or direction) and the *amplitude*. Alternating current (ac) is more complicated. There are additional variables: the *period* (and its reciprocal, the *frequency*), the *waveform,* and the *phase.*

Definition of Alternating Current

Direct current has a polarity, or direction, that stays the same over a long period of time. Although the amplitude can vary—the number of amperes, volts, or watts can fluctuate—the charge carriers always flow in the same direction through the circuit. In ac, the polarity reverses repeatedly.

PERIOD

In a *periodic ac wave,* the kind discussed in this chapter, the mathematical function of amplitude versus time repeats precisely and indefinitely; the same pattern recurs countless times. The period is the length of time between one repetition of the pattern, or one wave cycle, and the next. This is illustrated in Fig. 13-1 for a simple ac wave.

The period of a wave, in theory, can be anywhere from a minuscule fraction of a second to many centuries. Some electromagnetic (EM) fields have periods measured in quadrillionths of a second or smaller. The charged particles held captive by the magnetic field of the Sun reverse their direction

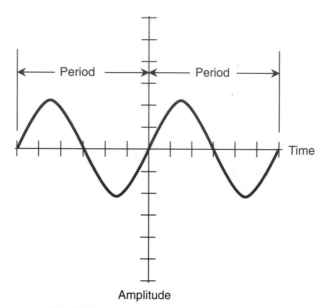

Fig. 13-1. A sine wave. The period is the length of time required for one cycle to be completed.

over periods measured in years. Period, when measured in seconds, is symbolized T.

FREQUENCY

The frequency, denoted f, of a wave is the reciprocal of the period. That is, $f = 1/T$, and $T = 1/f$. In the olden days (prior to the 1970s), frequency was specified in *cycles per second,* abbreviated *cps.* High frequencies were expressed in *kilocycles, megacycles,* or *gigacycles,* representing thousands, millions, or billions of cycles per second. Nowadays, the standard unit of frequency is known as the *hertz,* abbreviated *Hz.* Thus 1 Hz = 1 cps, 10 Hz = 10 cps, and so on.

Higher frequencies are given in *kilohertz* (kHz), *megahertz* (MHz), *gigahertz* (GHz), and *terahertz* (THz). The relationships are

$$1 \text{ kHz} = 1,000 \text{ Hz} = 10^3 \text{ Hz}$$

$$1 \text{ MHz} = 1,000 \text{ kHz} = 10^6 \text{ Hz}$$

$$1 \text{ GHz} = 1,000 \text{ MHz} = 10^9 \text{ Hz}$$

$$1 \text{ THz} = 1,000 \text{ GHz} = 10^{12} \text{ Hz}$$

PROBLEM 13-1
The period of an ac wave is 5.000×10^{-6} s. What is the frequency in hertz? In kilohertz? In megahertz?

SOLUTION 13-1
First, find the frequency f_{Hz} in hertz by taking the reciprocal of the period in seconds:

$$f_{Hz} = 1/(5.000 \times 10^{-6}) = 2.000 \times 10^5 \text{ Hz}$$

Next, divide f_{Hz} by 1,000 or 10^3 to get the frequency f_{kHz} in kilohertz:

$$f_{kHz} = f_{Hz}/10^3 = 2.000 \times 10^5/10^3 = 200.0 \text{ kHz}$$

Finally, divide f_{kHz} by 1,000 or 10^3 to get the frequency f_{MHz} in megahertz:

$$f_{MHz} = f_{kHz}/10^3 = 200.0/10^3 = 0.2000 \text{ MHz}$$

Waveforms

If you graph the instantaneous current or voltage in an ac system as a function of time, you get a waveform. Alternating currents can manifest themselves in an infinite variety of waveforms. Here are some of the simplest ones.

SINE WAVE

In its purest form, alternating current has a *sine-wave,* or *sinusoidal,* nature. The waveform in Fig. 13-1 is a sine wave. Any ac wave that consists of a single frequency has a perfect sine-wave shape. Any perfect sine-wave current contains one, and only one, component frequency.

In practice, a wave can be so close to a sine wave that it looks exactly like the sine function on an oscilloscope when in reality there are traces of other frequencies present. Imperfections are often too small to see. Utility ac in the United States has an almost perfect sine-wave shape, with a frequency of 60 Hz. However, there are slight aberrations.

SQUARE WAVE

On an oscilloscope, a theoretically perfect *square wave* would look like a pair of parallel dotted lines, one having positive polarity and the other having negative polarity (Fig. 13-2a). In real life, the transitions often can be seen as vertical lines (see Fig. 13-2b).

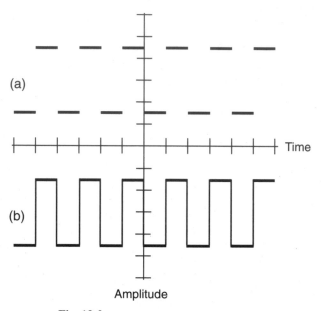

Fig. 13-2. (*a*) A theoretically perfect square wave. (*b*) The more common rendition.

A square wave might have equal negative and positive peaks. Then the absolute amplitude of the wave is constant at a certain voltage, current, or power level. Half the time the amplitude is +*x,* and the other half it is −*x* volts, amperes, or watts.

Some square waves are asymmetrical, with the positive and negative magnitudes differing. If the length of time for which the amplitude is positive differs from the length of time for which the amplitude is negative, the wave is not truly square but is described by the more general term *rectangular wave.*

SAWTOOTH WAVES

Some ac waves reverse their polarity at constant but not instantaneous rates. The slope of the amplitude-versus-time line indicates how fast the magnitude is changing. Such waves are called *sawtooth waves* because of their appearance.

In Fig. 13-3, one form of sawtooth wave is shown. The positive-going slope (rise) is extremely steep, as with a square wave, but the negative-

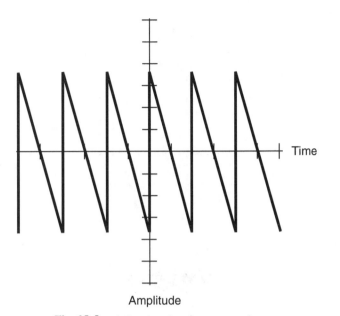

Fig. 13-3. A fast-rise, slow-decay sawtooth wave.

going slope (fall or decay) is gradual. The period of the wave is the time between points at identical positions on two successive pulses.

Another form of sawtooth wave is just the opposite, with a gradual positive-going slope and a vertical negative-going transition. This type of wave is sometimes called a *ramp* (Fig. 13-4). This waveform is used for scanning in *cathode-ray-tube* (CRT) television sets and oscilloscopes.

Sawtooth waves can have rise and decay slopes in an infinite number of different combinations. One example is shown in Fig. 13-5. In this case, the positive-going slope is the same as the negative-going slope. This is a *triangular wave.*

PROBLEM 13-2
Suppose that each horizontal division in Fig. 13-5 represents 1.0 microsecond (1.0 μs or 1.0×10^{-6} s). What is the period of this triangular wave? What is the frequency?

SOLUTION 13-2
The easiest way to look at this is to evaluate the wave from a point where it crosses the time axis going upward and then find the next point (to the right or left) where the wave crosses the time axis going upward. This is four horizontal divisions, at least within the limit of our ability to tell by looking at it. The period T is therefore 4.0 μs or 4.0×10^{-6} s. The frequency is the reciprocal of this: $f = 1/T = 1/(4.0 \times 10^{-6}) = 2.5 \times 10^{5}$ Hz.

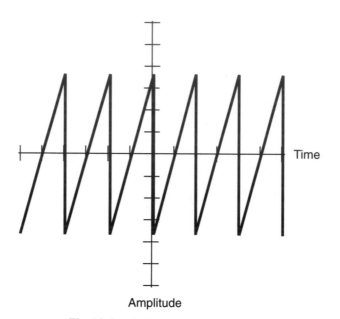

Time

Amplitude

Fig. 13-4. A slow-rise, fast-decay sawtooth
wave, also called a *ramp wave*.

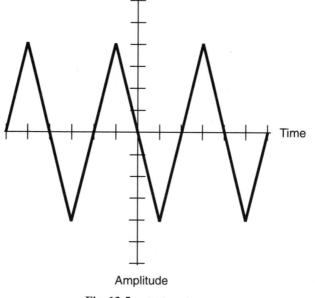

Time

Amplitude

Fig. 13-5. A triangular wave.

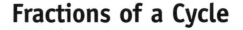

Fractions of a Cycle

Scientists and engineers break the ac cycle down into small parts for analysis and reference. One complete cycle can be likened to a single revolution around a circle.

SINE WAVES AS CIRCULAR MOTION

Suppose that you swing a glowing ball around and around at the end of a string at a rate of one revolution per second. The ball thus describes a circle in space (Fig. 13-6a). Imagine that you swing the ball around so that it is always at the same level; it takes a path that lies in a horizontal plane. Imagine that you do this in a pitch-dark gymnasium. If a friend stands some distance away with his or her eyes in the plane of the ball's path, what does your friend see? Only the glowing ball, oscillating back and forth. The ball seems to move toward the right, slow down, and then reverse its direction,

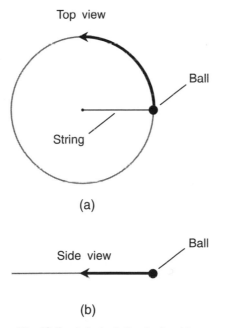

(a)

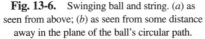

(b)

Fig. 13-6.   Swinging ball and string. (*a*) as seen from above; (*b*) as seen from some distance away in the plane of the ball's circular path.

going back toward the left (see Fig. 13-6*b*). Then it moves faster and faster and then slower again, reaching its left-most point, at which it turns around again. This goes on and on, with a frequency of 1 Hz, or a complete cycle per second, because you are swinging the ball around at one revolution per second.

If you graph the position of the ball as seen by your friend with respect to time, the result will be a sine wave (Fig. 13-7). This wave has the same characteristic shape as all sine waves. The standard, or basic, sine wave is described by the mathematical function $y = \sin x$ in the (x, y) coordinate plane. The general form is $y = a \sin bx,$ where a and b are real-number constants.

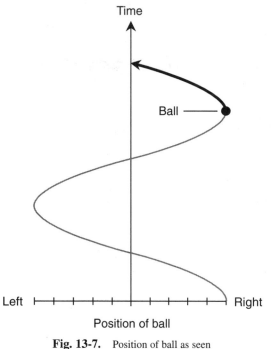

Fig. 13-7. Position of ball as seen edge-on as a function of time.

DEGREES

One method of specifying fractions of an ac cycle is to divide it into 360 equal increments called *degrees,* symbolized ° or deg (but it's okay to write out the whole word). The value 0° is assigned to the point in the cycle where the magnitude is zero and positive-going. The same point on the next cycle is given the value 360°. Halfway through the cycle is 180°; a quarter cycle is 90°; an eighth cycle is 45°. This is illustrated in Fig. 13-8.

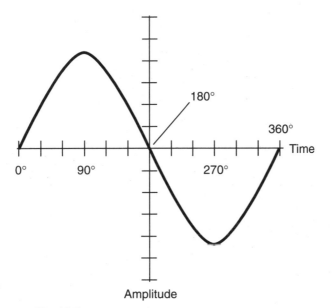

Fig. 13-8.   A wave cycle can be divided into 360 degrees.

RADIANS

The other method of specifying fractions of an ac cycle is to divide it into exactly 2π, or approximately 6.2832, equal parts. This is the number of radii of a circle that can be laid end to end around the circumference. One *radian,* symbolized rad (although you can write out the whole word), is equal to about 57.296°. Physicists use the radian more often than the degree when talking about fractional parts of an ac cycle.

Sometimes the frequency of an ac wave is measured in radians per second (rad/s) rather than in hertz (cycles per second). Because there are 2π radians in a complete cycle of 360°, the *angular frequency* of a wave, in radians per second, is equal to 2π times the frequency in hertz. Angular frequency is symbolized by the lowercase italicized Greek letter omega (ω).

PROBLEM 13-3
What is the angular frequency of household ac? Assume that the frequency of utility ac is 60.0 Hz.

SOLUTION 13-3
Multiply the frequency in hertz by 2π. If this value is taken as 6.2832, then the angular frequency is

$$\omega = 6.2832 \times 60.0 = 376.992 \text{ rad/s}$$

This should be rounded off to 377 rad/s because our input data are given only to three significant figures.

PROBLEM 13-4
A certain wave has an angular frequency of 3.8865×10^5 rad/s. What is the frequency in kilohertz? Express the answer to three significant figures.

SOLUTION 13-4
To solve this, first find the frequency in hertz. This requires that the angular frequency, in radians per second, be divided by 2π, which is approximately 6.2832. The frequency f_{Hz} is therefore

$$f_{Hz} = (3.8865 \times 10^5)/6.2832$$

$$= 6.1855 \times 10^4 \text{ Hz}$$

To obtain the frequency in kilohertz, divide by 10^3, and then round off to three significant figures:

$$f_{kHz} = 6.1855 \times 10^4/10^3$$

$$= 61.855 \text{ kHz} \approx 61.9 \text{ kHz}$$

Amplitude

Amplitude also can be called *magnitude, level, strength,* or *intensity.* Depending on the quantity being measured, the amplitude of an ac wave can be specified in amperes (for current), volts (for voltage), or watts (for power).

INSTANTANEOUS AMPLITUDE

The *instantaneous amplitude* of an ac wave is the voltage, current, or power at some precise moment in time. This constantly changes. The manner in which it varies depends on the waveform. Instantaneous amplitudes are represented by individual points on the wave curves.

AVERAGE AMPLITUDE

The *average amplitude* of an ac wave is the mathematical average (or mean) instantaneous voltage, current, or power evaluated over exactly one wave cycle or any exact whole number of wave cycles. A pure ac sine wave always has an average amplitude of zero. The same is true of a pure ac square wave or triangular wave. It is not generally the case for sawtooth waves. You can get an idea of why these things are true by carefully looking at the waveforms illustrated by Figs. 13-1 through 13-5. If you know calculus, you know that the average amplitude is the integral of the waveform evaluated over one full cycle.

PEAK AMPLITUDE

The *peak amplitude* of an ac wave is the maximum extent, either positive or negative, that the instantaneous amplitude attains. In many waves, the positive and negative peak amplitudes are the same. Sometimes they differ, however. Figure 13-9 is an example of a wave in which the positive peak amplitude is the same as the negative peak amplitude. Figure 13-10 is an illustration of a wave that has different positive and negative peak amplitudes.

PEAK-TO-PEAK AMPLITUDE

The *peak-to-peak (pk-pk) amplitude* of a wave is the net difference between the positive peak amplitude and the negative peak amplitude (Fig. 13-11). Another way of saying this is that the pk-pk amplitude is equal to the positive peak amplitude plus the absolute value of the negative peak amplitude.

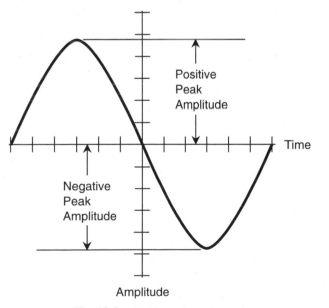

Fig. 13-9.   Positive and negative peak amplitudes. In this case, they are equal.

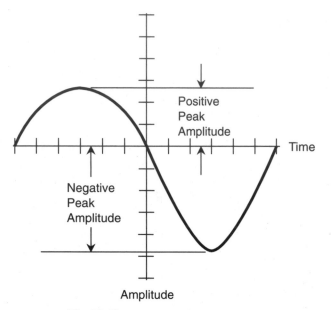

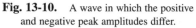

Fig. 13-10.   A wave in which the positive
and negative peak amplitudes differ.

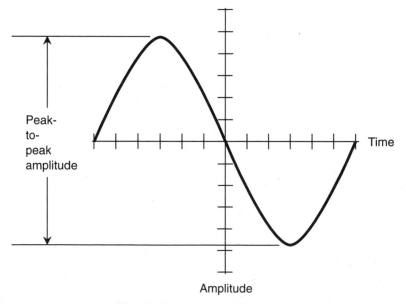

Fig. 13-11.   Peak-to-peak amplitude.

Peak to peak is a way of expressing how much the wave level "swings" during the cycle.

In many waves, the pk-pk amplitude is twice the peak amplitude. This is the case when the positive and negative peak amplitudes are the same.

ROOT-MEAN-SQUARE AMPLITUDE

Often it is necessary to express the *effective amplitude* of an ac wave. This is the voltage, current, or power that a dc source would produce to have the same general effect in a real circuit or system. When you say a wall outlet has 117 V, you mean 117 effective volts. The most common figure for effective ac levels is called the *root-mean-square,* or *rms, value.*

The expression *root mean square* means that the waveform is mathematically "operated on" by taking the square root of the mean of the square of all its instantaneous values. The rms amplitude is not the same thing as the average amplitude. For a perfect sine wave, the rms value is equal to 0.707 times the peak value, or 0.354 times the pk-pk value. Conversely, the peak value is 1.414 times the rms value, and the pk-pk value is 2.828 times the rms value. The rms figures often are quoted for perfect sine-wave sources of voltage, such as the utility voltage or the effective voltage of a radio signal.

For a perfect square wave, the rms value is the same as the peak value, and the pk-pk value is twice the rms value and twice the peak value. For sawtooth and irregular waves, the relationship between the rms value and the peak value depends on the exact shape of the wave. The rms value is never more than the peak value for any waveshape.

SUPERIMPOSED DC

Sometimes a wave can have components of both ac and dc. The simplest example of an ac/dc combination is illustrated by the connection of a dc source, such as a battery, in series with an ac source, such as the utility main.

Any ac wave can have a dc component along with it. If the dc component exceeds the peak value of the ac wave, then fluctuating or pulsating dc will result. This would happen, for example, if a 200-V dc source were connected in series with the utility output. Pulsating dc would appear, with an average value of 200 V but with instantaneous values much higher and lower. The waveshape in this case is illustrated by Fig. 13-12.

PROBLEM 13-5
An ac sine wave measures 60 V pk-pk. There is no dc component. What is the peak voltage?

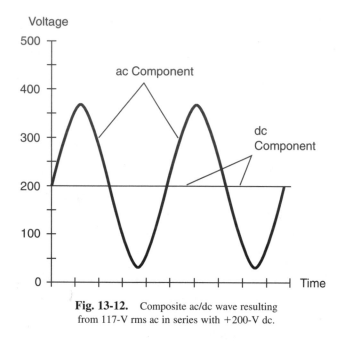

Fig. 13-12. Composite ac/dc wave resulting
from 117-V rms ac in series with +200-V dc.

SOLUTION 13-5
In this case, the peak voltage is exactly half the peak-to-peak value, or 30 V pk.
Half the peaks are +30 V; half are −30 V.

PROBLEM 13-6
Suppose that a dc component of +10 V is superimposed on the sine wave
described in Problem 13-5. What is the peak voltage?

SOLUTION 13-6
This can't be answered simply, because the absolute values of the positive
peak and negative peak voltages differ. In the case of Problem 13-5, the positive peak is +30 V and the negative peak is −30 V, so their absolute values
are the same. However, when a dc component of +10 V is superimposed on
the wave, both the positive peak and the negative peak voltages change by
+10 V. The positive peak voltage thus becomes +40 V, and the negative
peak voltage becomes −20 V.

Phase Angle

Phase angle is an expression of the displacement between two waves having
identical frequencies. There are various ways of defining this. Phase angles
are usually expressed as values ϕ such that $0° \leq \phi < 360°$. In radians, this

range is $0 \leq \phi < 2\pi$. Once in awhile you will hear about phase angles specified over a range of $-180° < \phi \leq +180°$. In radians, this range is $-\pi < \phi \leq +\pi$. Phase angle figures can be defined only for pairs of waves whose frequencies are the same. If the frequencies differ, the phase changes from moment to moment and cannot be denoted as a specific number.

PHASE COINCIDENCE

Phase coincidence means that two waves begin at exactly the same moment. They are "lined up." This is shown in Fig. 13-13 for two waves having different amplitudes. (If the amplitudes were the same, you would see only one wave.) The phase difference in this case is 0°.

If two sine waves are in phase coincidence, the peak amplitude of the resulting wave, which also will be a sine wave, is equal to the sum of the peak amplitudes of the two composite waves. The phase of the resultant is the same as that of the composite waves.

PHASE OPPOSITION

When two sine waves begin exactly one-half cycle, or 180°, apart, they are said to be in *phase opposition*. This is illustrated by the drawing of Fig. 13-14.

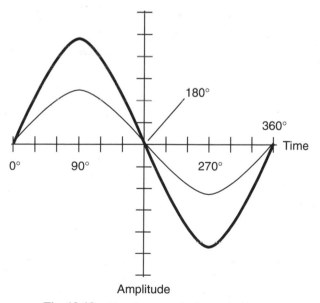

Fig. 13-13. Two sine waves in phase coincidence.

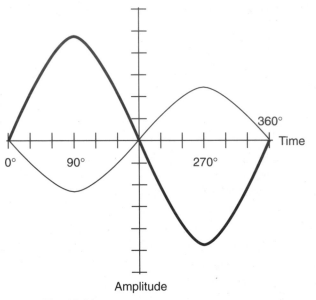

Fig. 13-14.   Two sine waves in phase opposition.

If two sine waves have the same amplitude and are in phase opposition, they cancel each other out because the instantaneous amplitudes of the two waves are equal and opposite at every moment in time.

If two sine waves have different amplitudes and are in phase opposition, the peak value of the resulting wave, which is a sine wave, is equal to the difference between the peak values of the two composite waves. The phase of the resultant is the same as the phase of the stronger of the two composite waves.

LEADING PHASE

Suppose that there are two sine waves, wave X and wave Y, with identical frequencies. If wave X begins a fraction of a cycle earlier than wave Y, then wave X is said to be *leading* wave Y in phase. For this to be true, X must begin its cycle less than 180° before Y. Figure 13-15 shows wave X leading wave Y by 90°. The difference can be anything greater than 0°, up to but not including 180°.

Leading phase is sometimes expressed as a phase angle ϕ such that $0° < \phi < +180°$. In radians, this is $0 < \phi < +\pi$. If we say that wave X has a phase of $+\pi/2$ rad relative to wave Y, we mean that wave X leads wave Y by $\pi/2$ rad.

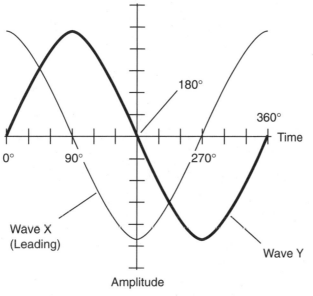

Fig. 13-15.   Wave *X* leads wave *Y* by 90°.

LAGGING PHASE

Suppose that wave *X* begins its cycle more than 180° but less than 360° ahead of wave *Y*. In this situation, it is easier to imagine that wave *X* starts its cycle later than wave *Y* by some value between but not including 0° and 180°. Then wave *X* is *lagging* wave *Y*. Figure 13-16 shows wave *X* lagging wave *Y* by 90°. The difference can be anything between but not including 0° and 180°.

   Lagging phase is sometimes expressed as a negative angle ϕ such that $-180° < \phi < 0°$. In radians, this is $-\pi < \phi < 0$. If we say that wave *X* has a phase of $-45°$ relative to wave *Y*, we mean that wave *X* lags wave *Y* by 45°.

VECTOR REPRESENTATIONS OF PHASE

If a sine wave *X* is leading a sine wave *Y* by *x* degrees, then the two waves can be drawn as vectors, with vector *X* oriented *x* degrees counterclockwise from vector *Y*. If wave *X* lags *Y* by *y* degrees, then *X* is oriented *y* degrees clockwise from *Y*. If two waves are in phase, their vectors overlap (line up). If they are in phase opposition, they point in exactly opposite directions.

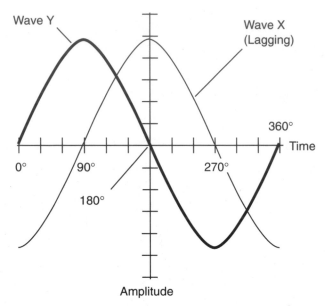

Fig. 13-16. Wave *X* lags wave *Y* by 90°.

Figure 13-17 shows four phase relationships between waves *X* and *Y*. Wave *X* always has twice the amplitude of wave *Y*, so vector *X* is always twice as long as vector *Y*. In part *a*, wave *X* is in phase with wave *Y*. In part *b*, wave *X* leads wave *Y* by 90°. In part *c*, waves *X* and *Y* are 180° opposite in phase. In part *d*, wave *X* lags wave *Y* by 90°.

In all cases, the vectors rotate counterclockwise at the rate of one complete circle per wave cycle. Mathematically, a sine wave is a vector that goes around and around, just like the ball goes around and around your head when you put it on a string and whirl it.

In a sine wave, the vector magnitude stays the same at all times. If the waveform is not sinusoidal, the vector magnitude is greater in some directions than in others. As you can guess, there exist an infinite number of variations on this theme, and some of them can get complicated.

PROBLEM 13-7
Suppose that there are three waves, called *X, Y,* and *Z*. Wave *X* leads wave *Y* by 0.5000 rad; wave *Y* leads wave *Z* by precisely one-eighth cycle. By how many degrees does wave *X* lead or lag wave *Z*?

SOLUTION 13-7
To solve this, let's convert all phase-angle measures to degrees. One radian is approximately equal to 57.296°; therefore, 0.5000 rad = 57.296° × 0.5000

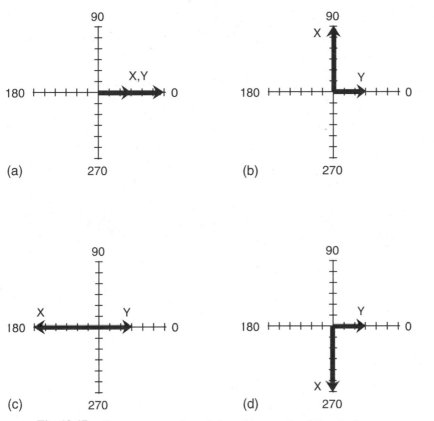

Fig. 13-17. Vector representations of phase. (*a*) waves *X* and *Y* are in phase;
(*b*) wave *X* leads wave *Y* by 90 degrees; (*c*) waves *X* and *Y* are
in phase opposition; (*d*) wave *X* lags wave *Y* by 90 degrees.

= 28.65° (to four significant figures). One-eighth of a cycle is equal to 45.00°
(that is 360°/8.000). The phase angles therefore add up, so wave *X* leads
wave *Z* by 28.65° + 45.00°, or 73.65°.

PROBLEM 13-8
Suppose that there are three waves *X, Y,* and *Z*. Wave *X* leads wave *Y* by
0.5000 rad; wave *Y* lags wave *Z* by precisely one-eighth cycle. By how many
degrees does wave *X* lead or lag wave *Z*?

SOLUTION 13-8
The difference in phase between *X* and *Y* in this problem is the same as that
in the preceding problem, namely, 28.65°. The difference between *Y* and *Z* is
also the same, but in the opposite sense. Wave *Y* lags wave *Z* by 45.00°. This
is the same as saying that wave *Y* leads wave *Z* by −45.00°. Thus wave *X*

leads wave Z by $28.65° + (−45.00°)$, which is equivalent to $28.65° − 45.00°$, or $−16.35°$. It is better in this case to say that wave X lags wave Z by $16.35°$ or that wave Z leads wave X by $16.35°$.

As you can see, phase relationships can get confusing. It's the same sort of thing that happens when you talk about negative numbers. Which number is larger than which? It depends on point of view. If it helps you to draw pictures of waves when thinking about phase, then by all means go ahead.

Quiz

Refer to the text in this chapter if necessary. A good score is eight correct. Answers are in the back of the book.

1. Approximately how many radians are in a quarter of a cycle?
   (a) 0.7854
   (b) 1.571
   (c) 3.142
   (d) 6.284

2. Refer to Fig. 13-18. Suppose that each horizontal division represents 1.0 ns (1.0×10^{-9} s) and that each vertical division represents 1 mV (1.0×10^{-3} V). What is the approximate rms voltage? Assume the wave is sinusoidal.
   (a) 4.8 mV
   (b) 9.6 mV
   (c) 3.4 mV
   (d) 6.8 mV

3. In the wave illustrated by Fig. 13-18, given the same specifications as those for the previous problem, what is the approximate frequency of this wave?
   (a) 330 MHz
   (b) 660 MHz
   (c) 4.1×10^{9} rad/s
   (d) It cannot be determined from this information.

4. In the wave illustrated by Fig. 13-18, what fraction of a cycle, in degrees, is represented by one horizontal division?
   (a) 60
   (b) 90

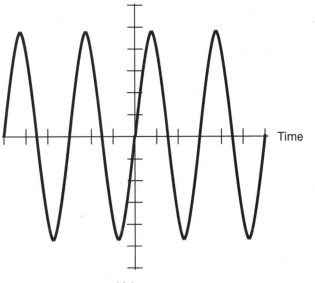

Fig. 13-18.   Illustration for quiz questions 2, 3, and 4.

(c) 120
(d) 180

5. The maximum instantaneous current in a fluctuating dc wave is +543 mA over several cycles. The minimum instantaneous current is +105 mA, also over several cycles. What is the peak-to-peak current in this wave?
(a) 438 mA
(b) 648 mA
(c) 543 mA
(d) It cannot be calculated from this information.

6. The pk-pk voltage in a square wave is 5.50 V. The wave is ac, but it has a dc component of +1.00 V. What is the instantaneous voltage?
(a) More information is needed to answer this question.
(b) +3.25 V
(c) −1.25 V
(d) +1.00 V

7. Given the situation in the preceding question, what is the average voltage?
(a) More information is needed to answer this question.
(b) +3.25 V
(c) −1.25 V
(d) +1.00 V

8. Suppose that there are two sine waves having identical frequency and that their vector representations are at right angles to each other. What is the difference in phase?
    (a) More information is needed to answer this question.
    (b) 90°
    (c) 180°
    (d) 2π rad

9. A square wave is a special form of
    (a) sine wave.
    (b) sawtooth wave.
    (c) ramp wave.
    (d) rectangular wave.

10. An ac wave has a constant frequency f. Its peak voltage V_{pk} is doubled. What happens to the period T?
    (a) It doubles to $2T$.
    (b) It is reduced to $T/2$.
    (c) It is reduced to $0.707T$.
    (d) It remains at T.

CHAPTER 14

Magnetism

The study of magnetism is a science in itself. Electrical and magnetic phenomena interact; a detailed study of magnetism and electromagnetism could easily fill a book. Magnetism exists whenever electric charges move relative to other objects or relative to a frame of reference.

Geomagnetism

The Earth has a core made up largely of iron heated to the extent that some of it is liquid. As the Earth rotates, the iron flows in complex ways. This flow gives rise to a huge magnetic field, called the *geomagnetic field,* that surrounds the Earth.

EARTH'S MAGNETIC POLES AND AXIS

The geomagnetic field has poles, as a bar magnet does. These poles are near, but not at, the geographic poles. The *north geomagnetic pole* is located in the frozen island region of northern Canada. The *south geomagnetic pole* is in the ocean near the coast of Antarctica. The *geomagnetic axis* is thus somewhat tilted relative to the axis on which the Earth rotates. Not only this, but the geomagnetic axis does not exactly run through the center of the Earth. It's like an apple core that's off center.

SOLAR WIND

Charged particles from the Sun, constantly streaming outward through the solar system, distort the geomagnetic field. This *solar wind* in effect "blows" the field out of shape. On the side of the Earth facing the Sun, the field is compressed; on the side of the Earth opposite the Sun, the field is stretched out. This effect occurs with the magnetic fields around the other planets, too, notably Jupiter.

As the Earth rotates, the geomagnetic field does a complex twist-and-turn dance into space in the direction facing away from the Sun. At and near the Earth's surface, the field is nearly symmetrical with respect to the geomagnetic poles. As the distance from the Earth increases, the extent of geomagnetic-field distortion increases.

THE MAGNETIC COMPASS

The presence of the Earth's magnetic field was noticed in ancient times. Certain rocks, called *lodestones,* when hung by strings, always orient themselves in a generally north-south direction. Long ago this was correctly attributed to the presence of a "force" in the air. It was some time before the reasons for this phenomenon were known, but the effect was put to use by seafarers and land explorers. Today, a *magnetic compass* can still be a valuable navigation aid, used by mariners, backpackers, and others who travel far from familiar landmarks. It can work when more sophisticated navigational devices fail.

The geomagnetic field and the magnetic field around a compass needle interact so that a force is exerted on the little magnet inside the compass. This force works not only in a horizontal plane (parallel to the Earth's surface) but vertically, too, in most locations. The vertical component is zero at the *geomagnetic equator,* a line running around the globe equidistant from both geomagnetic poles. As the *geomagnetic latitude* increases toward either the north or the south geomagnetic pole, the magnetic force pulls up and down on the compass needle more and more. The extent of this vertical component at any particular location is called the *inclination* of the geomagnetic field at that location. You have noticed this when you hold a compass. One end of the needle seems to insist on touching the compass face, whereas the other end tilts up toward the glass.

Magnetic Force

As children, most of us discovered that magnets "stick" to some metals. Iron, nickel, and alloys containing either or both of these elements are known as *ferromagnetic materials*. Magnets exert force on these metals. Magnets generally do not exert force on other metals unless those metals carry electric currents. Electrically insulating substances never attract magnets under normal conditions.

CAUSE AND STRENGTH

When a magnet is brought near a piece of ferromagnetic material, the atoms in the material become lined up so that the metal is temporarily magnetized. This produces a *magnetic force* between the atoms of the ferromagnetic substance and those in the magnet.

If a magnet is near another magnet, the force is even stronger than it is when the same magnet is near a ferromagnetic substance. In addition, the force can be either repulsive (the magnets repel, or push away from each other) or attractive (the magnets attract, or pull toward each other) depending on the way the magnets are turned. The force gets stronger as the magnets are brought closer and closer together.

Some magnets are so strong that no human being can pull them apart if they get "stuck" together, and no person can bring them all the way together against their mutual repulsive force. This is especially true of *electromagnets,* discussed later in this chapter. The tremendous forces available are of use in industry. A huge electromagnet can be used to carry heavy pieces of scrap iron or steel from place to place. Other electromagnets can provide sufficient repulsion to suspend one object above another. This is called *magnetic levitation.*

ELECTRIC CHARGE CARRIERS IN MOTION

Whenever the atoms in a ferromagnetic material are aligned, a magnetic field exists. A magnetic field also can be caused by the motion of electric *charge carriers* either in a wire or in free space.

The magnetic field around a *permanent magnet* arises from the same cause as the field around a wire that carries an electric current. The responsible factor in either case is the motion of electrically charged particles. In a wire, the electrons move along the conductor, being passed from atom to atom. In a permanent magnet, the movement of orbiting electrons occurs in such a manner that an "effective current" is produced by the way the electrons move within individual atoms.

Magnetic fields can be produced by the motion of charged particles through space. The Sun is constantly ejecting protons and helium nuclei. These particles carry a positive electric charge. Because of this, they produce "effective currents" as they travel through space. These currents in turn generate magnetic fields. When these fields interact with the Earth's geomagnetic field, the particles are forced to change direction, and they are accelerated toward the geomagnetic poles.

If there is an eruption on the Sun called a *solar flare,* the Sun ejects more charged particles than normal. When these arrive at the Earth's geomagnetic poles, their magnetic fields, collectively working together, can disrupt the Earth's geomagnetic field. Then there is a *geomagnetic storm.* Such an event causes changes in the Earth's ionosphere, affecting long-distance radio communications at certain frequencies. If the fluctuations are intense enough, even wire communications and electrical power transmission can be interfered with. Microwave transmissions generally are immune to the effects of geomagnetic storms. Fiberoptic cable links and free-space laser communications are not affected. Aurora (northern or southern lights) are frequently observed at night during geomagnetic storms.

LINES OF FLUX

Physicists consider magnetic fields to be comprised of *flux lines,* or *lines of flux.* The intensity of the field is determined according to the number of flux lines passing through a certain cross section, such as a centimeter squared (cm^2) or a meter squared (m^2). The lines are not actual threads in space, but it is intuitively appealing to imagine them this way, and their presence can be shown by simple experimentation.

Have you seen the classical demonstration in which iron filings are placed on a sheet of paper, and then a magnet is placed underneath the paper? The filings arrange themselves in a pattern that shows, roughly, the "shape" of the magnetic field in the vicinity of the magnet. A bar magnet has a field whose lines of flux have a characteristic pattern (Fig. 14-1).

Another experiment involves passing a current-carrying wire through the paper at a right angle. The iron filings become grouped along circles

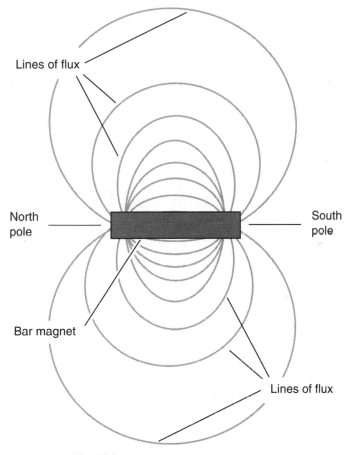

Fig. 14-1. Magnetic flux around a bar magnet.

centered at the point where the wire passes through the paper. This shows that the lines of flux are circular as viewed through any plane passing through the wire at a right angle. The flux circles are centered on the axis of the wire, or the axis along which the charge carriers move (Fig. 14-2).

POLARITY

A magnetic field has a direction, or orientation, at any point in space near a current-carrying wire or a permanent magnet. The flux lines run parallel to the direction of the field. A magnetic field is considered to begin, or originate,

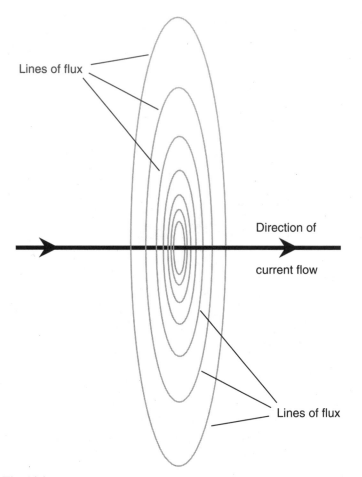

Fig. 14-2. Magnetic flux produced by charge carriers traveling in a straight line.

at a *north pole* and to end, or terminate, at a *south pole.* These poles are not the same as the geomagnetic poles; in fact, they are precisely the opposite! The north geomagnetic pole is in reality a south pole because it attracts the north poles of magnetic compasses. Similarly, the south geomagnetic pole is a north pole because it attracts the south poles of compasses. In the case of a permanent magnet, it is usually, but not always, apparent where the magnetic poles are located. With a current-carrying wire, the magnetic field goes around and around endlessly, like a dog chasing its own tail.

A charged electric particle, such as a proton, hovering in space, is an *electric monopole,* and the electrical flux lines around it aren't closed. A positive charge does not have to be mated with a negative charge. The electrical flux lines around any stationary charged particle run outward in all directions for a theoretically infinite distance. However, a magnetic field is different. Under normal circumstances, all magnetic flux lines are closed loops. With permanent magnets, there is always a starting point (the north pole) and an ending point (the south pole). Around the current-carrying wire, the loops are circles. This can be seen plainly in experiments with iron filings on paper.

DIPOLES AND MONOPOLES

You might at first think that the magnetic field around a current-carrying wire is caused by a monopole or that there aren't any poles at all because the concentric circles apparently don't originate or terminate anywhere. However, think of any geometric plane containing the wire. A *magnetic dipole,* or pair of opposite magnetic poles, is formed by the lines of flux going halfway around on either side. There in effect are two such "magnets" stuck together. The north poles and the south poles are thus not points but rather faces of the plane backed right up against each other.

The lines of flux in the vicinity of a magnetic dipole always connect the two poles. Some flux lines are straight in a local sense, but in a larger sense they are always curves. The greatest magnetic field strength around a bar magnet is near the poles, where the flux lines converge. Around a current-carrying wire, the greatest field strength is near the wire.

Magnetic Field Strength

The overall magnitude of a magnetic field is measured in units called *webers,* symbolized Wb. A smaller unit, the *maxwell* (Mx), is sometimes used if a magnetic field is very weak. One weber is equivalent to 100 million maxwells. Thus 1 Wb $= 10^8$ Mx, and 1 Mx $= 10^{-8}$ Wb.

THE TESLA AND THE GAUSS

If you have a permanent magnet or electromagnet, you might see its "strength" expressed in terms of webers or maxwells. More often, though, you'll hear or read about units called *teslas* (T) or *gauss* (G). These units are expressions of the concentration, or intensity, of the magnetic field within a certain cross section. The *flux density,* or number of "flux lines per unit cross-sectional area," is a more useful expressions for magnetic effects than the overall quantity of magnetism. Flux density is customarily denoted B in equations. A flux density of 1 tesla is equal to 1 weber per meter squared (1 Wb/m^2). A flux density of 1 gauss is equal to 1 maxwell per centimeter squared (1 Mx/cm^2). It turns out that the gauss is equivalent to exactly 0.0001 tesla. That is, $1 \text{ G} = 10^{-4} \text{ T}$, and $1 \text{ T} = 10^{4} \text{ G}$. To convert from teslas to gauss (not gausses!), multiply by 10^4; to convert from gauss to teslas, multiply by 10^{-4}.

If you are confused by the distinctions between webers and teslas or between maxwells and gauss, think of a light bulb. Suppose that a lamp emits 20 W of visible-light power. If you enclose the bulb completely, then 20 W of visible light strike the interior walls of the chamber, no matter how large or small the chamber. However, this is not a very useful notion of the brightness of the light. You know that a single bulb gives plenty of light for a small walk-in closet but is nowhere near adequate to illuminate a gymnasium. The important consideration is the number of watts *per unit area.* When we say the bulb gives off a certain number of watts of visible light, it's like saying a magnet has an overall magnetism of so many webers or maxwells. When we say that the bulb produces a certain number of watts per unit area, it's analogous to saying that a magnetic field has a flux density of so many teslas or gauss.

THE AMPERE-TURN AND THE GILBERT

When working with electromagnets, another unit is employed. This is the *ampere-turn* (At). It is a unit of *magnetomotive force.* A wire bent into a circle and carrying 1 A of current produces 1 At of magnetomotive force. If the wire is bent into a loop having 50 turns, and the current stays the same, the resulting magnetomotive force becomes 50 times as great, that is, 50 At. If the current in the 50-turn loop is reduced to 1/50 A or 20 mA, the magnetomotive force goes back down to 1 At.

A unit called the *gilbert* is sometimes used to express magnetomotive force. This unit is equal to about 1.256 At. To approximate ampere-turns when the number of gilberts is known, multiply by 1.256. To approximate gilberts when the number of ampere-turns is known, multiply by 0.796.

FLUX DENSITY VERSUS CURRENT

In a straight wire carrying a steady direct current surrounded by air or by free space (a vacuum), the flux density is greatest near the wire and diminishes with increasing distance from the wire. You ask, "Is there a formula that expresses flux density as a function of distance from the wire?" The answer is yes. Like all formulas in physics, it is perfectly accurate only under idealized circumstances.

Consider a wire that is perfectly thin, as well as perfectly straight. Suppose that it carries a current of I amperes. Let the flux density (in teslas) be denoted B. Consider a point P at a distance r (in meters) from the wire, as measured along the shortest possible route (that is, within a plane perpendicular to the wire). This is illustrated in Fig. 14-3. The following formula applies:

$$B = 2 \times 10^{-7} \, (I/r)$$

In this formula, the value 2 can be considered mathematically exact to any desired number of significant figures.

As long as the thickness of the wire is small compared with the distance r from it, and as long as the wire is reasonably straight in the vicinity of the point P at which the flux density is measured, this formula is a good indicator of what happens in real life.

PROBLEM 14-1
What is the flux density in teslas at a distance of 20 cm from a straight, thin wire carrying 400 mA of direct current?

SOLUTION 14-1
First, convert everything to units in the International System (SI). This means that $r = 0.20$ m and $I = 0.400$ A. Knowing these values, plug them directly into the formula:

$$B = 2 \times 10^{-7} \, (I/r)$$
$$= 2.00 \times 10^{-7} \, (0.400/0.20)$$
$$= 4.0 \times 10^{-7} \, \text{T}$$

PROBLEM 14-2
In the preceding scenario, what is the flux density B_{gauss} (in gauss) at point P?

SOLUTION 14-2
To figure this out, we must convert from teslas to gauss. This means that we must multiply the answer from the preceding problem by 10^4:

$$B_{gauss} = 4.0 \times 10^{-7} \times 10^4$$
$$= 4.0 \times 10^{-3} \, \text{G}$$

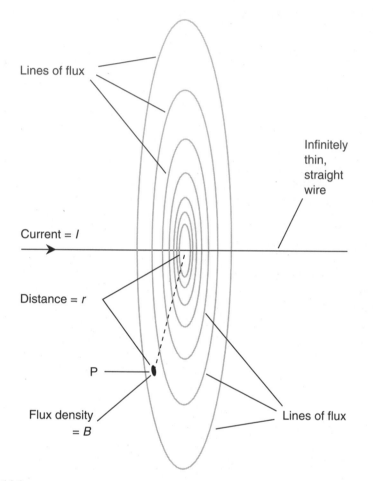

Lines of flux

Infinitely thin, straight wire

Current = *I*

Distance = *r*

P

Flux density = *B*

Lines of flux

Fig. 14-3. Flux density varies inversely with the distance from a wire carrying direct current.

Electromagnets

Any electric current, or movement of charge carriers, produces a magnetic field. This field can become intense in a tightly coiled wire having many turns and carrying a large electric current. When a ferromagnetic rod, called a *core,* is placed inside the coil, the magnetic lines of flux are concentrated in the core, and the field strength in and near the core becomes tremendous. This is the principle of an electromagnet (Fig. 14-4).

Electromagnets are almost always cylindrical in shape. Sometimes the cylinder is long and thin; in other cases it is short and fat. Whatever the

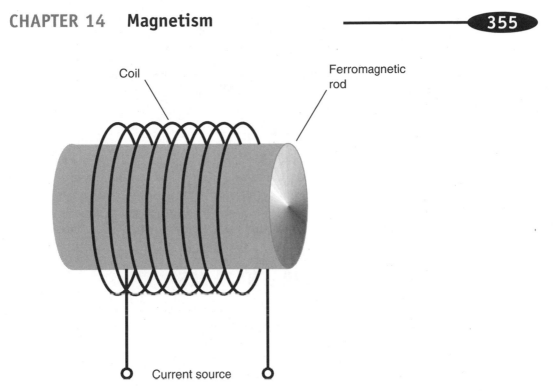

Coil   Ferromagnetic
rod

Fig. 14-4. A simple electromagnet.

Current source

ratio of diameter to length for the core, however, the principle is always the same: The flux produced by the current temporarily magnetizes the core.

DIRECT-CURRENT TYPES

You can build a dc electromagnet by taking a large iron or steel bolt (such as a stove bolt) and wrapping a couple of hundred turns of wire around it. These items are available in almost any hardware store. Be sure the bolt is made of ferromagnetic material. (If a permanent magnet "sticks" to the bolt, the bolt is ferromagnetic.) Ideally, the bolt should be at least $^{3}/_{8}$ inch in diameter and several inches long. You must use insulated or enameled wire, preferably made of solid, soft copper. "Bell wire" works well.

Be sure that all the wire turns go in the same direction. A large 6-V "lantern battery" can provide plenty of dc to operate the electromagnet. Never leave the coil connected to the battery for more than a few seconds at a time. And do not—repeat, *do not*—use an automotive battery for this experiment. The near-short-circuit produced by an electromagnet can cause the acid from such a battery to violently boil out, and this acid is dangerous stuff.

Direct-current electromagnets have defined north and south poles, just like permanent magnets. The main difference is that an electromagnet can get much stronger than any permanent magnet. You should see evidence of this if you do the preceding experiment with a large enough bolt and enough turns of wire. Another difference between an electromagnet and a permanent magnet is the fact that in an electromagnet, the magnetic field exists only as long as the coil carries current. When the power source is removed, the magnetic field collapses. In some cases, a small amount of *residual magnetism* remains in the core, but this is much weaker than the magnetism generated when current flows in the coil.

ALTERNATING-CURRENT TYPES

You might get the idea that the electromagnet can be made far stronger if, rather than using a lantern battery for the current source, you plug the wires into a wall outlet. In theory, this is true. In practice, you'll blow the fuse or circuit breaker. Do not try this. The electrical circuits in some buildings are not adequately protected, and a short circuit can create a fire hazard. Also, you can get a lethal shock from the 117-V utility mains. (Do this experiment in your mind, and leave it at that.)

Some electromagnets use 60-Hz ac. These magnets "stick" to ferromagnetic objects. The polarity of the magnetic field reverses every time the direction of the current reverses; there are 120 fluctuations, or 60 complete north-to-south-to-north polarity changes, every second (Fig. 14-5). If a permanent magnet is brought near either "pole" of an ac electromagnet of the same strength, there is no net force resulting from the ac electromagnetism because there is an equal amount of attractive and repulsive force between the alternating magnetic field and the steady external field. However, there is an attractive force between the core material and the nearby magnet produced independently of the alternating magnetic field resulting from the ac in the coil.

PROBLEM 14-3
Suppose that the frequency of the ac applied to an electromagnet is 600 Hz instead of 60 Hz. What will happen to the interaction between the alternating magnetic field and a nearby permanent magnet of the same strength?

SOLUTION 14-3
Assuming that no change occurs in the behavior of the core material, the situation will be the same as is the case at 60 Hz or at any other ac frequency.

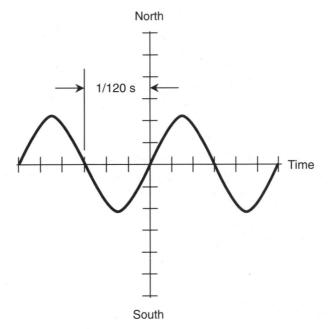

Fig. 14-5. Polarity change in an ac electromagnet.

Magnetic Materials

Some substances cause magnetic lines of flux to bunch closer together than they are in the air; other materials cause the lines of flux to spread farther apart. The first kind of material is ferromagnetic. Substances of this type are, as we have discussed already, "magnetizable." The other kind of material is called *diamagnetic*. Wax, dry wood, bismuth, and silver are examples of substances that decrease magnetic flux density. No diamagnetic material reduces the strength of a magnetic field by anywhere near the factor that ferromagnetic substances can increase it.

The magnetic characteristics of a substance or medium can be quantified in two important but independent ways: *permeability* and *retentivity*.

PERMEABILITY

Permeability, symbolized by the lowercase Greek mu (μ), is measured on a scale relative to a vacuum, or free space. A perfect vacuum is assigned, by convention, a permeability figure of exactly 1. If current is forced through a wire loop or coil in air, then the flux density in and around the coil is about the same as it would be in a vacuum. Therefore, the permeability of pure air is about equal to 1. If you place an iron core in the coil, the flux density increases by a factor ranging from a few dozen to several thousand times, depending on the purity of the iron. The permeability of iron can be as low as about 60 (impure) to as high as about 8,000 (highly refined).

If you use special metallic alloys called *permalloys* as the core material in electromagnets, you can increase the flux density, and therefore the local strength of the field, by as much as 1 million (10^6) times. Such substances thus have permeability as great as 10^6.

If, for some reason, you feel compelled to make an electromagnet that is as weak as possible, you can use dry wood or wax for the core material. Usually, however, diamagnetic substances are used to keep magnetic objects apart while minimizing the interaction between them.

RETENTIVITY

Certain ferromagnetic materials stay magnetized better than others. When a substance such as iron is subjected to a magnetic field as intense as it can handle, say, by enclosing it in a wire coil carrying a high current, there will be some residual magnetism left when the current stops flowing in the coil. Retentivity, also sometimes called *remanence,* is a measure of how well a substance can "memorize" a magnetic field imposed on it and thereby become a permanent magnet.

Retentivity is expressed as a percentage. If the maximum possible flux density in a material is x teslas or gauss and then goes down to y teslas or gauss when the current is removed, the retentivity B_r of that material is given by the following formula:

$$B_r = 100y/x$$

What is meant by *maximum possible flux density* in the foregoing definition? This is an astute question. In the real world, if you make an electromagnet with a core material, there is a limit to the flux density that can be generated in that core. As the current in the coil increases, the flux density inside the core goes up in proportion—for awhile. Beyond a certain point, however, the flux density levels off, and further increases in current do not produce any further increase in the flux density. This condition is called *core saturation.* When we determine retentivity for a material, we are referring to the ratio of the flux density when it is saturated and the flux density when there is no magnetomotive force acting on it.

As an example, suppose that a metal rod can be magnetized to 135 G when it is enclosed by a coil carrying an electric current. Imagine that this is the maximum possible flux density that the rod can be forced to have. For any substance, there is always such a maximum; further increasing the current in the wire will not make the rod any more magnetic. Now suppose that the current is shut off and that 19 G remain in the rod. Then the retentivity B_r is

$$B_r = 100 \times 19/135 = 100 \times 0.14 = 14 \text{ percent}$$

Certain ferromagnetic substances have good retentivity and are excellent for making permanent magnets. Other ferromagnetic materials have poor retentivity. They can work well as the cores of electromagnets, but they do not make good permanent magnets. Sometimes it is desirable to have a substance with good ferromagnetic properties but poor retentivity. This is the case when you want to have an electromagnet that will operate from dc so that it maintains a constant polarity but that will lose its magnetism when the current is shut off.

If a ferromagnetic substance has poor retentivity, it's easy to make it work as the core for an ac electromagnet because the polarity is easy to switch. However, if the retentivity is high, the material is "magnetically sluggish" and has trouble following the current reversals in the coil. This sort of stuff doesn't function well as the core of an ac electromagnet.

PROBLEM 14-4
Suppose that a metal rod is surrounded by a coil and that the magnetic flux density can be made as great as 0.500 T; further increases in current cause no further increase in the flux density inside the core. Then the current is removed; the flux density drops to 500 G. What is the retentivity of this core material?

SOLUTION 14-4
First, convert both flux density figures to the same units. Remember that 1 T = 10^4 G. Thus the flux density is $0.500 \times 10^4 = 5,000$ G with the current and 500 G without the current. "Plugging in" these numbers gives us this:

$$B_r = 100 \times 500/5,000 = 100 \times 0.100 = 10.0 \text{ percent}$$

PERMANENT MAGNETS

Any ferromagnetic material, or substance whose atoms can be aligned permanently, can be made into a permanent magnet. These are the magnets you played with as a child (and maybe still play with when you use them to stick notes to your refrigerator door). Some alloys can be made into stronger permanent magnets than others.

One alloy that is especially suited to making strong permanent magnets is known by the trade name *Alnico*. This word derives from the chemical symbols of the metals that comprise it: aluminum (Al), nickel (Ni), and cobalt (Co). Other elements are sometimes added, including copper and titanium. However, any piece of iron or steel can be magnetized to some extent. Many technicians use screwdrivers that are slightly magnetized so that they can hold onto screws when installing or removing them from hard-to-reach places.

Permanent magnets are best made from materials with high retentivity. They are made by using the material as the core of an electromagnet for an extended period of time. If you want to magnetize a screwdriver a little bit so that it will hold onto screws, stroke the shaft of the screwdriver with the end of a bar magnet several dozen times. However, take note: Once you have magnetized a tool, it is practically impossible to completely demagnetize it.

FLUX DENSITY INSIDE A LONG COIL

Suppose that you have a long coil of wire, commonly known as a *solenoid,* with n turns and whose length in meters is s. Suppose that this coil carries a direct current of I amperes and has a core whose permeability is μ. The flux density B in teslas inside the core, assuming that it is not in a state of saturation, can be found using this formula:

$$B = 4\pi \times 10^{-7} \, (\mu n I / s)$$

A good approximation is

$$B = 1.2566 \times 10^{-6} \, (\mu n I / s)$$

PROBLEM 14-5
Consider a dc electromagnet that carries a certain current. It measures 20 cm long and has 100 turns of wire. The flux density in the core, which is known not to be in a state of saturation, is 20 G. The permeability of the core material is 100. What is the current in the wire?

SOLUTION 14-5
As always, start by making sure that all units are correct for the formula that will be used. The length s is 20 cm, that is, 0.20 m. The flux density B is 20 G, which is 0.0020 T. Rearrange the preceding formula so it solves for I:

$$B = 1.2566 \times 10^{-6} \ (\mu nI/s)$$

$$B/I = 1.2566 \times 10^{-6} \ (\mu n/s)$$

$$I^{-1} = 1.2566 \times 10^{-6} \ (\mu n/sB)$$

$$I = 7.9580 \times 10^{5} \ (sB/\mu n)$$

This is an exercise, but it is straightforward. Derivations such as this are subject to the constraint that we not divide by any quantity that can attain a value of zero in a practical situation. (This is not a problem here. We aren't concerned with scenarios involving zero current, zero turns of wire, permeability of zero, or coils having zero length.) Let's "plug in the numbers":

$$I = 7.9580 \times 10^{5} \ (0.20 \times 0.0020)/(100 \times 100)$$

$$= 7.9580 \times 10^{5} \times 4.0 \times 10^{-8}$$

$$= 0.031832 \ A = 31.832 \ mA$$

This must be rounded off to 32 mA because we are only entitled to claim two significant figures.

Magnetic Machines

A solenoid, having a movable ferromagnetic core, can do various things. Electrical relays, bell ringers, electric "hammers," and other mechanical devices make use of the principle of the solenoid. More sophisticated electromagnets, sometimes in conjunction with permanent magnets, can be used to build motors, meters, generators, and other devices.

A RINGER DEVICE

Figure 14-6 is a simplified diagram of a bell ringer. Its solenoid is an electromagnet. The core has a hollow region in the center, along its axis, through which a steel rod passes. The coil has many turns of wire, so the electromagnet is powerful if a substantial current passes through the coil.

When there is no current flowing in the coil, the rod is held down by the force of gravity. When a pulse of current passes through the coil, the rod is pulled forcibly upward. The magnetic force "wants" the ends of the rod,

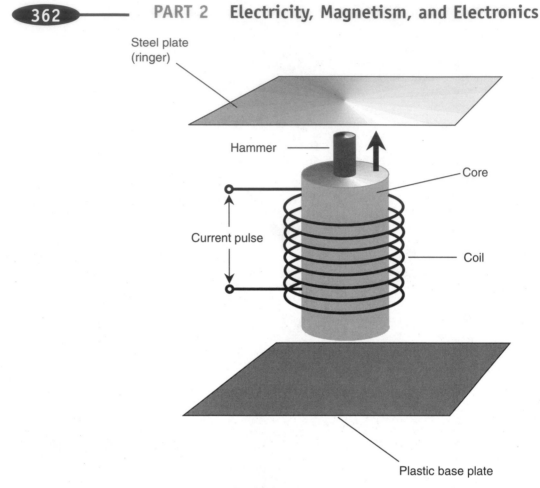

Fig. 14-6. A bell ringer using a solenoid.

which is the same length as the core, to be aligned with the ends of the core. However, the pulse is brief, and the upward momentum is such that the rod passes all the way through the core and strikes the ringer plate. Then the steel rod falls back down again to its resting position, allowing the plate to reverberate. Some office telephones are equipped with ringers that produce this noise rather than conventional ringing, buzzing, beeping, or chirping emitted by most phone sets. The "gong" sound is less irritating to some people than other attention-demanding signals.

A RELAY

In some electronic devices, it is inconvenient to place a switch exactly where it should be. For example, you might want to switch a communications line from one branch to another from a long distance away. In wireless transmitters, some of the wiring carries high-frequency alternating currents that must be kept within certain parts of the circuit and not routed out to the front panel for switching. A *relay* makes use of a solenoid to allow remote-control switching.

A drawing and a diagram of a relay are shown in Fig. 14-7. The movable lever, called the *armature,* is held to one side by a spring when there is no current flowing through the electromagnet. Under these conditions, terminal X is connected to terminal Y but not to terminal Z. When a sufficient

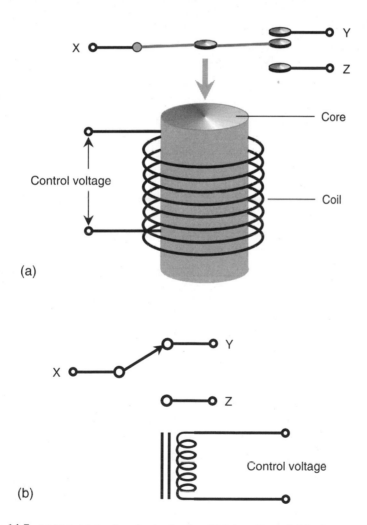

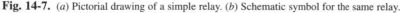

Fig. 14-7. (*a*) Pictorial drawing of a simple relay. (*b*) Schematic symbol for the same relay.

current is applied, the armature is pulled over to the other side. This disconnects terminal *X* from terminal *Y* and connects *X* to *Z*.

There are numerous types of relays, each used for a different purpose. Some are meant for use with dc, and others are for ac; some will work with either ac or dc. A *normally closed relay* completes a circuit when there is no current flowing in its electromagnet and breaks the circuit when current flows. A *normally open relay* is just the opposite. (*Normal* in this sense means "no current in the coil.") The relay shown in Fig. 14-7 can be used either as a normally open or normally closed relay depending on which contacts are selected. It also can be used to switch a line between two different circuits.

These days, relays are used only in circuits and systems carrying extreme currents or voltages. In most ordinary applications, electronic semiconductor switches, which have no moving parts and can last far longer than relays, are preferred.

THE DC MOTOR

Magnetic fields can produce considerable mechanical forces. These forces can be harnessed to do work. The device that converts dc energy into rotating mechanical energy is a *dc motor.* In this sense, a dc motor is a form of *transducer.* Motors can be microscopic in size or as big as a house. Some tiny motors are being considered for use in medical devices that actually can circulate in the bloodstream or be installed in body organs. Others can pull a train at freeway speeds.

In a dc motor, the source of electricity is connected to a set of coils producing magnetic fields. The attraction of opposite poles, and the repulsion of like poles, is switched in such a way that a constant torque, or rotational force, results. The greater the current that flows in the coils, the stronger is the torque, and the more electrical energy is needed. One set of coils, called the *armature coil,* goes around with the motor shaft. The other set of coils, called the *field coil,* is stationary (Fig. 14-8). In some motors, the field coils are replaced by a pair of permanent magnets. The current direction in the armature coil is reversed every half-rotation by the *commutator.* This keeps the force going in the same angular direction. The shaft is carried along by its own angular momentum so that it doesn't come to a stop during those instants when the current is being switched in polarity.

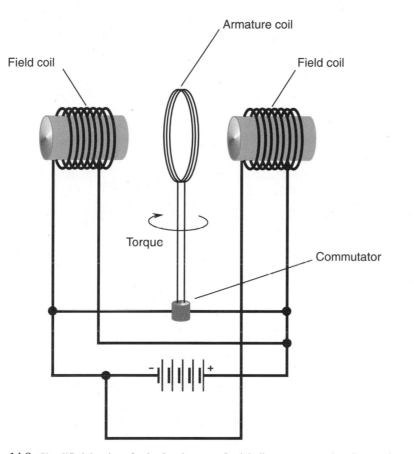

Fig. 14-8. Simplified drawing of a dc electric motor. Straight lines represent wires. Intersecting lines indicate connections only when there is a dot at the point where the lines cross.

THE ELECTRIC GENERATOR

An *electric generator* is constructed somewhat like a conventional motor, although it functions in the opposite sense. Some generators also can operate as motors; they are called *motor/generators*. Generators, like motors, are energy transducers of a special sort.

A typical generator produces ac when a coil is rotated rapidly in a strong magnetic field. The magnetic field can be provided by a pair of permanent magnets (Fig. 14-9). The rotating shaft is driven by a gasoline-powered motor, a turbine, or some other source of mechanical energy. A commutator can be used with a generator to produce pulsating dc output, which can be filtered to obtain pure dc for use with precision equipment.

Magnetic Data Storage

Magnetic fields can be used to store data in various forms. Common media for data storage include *magnetic tape* and the *magnetic disk.*

MAGNETIC TAPE

Recording tape is the stuff you find in cassette players. These days, magnetic tape is largely obsolete, but it is still sometimes used for home enter-

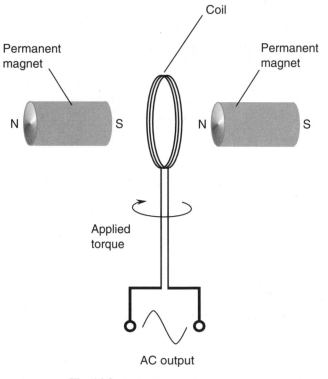

Fig. 14-9. A simple type of ac generator.

tainment, especially high-fidelity (hi-fi) music and home video. It also can be found in some high-capacity computer data storage systems.

The tape consists of millions of particles of iron oxide attached to a plastic or nonferromagnetic metal strip. A fluctuating magnetic field, produced by the *recording head,* polarizes these particles. As the field changes in strength next to the recording head, the tape passes by at a constant, controlled speed. This produces regions in which the iron oxide particles are polarized in either direction. When the tape is run at the same speed through the recorder in the playback mode, the magnetic fields around the individual particles cause a fluctuating field that is detected by a *pickup head.* This field has the same pattern of variations as the original field from the recording head.

Magnetic tape is available in various widths and thicknesses for different applications. Thick-tape cassettes don't play as long as thin-tape ones, but the thicker tape is more resistant to stretching. The speed of the tape determines the fidelity of the recording. Higher speeds are preferred for music and video and lower speeds for voice.

The data on a magnetic tape can be distorted or erased by external magnetic fields. Therefore, tapes should be protected from such fields. Keep magnetic tape away from permanent magnets or electromagnets. Extreme heat also can damage the data on magnetic tape, and if the temperature is high enough, physical damage occurs as well.

MAGNETIC DISK

The era of the personal computer has seen the development of ever-more-compact data storage systems. One of the most versatile is the magnetic disk. Such a disk can be either rigid or flexible. Disks are available in various sizes. *Hard disks* (also called *hard drives*) store the most data and generally are found inside computer units. *Diskettes* are usually 3.5 inches (8.9 cm) in diameter and can be inserted and removed from digital recording/playback machines called *disk drives.*

The principle of the magnetic disk, on the micro scale, is the same as that of magnetic tape. But disk data is stored in binary form; that is, there are only two different ways that the particles are magnetized. This results in almost perfect, error-free storage. On a larger scale, the disk works differently than tape because of the difference in geometry. On a tape, the information is spread out over a long span, and some bits of data are far away from others. On a disk, no two bits are ever farther apart than the diameter of the disk. Therefore, data can be transferred to or from a disk more rapidly than is possible with tape.

A typical diskette can store an amount of digital information equivalent to a short novel. Specialized high-capacity diskettes can store the equivalent of hundreds of long novels or even a complete encyclopedia.

The same precautions should be observed when handling and storing magnetic disks as are necessary with magnetic tape.

Quiz

Refer to the text in this chapter if necessary. A good score is eight correct. Answers are in the back of the book.

1. The geomagnetic field
   (a) makes the Earth like a huge horseshoe magnet.
   (b) runs exactly through the geographic poles.
   (c) makes a compass work.
   (d) makes an electromagnet work.

2. A material that can be permanently magnetized is generally said to be
   (a) magnetic.
   (b) electromagnetic.
   (c) permanently magnetic.
   (d) ferromagnetic.

3. The magnetic flux around a straight current-carrying wire
   (a) gets stronger with increasing distance from the wire.
   (b) is strongest near the wire.
   (c) does not vary in strength with distance from the wire.
   (d) consists of straight lines parallel to the wire.

4. The gauss is a unit of
   (a) overall magnetic field strength.
   (b) ampere-turns.
   (c) magnetic flux density.
   (d) magnetic power.

5. If a wire coil has 10 turns and carries 500 mA of current, what is the magnetomotive force in ampere-turns?
   (a) 5,000
   (b) 50
   (c) 5.0
   (d) 0.02

6. Which of the following is not generally observed in a geomagnetic storm?
    (a) Charged particles streaming out from the Sun
    (b) Fluctuations in the Earth's magnetic field
    (c) Disruption of electrical power transmission
    (d) Disruption of microwave propagation

7. An ac electromagnet
    (a) will attract only other magnetized objects.
    (b) will attract iron filings.
    (c) will repel other magnetized objects.
    (d) will either attract or repel permanent magnets depending on the polarity.

8. A substance with high retentivity is best suited for making
    (a) an ac electromagnet.
    (b) a dc electromagnet.
    (c) an electrostatic shield.
    (d) a permanent magnet.

9. A device that reverses magnetic field polarity to keep a dc motor rotating is
    (a) a solenoid.
    (b) an armature coil.
    (c) a commutator.
    (d) a field coil.

10. An advantage of a magnetic disk, as compared with magnetic tape, for data
    storage and retrieval is that
    (a) a disk lasts longer.
    (b) data can be stored and retrieved more quickly with disks than with tapes.
    (c) disks look better.
    (d) disks are less susceptible to magnetic fields.

CHAPTER 15

More About Alternating Current

In dc electrical circuits, the relationship among current, voltage, resistance, and power is simple. The same is true for ac circuits as long as those circuits do not store or release any energy during the course of each current cycle. When energy is stored and released during each cycle, an ac circuit is said to contain *reactance*. This can be caused by *inductance, capacitance,* or both.

Inductance

Inductance is opposition to ac by temporarily storing some of the electrical energy as a magnetic field. Components that do this are called *inductors*. Inductors often, but not always, consist of wire coils.

THE PROPERTY OF INDUCTANCE

Suppose that you have a wire 1 million (10^6) km long. What happens if you make this wire into a huge loop and connect its ends to the terminals of a battery? A current flows through the loop, and this produces a magnetic field. The field is small at first because current flows in only part of the loop. The magnetic flux increases over a period of a few seconds as the motion of charge carriers (mainly electrons) makes its way around the loop. A certain amount of energy is stored in this magnetic field. The ability of the loop to store energy in this way is the property of inductance, symbolized in equations by an italicized uppercase letter *L*.

PRACTICAL INDUCTORS

In practice, you cannot make wire loops anywhere near 10^6 km in circumference. But lengths of wire can be coiled up. When this is done, the magnetic flux is increased many times for a given length of wire compared with the flux produced by a single-turn loop.

For any coil, the magnetic flux density is multiplied when a ferromagnetic core is placed within the coil of wire. You may remember this from the study of magnetism. The increase in flux density has the effect of multiplying the inductance of a coil so that it is many times greater with a ferromagnetic core than with an air core. The inductance also depends on the number of turns in the coil, the diameter of the coil, and the overall shape of the coil.

In general, inductance of a coil is directly proportional to the number of turns of wire. Inductance is directly proportional to the diameter of the coil. The length of a coil, given a certain number of turns and a certain diameter, has an effect such that the longer the coil is made, the lower the inductance becomes.

THE UNIT OF INDUCTANCE

When a source of dc is connected across an inductor, it takes awhile for the current flow to establish itself throughout the inductor. The current changes at a rate that depends on the inductance. The greater the inductance, the slower is the rate of change of current for a given dc voltage. The unit of inductance is an expression of the ratio between the rate of current change and the voltage across an inductor. An inductance of 1 *henry* (1 H) represents a potential difference of 1 volt (1 V) across an inductor within which the current is increasing or decreasing at 1 ampere per second (1 A/s).

The henry is an extremely large unit of inductance. Rarely will you see an inductor this large, although some power-supply filter chokes have inductances up to several henrys. Usually inductances are expressed in *millihenrys* (mH), *microhenrys* (μH), or *nanohenrys* (nH). You should know your prefix multipliers fairly well by now, but in case you've forgotten, 1 mH = 0.001 H = 10^{-3} H, 1 μH = 0.001 mH = 0.000001 H = 10^{-6} H, and 1 nH = 0.001 μH = 0.000000001 H = 10^{-9} H.

Small coils with few turns of wire have small inductances, in which the current changes quickly and the voltages are small. Huge coils with ferromagnetic cores and many turns of wire have large inductances, in which the current changes slowly and the voltages are large.

INDUCTORS IN SERIES

As long as the magnetic fields around inductors do not interact, inductances in series add like resistances in series. The total value is the sum of the individual values. It is important to be sure that you are using the same size units for all the inductors when you add their values.

Suppose that you have inductances L_1, L_2, L_3,..., L_n connected in series (Fig. 15-1). As long as the magnetic fields of the inductors do not interact—

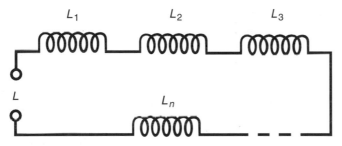

Fig. 15-1.   Inductances in series add like resistances in series.

that is, as long as there is no *mutual inductance* between or among the components—the total inductance L is given by this formula:

$$L = L_1 + L_2 + L_3 + \cdots + L_n$$

INDUCTORS IN PARALLEL

If there is no mutual inductance among two or more parallel-connected inductors, their values add up like the values of resistors in parallel. Suppose that you have inductances L_1, L_2, L_3,..., L_n connected in parallel (Fig. 15-2). Then you can find the reciprocal of the total inductance $1/L$ using the following formula:

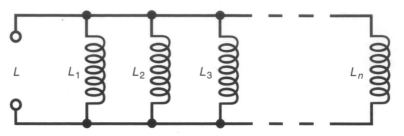

Fig. 15-2.   Inductances in parallel add like resistances in parallel.

$$1/L = 1/L_1 + 1/L_2 + 1/L_3 + \cdots + 1/L_n$$

The total inductance L is found by taking the reciprocal of the number you get for $1/L$. That is:

$$L = 1/ (1/L_1 + 1/L_2 + 1/L_3 + \cdots + 1/L_n)$$
$$= (1/L_1 + 1/L_2 + 1/L_3 + \cdots + 1/L_n)^{-1}$$

Again, as with inductances in series, it is important to remember that all the units have to agree. Don't mix microhenrys with millihenrys or henrys with nanohenrys. The units you use for the individual component values will be the units you get for the final answer.

Let's not concern ourselves with what happens when there is mutual inductance. Sometimes mutual inductance increases the net inductance of a combination to values greater than the formulas indicate, and in other cases mutual inductance reduces the net inductance of a combination. Engineers sometimes must worry about mutual inductance when building radios or other sophisticated electronic circuits, especially at high frequencies.

PROBLEM 15-1
Suppose that there are three inductors in series with no mutual inductance. Their values are 1.50 mH, 150 μH, and 120 μH. What is the net inductance of the combination?

SOLUTION 15-1
Convert all the inductances to the same units and then add them up. Let's use millihenrys (mH). The second and third values must be multiplied by 0.001 (10^{-3}) to convert from microhenrys to millihenrys. Therefore, the net series inductance L_s is:

$$L_s = (1.50 + 0.150 + 0.120) \text{ mH} = 1.77 \text{ mH}$$

PROBLEM 15-2
What is the total inductance of the same three inductors connected in parallel, still assuming that there is no mutual inductance?

SOLUTION 15-2

First, convert all the inductances to the same units. Let's use millihenrys again. Then take the reciprocals of these numbers. The value of the first inductance is 1.50 mH, so the reciprocal of this is 0.667 mH^{-1}. Similarly, the reciprocals of the second and third inductances are 6.667 mH^{-1} and 8.333 mH^{-1}. (The units of "reciprocal millihenrys" don't mean much in real life, but they are useful for keeping track of what we're doing in the process of the calculation.) Now add these to get the reciprocal of the net parallel inductance L_p:

$$L_p^{-1} = (0.667 + 6.667 + 8.333) \text{ mH}^{-1} = 15.667 \text{ mH}^{-1}$$

Finally, take the reciprocal of L_p^{-1} to obtain L_p:

$$L_p^{-1} = (15.667 \text{ mH}^{-1})^{-1} = 0.0638 \text{ mH}$$

This might better be expressed as 63.8 μH.

Inductive Reactance

In dc circuits, resistance is a simple thing. It can be expressed as a number ranging from zero (a perfect conductor) to extremely large values, increasing without limit through thousands, millions, and billions of ohms. Physicists call resistance a *scalar quantity* because it can be expressed on a one-dimensional scale. In fact, dc resistance can be represented along a half-line (also called a *ray*).

Given a certain dc voltage, the current decreases as the resistance increases in accordance with Ohm's law. The same law holds for ac through a resistance if the ac voltage and current are both specified as peak, pk-pk, or rms values.

INDUCTORS AND DC

Suppose that you have some wire that conducts electricity very well. If you wind a length of the wire into a coil and connect it to a source of dc, the wire draws a small amount of current at first, but the current quickly becomes large, possibly blowing a fuse or overstressing a battery. It does not matter whether the wire is a single-turn loop, lying haphazardly on the floor, or wrapped around a stick. The current is large. In amperes, it is equal to $I = E/R$, where I is the current, E is the dc voltage, and R is the resistance of the wire (a low resistance).

You can make an electromagnet by passing dc through a coil wound around an iron rod. However, there is still a large, constant current in the coil. In a practical electromagnet, the coil heats up as energy is dissipated in the resistance of the wire; not all the electrical energy goes into the magnetic field. If the voltage of the battery or power supply is increased, the wire in the coil, iron core or not, gets hotter. Ultimately, if the supply can deliver the necessary current, the wire will melt.

INDUCTORS AND AC

Suppose that you change the voltage source connected across a coil from dc to ac. Imagine that you can vary the frequency of the ac from a few hertz to hundreds of hertz, then kilohertz, and then megahertz.

At first, the ac will be high, just as is the case with dc. However, the coil has a certain amount of inductance, and it takes a little time for current to establish itself in the coil. Depending on how many turns there are and on whether the core is air or a ferromagnetic material, you'll reach a point, as the ac frequency increases, when the coil starts to get sluggish. That is, the current won't have time to get established in the coil before the polarity reverses. At high ac frequencies, the current through the coil has difficulty following the voltage placed across the coil. Just as the coil starts to "think" that it's making a good short circuit, the ac voltage wave passes its peak, goes back to zero, and then tries to pull the electrons the other way. This sluggishness in a coil for ac is, in effect, similar to dc resistance. As the frequency is raised, the effect gets more pronounced. Eventually, if you keep on increasing the frequency of the ac source, the coil does not even come near establishing a current with each cycle. It then acts like a large resistance. Hardly any ac current flows through it.

The opposition that the coil offers to ac is called *inductive reactance.* It, like resistance, is measured in ohms (Ω). It can vary just as resistance does, from near zero (a short piece of wire) to a few ohms (a small coil) to kilohms or megohms (bigger and bigger coils or coils with ferromagnetic cores at high frequencies). Inductive reactance can be depicted on a ray, just like resistance, as shown in Fig. 15-3.

REACTANCE AND FREQUENCY

Inductive reactance is one of two kinds of reactance. (The other kind will be dealt with in a little while.) In mathematical expressions, reactance is symbolized X. Inductive reactance is denoted X_L.

Ohms, kilohms, megohms, or whatever

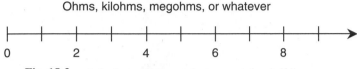

Fig. 15-3.   Inductive reactance can be represented on half-line or ray.
There is no limit to how large it can get, but it can never be negative.

If the frequency of an ac source is f (in hertz) and the inductance of a coil is L (in henrys), then the inductive reactance X_L (in ohms) is

$$X_L = 2\pi fL \approx 6.2832 fL$$

This same formula applies if the frequency f is in kilohertz and the inductance L is in millihenrys. It also applies if f is in megahertz and L is in microhenrys. Remember that if frequency is in thousands, inductance must be in thousandths, and if frequency is in millions, inductance must be in millionths.

Inductive reactance increases linearly with increasing ac frequency. This means that the function of X_L versus f is a straight line when graphed. Inductive reactance also increases linearly with inductance. Therefore, the function of X_L versus L also appears as a straight line on a graph. The value of X_L is directly proportional to f; X_L is also directly proportional to L. These relationships are graphed, in relative form, in Fig. 15-4.

PROBLEM 15-3
An inductor has a value of 10.0 mH. What is the inductive reactance at a frequency of 100 kHz?

SOLUTION 15-3
We are dealing in millihenrys (thousandths of henrys) and kilohertz (thousands of hertz), so we can apply the preceding formula directly. Using 6.2832 to represent an approximation of 2π, we get

$$X_L = 6.2832 \times 100 \times 10.0 = 6283.2 \text{ ohms}$$

Because our input data are given to only three significant figures, we must round this to 6,280 ohms or 6.28 kilohms (kΩ).

POINTS IN THE *RL* QUARTER-PLANE

In a circuit containing both resistance and inductance, the characteristics become two-dimensional. You can orient the resistance and reactance half-lines perpendicular to each other to make a quarter-plane coordinate system,

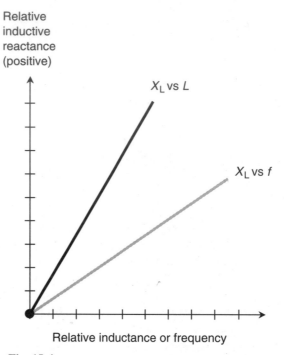

Relative
inductive
reactance
(positive)

X_L vs L

X_L vs f

Relative inductance or frequency

Fig. 15-4. Inductive reactance is directly proportional to
inductance, as well as to frequency.

as shown in Fig. 15-5. Resistance is shown on the horizontal axis, and
inductive reactance is plotted vertically, going upward.

Each point on the *RL quarter-plane* corresponds to a unique *complex-
number impedance*. Conversely, each complex-number impedance value
corresponds to a unique point on the quarter-plane. Impedances on the *RL*
quarter-plane are written in the form $R + jX_L$, where R is the resistance in
ohms, X_L is the inductive reactance in ohms, and j is the unit imaginary
number, that is, the positive square root of -1. The value j in this application is known as the *j operator.* (If you're uncomfortable with imaginary
and complex numbers, go back and review that material in Chap. 1.)

Suppose that you have a pure resistance, say, $R = 5$ ohms. Then the
complex-number impedance is $5 + j0$ and is at the point $(5, 0)$ on the *RL*
quarter-plane. If you have a pure inductive reactance, such as $X_L = 3$
ohms, then the complex-number impedance is $0 + j3$ and is at the point
$(0, 3)$ on the *RL* quarter-plane. These points, and a few others, are shown
in Fig. 15-5.

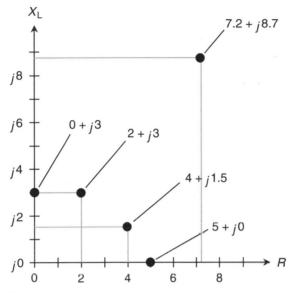

Fig. 15-5.   The *RL* impedance quarter-plane showing five points
for specific complex-number impedances.

In the real world, all coil inductors have some resistance because no wire is a perfect conductor. All resistors have at least a tiny bit of inductive reactance because they have wire leads at each end and they have a measurable physical length. There is no such thing as a mathematically perfect pure resistance like $5 + j0$ or a mathematically perfect pure reactance like $0 + j3$. Sometimes you can get close to such ideals, but absolutely pure resistances or reactances never exist (except occasionally in quiz and test problems, of course!).

In electronic circuits, resistance and inductive reactance are sometimes both incorporated deliberately. Then you get impedances values such as $2 + j3$ or $4 + j1.5$.

Remember that the values for X_L are reactances (expressed in ohms) and not inductances (which are expressed in henrys). Reactances vary with the frequency in an *RL* circuit. Changing the frequency has the effect of making the points move in the *RL* quarter-plane. They move vertically, going upward as the ac frequency increases and downward as the ac frequency decreases. If the ac frequency goes down to zero, thereby resulting in dc, the inductive reactance vanishes. Then $X_L = 0$, and the point is along the resistance axis of the *RL* quarter-plane.

Capacitance

Capacitance impedes the flow of ac charge carriers by temporarily storing the energy as an electrical field. This energy is given back later. Capacitance generally is not important in pure dc circuits, but it can have significance in circuits where dc is pulsating and not steady. Capacitance, like inductance, can appear when it is not wanted or intended. Capacitive effects become more evident as the frequency increases.

THE PROPERTY OF CAPACITANCE

Imagine two huge flat sheets of metal that are excellent electrical conductors. Suppose that they are each the size of the state of Nebraska and are placed one over the other, separated uniformly by a few centimeters of air. If these two sheets of metal are connected to the terminals of a battery, they will become charged, one positively and the other negatively. This will take a little while because the sheets are so big.

If the plates were small, they would both become charged almost instantly, attaining a potential difference equal to the voltage of the battery. However, because the plates are gigantic, it takes awhile for the negative plate to "fill up" with electrons, and it also takes some time for the positive plate to get electrons "sucked out."

Ultimately, the potential difference between the two plates becomes equal to the battery voltage, and an electrical field exists in the space between the plates. This electrical field is small at first; the plates don't charge right away. However, the field increases over a period of time, depending on how large the plates are, as well as on how far apart they are. Energy is stored in this electrical field. Capacitance is a manifestation of the ability of the plates, and of the space between them, to store this energy. In formulas, capacitance is symbolized by the italicized uppercase letter C.

PRACTICAL CAPACITORS

It is out of the question to make a capacitor of the preceding dimensions. However, two sheets or strips of foil can be placed one atop the other, separated by a thin, nonconducting sheet such as paper, and then the whole assembly can be rolled up to get a large effective surface area. When this is done, the electrical flux becomes great enough that the device exhibits significant capacitance. Two sets of several plates can be meshed together, with air in between them, and the resulting capacitance is significant at high ac frequencies.

In a capacitor, the electrical flux concentration is multiplied when a *dielectric* of a certain type is placed between the plates. Some plastics work well for this purpose. The dielectric increases the effective surface area of the plates so that a physically small component can be made to have a large capacitance. Capacitance is directly proportional to the surface area of the conducting plates or sheets. Capacitance is inversely proportional to the separation between conducting sheets; the closer the sheets are to each other, the greater is the capacitance. The capacitance also depends on the *dielectric constant* of the material between the plates. This is the electrostatic equivalent of magnetic permeability. A vacuum has a dielectric constant of 1. Dry air is about the same as a vacuum. Some substances have high dielectric constants that multiply the effective capacitance many times.

In theory, if the dielectric constant of a material is x, then placing that material between the plates of a capacitor will increase the capacitance by a factor of x compared with the capacitance when there is only dry air or a vacuum between the plates. In practice, this is true only if the dielectric is 100 percent efficient—if it does not turn any of the energy contained in the electrical field into heat. It is also true only if all the electrical lines of flux between the plates are forced to pass through the dielectric material. These are ideal scenarios, and while they can never be attained absolutely, many manufactured capacitors come close.

THE UNIT OF CAPACITANCE

When a battery is connected between the plates of a capacitor, it takes some time before the electrical field reaches its full intensity. The voltage builds up at a rate that depends on the capacitance. The greater the capacitance, the slower is the rate of change of voltage in the plates.

The unit of capacitance is an expression of the ratio between the amount of current flowing and the rate of voltage change across the plates of a capacitor. A capacitance of 1 *farad,* abbreviated F, represents a current flow of 1 ampere (1 A) while there is a potential-difference increase or decrease of 1 volt per second (1 V/s). A capacitance of 1 F also results in 1 V of potential difference for an electric charge of 1 coulomb (1 C).

The farad is a huge unit of capacitance. You'll almost never see a capacitor with a value of 1 F. Commonly employed units of capacitance are the *microfarad* (μF) and the *picofarad* (pF). A capacitance of 1 μF represents 0.000001 (10^{-6}) F, and 1 pF is 0.000001 μF, or 10^{-12} F.

CAPACITORS IN SERIES

With capacitors, there is rarely any significant mutual interaction. At very high ac frequencies, however, *interelectrode capacitance* can sometimes be a problem for engineers. This effect, which shows up as an inherent tiny capacitance between wires that run near and parallel to each other, is almost always undesirable in practical circuits.

Capacitors in series add together like resistors or inductors in parallel. If you connect two capacitors of the same value in series, the result is half the capacitance of either component alone. In general, if there are several capacitors in series, the composite value is less than that of any of the single components. It is important that you always use the same size units when determining the capacitance of any combination. Don't mix microfarads with picofarads. The answer you get will be in whichever size units you use for the individual components.

Suppose that you have several capacitors with values C_1, C_2, C_3,..., C_n connected in series (Fig. 15-6). You can find the reciprocal of the total capacitance $1/C$ using the following formula:

$$1/C = 1/C_1 + 1/C_2 + 1/C_3 + \cdots + 1/C_n$$

The total capacitance C is found by taking the reciprocal of the number you get for $1/C$.

PROBLEM 15-4
Two capacitors, with values of $C_1 = 0.10$ μF and $C_2 = 0.050$ μF, are connected in series. What is the total capacitance?

SOLUTION 15-4
Using the preceding formula, first find the reciprocals of the values. They are $1/C_1 = 10$ μF^{-1} and $1/C_2 = 20$ μF^{-1}. ("Reciprocal microfarads" don't have any practical meaning, but using them can help us remember that we must take the reciprocal of the sum of the numbers before we come up with capacitance.) Then

$$1/C = 10 \text{ μF}^{-1} + 20 \text{ μF}^{-1} = 30 \text{ μF}^{-1}$$
$$C = 1/30 \text{ μF}^{-1} = 0.033 \text{ μF}$$

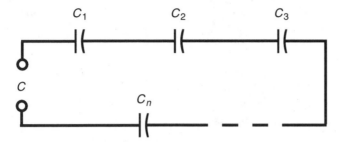

Fig. 15-6. Capacitances in series add like resistances or inductances in parallel.

PROBLEM 15-5
Two capacitors with values of 0.0010 μF and 100 pF are connected in series. What is the total capacitance?

SOLUTION 15-5
Convert to the same size units. A value of 100 pF represents 0.000100 μF. Then you can say that C_1 = 0.0010 μF and C_2 – 0.000100 μF. The reciprocals are $1/C_1$ = 1000 μF^{-1} and $1/C_2$ = 10,000 μF^{-1}. Therefore:

$$1/C = 1000\ \mu F^{-1} + 10,000\ \mu F^{-1} = 11,000\ \mu F^{-1}$$

$$C = 0.000091\ \mu F$$

This number is a little awkward, and you might rather say that it is 91 pF.

In the preceding problem, you can choose picofarads to work with rather than microfarads. In either case, there is some tricky decimal placement involved. It's important to double-check calculations when numbers get like this. Calculators will take care of the decimal placement problem, sometimes using exponent notation and sometimes not, but a calculator can only work with what you put into it. If you enter a wrong number, you will get a wrong answer, and if you miss a digit, you'll be off by a factor of 10 (an order of magnitude).

CAPACITORS IN PARALLEL

Capacitances in parallel add like resistances or inductances in series (Fig. 15-7). That is, the total capacitance is the sum of the individual component values. Again, you need to be sure that you use the same size units all the way through.

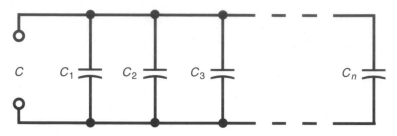

Fig. 15-7. Capacitances in parallel add like resistances or inductances in series.

PROBLEM 15-6
Three capacitors are set up in parallel, having values of $C_1 = 0.100$ μF, $C_2 = 0.0100$ μF, and $C_3 = 0.00100$ μF. What is C, the total capacitance?

SOLUTION 15-6
Add them up: $C = 0.100$ μF $+ 0.0100$ μF $+$ μF $0.00100 = 0.11100$ μF. Because the values are given to three significant figures, the final answer should be stated as $C = 0.111$ μF.

Capacitive Reactance

Inductive reactance has its counterpart in the form of *capacitive reactance.* This, too, can be represented as a ray starting at the same zero point as inductive reactance but running off in the opposite direction, having negative ohmic values (Fig. 15-8). When the ray for capacitive reactance is combined with the ray for inductive reactance, a complete real-number line is the result, with ohmic values that range from the huge negative numbers, through zero, to huge positive numbers.

CAPACITORS AND DC

Imagine two large parallel metal plates, as described earlier. If you connect them to a source of dc, they draw a large amount of current at first as they become electrically charged. However, as the plates reach equilibrium, this current diminishes, and when the two plates attain the same potential difference throughout, the current is zero.

If the voltage of the battery or power supply is increased, a point is eventually reached at which sparks begin to jump between them. Ultimately, if

Fig. 15-8.   Capacitive reactance can be represented on half-line or ray. There is no limit to how large it can get negatively, but it can never be positive.

the power supply can deliver the necessary voltage, this sparking, or *arcing,* becomes continuous. Then the pair of plates no longer acts like a capacitor. When the voltage across a capacitor is too great, the dielectric (whatever it is) no longer functions properly. This condition is known as *dielectric breakdown.*

In air-dielectric and vacuum-dielectric capacitors, dielectric breakdown is a temporary affair; it does not cause permanent damage. The device operates normally when the voltage is reduced, so the arcing stops. However, in capacitors with solid dielectric such as mica, paper, or tantalum, dielectric breakdown can burn or crack the dielectric, causing the component to conduct current even when the voltage is reduced below the arcing threshold. In such instances, the component is ruined.

CAPACITORS AND AC

Suppose that the power source connected to a capacitor is changed from dc to ac. Imagine that you can adjust the frequency of this ac from a low initial value of a few hertz up to hundreds of hertz, then to many kilohertz, and finally to many megahertz or gigahertz.

At first, the voltage between the plates follows along with the voltage of the power source as the source polarity reverses over and over. However, the set of plates has a certain amount of capacitance. The plates can charge up fast if they are small and if the space between them is large, but they can't charge instantaneously. As you increase the frequency of the ac source, there comes a point at which the plates do not get charged up very much before the source polarity reverses. The set of plates becomes sluggish. The charge does not have time to get established with each ac cycle.

At high ac frequencies, the voltage between the plates has trouble following the current that is charging and discharging them. Just as the plates begin to get a good charge, the ac current passes its peak and starts to discharge them, pulling electrons out of the negative plate and pumping electrons into the positive plate. As the frequency is raised, the set of plates starts to act more and more like a short circuit. Eventually, if you keep on increasing the frequency, the period of the wave is much shorter than the charging-discharging time, and current flows in and out of the plates in the same way as it would flow if the plates were shorted out.

Capacitive reactance is a quantitative measure of the opposition that the set of plates offers to ac. It is measured in ohms, just like inductive reactance and just like resistance. However, by convention, it is assigned negative values rather than positive ones. Capacitive reactance, denoted X_C in mathematical formulas, can vary from near zero (when the plates are huge and close together and/or the frequency is very high) to a few negative ohms to many negative kilohms or megohms.

Capacitive reactance varies with frequency. It gets larger negatively as the frequency goes down and smaller negatively as the frequency goes up. This is the opposite of what happens with inductive reactance, which gets larger (positively) as the frequency goes up. Sometimes capacitive reactance is talked about in terms of its absolute value, with the minus sign removed. Then you might say that X_C increases as the frequency decreases or that X_C diminishes as the frequency is raised. However, it is best if you learn to work with negative X_C values from the start.

REACTANCE AND FREQUENCY

Capacitive reactance behaves in many ways like a mirror image of inductive reactance. In another sense, however, X_C is an extension of X_L into negative values—below zero—with its own peculiar set of characteristics.

If the frequency of an ac source is given in hertz as f and the capacitance of a capacitor in farads is given as C, then the capacitive reactance is

$$X_C = -1/(2\pi fC) = -(2\pi fC)^{-1} \approx -(6.2832fC)^{-1}$$

This same formula applies if the frequency is in megahertz (MHz) and the capacitance is in microfarads (μF). Remember that if the frequency is in millions, the capacitance must be in millionths. This formula also would apply for frequencies in kilohertz (kHz) and millifarads (mF), but for some reason, you'll almost never see millifarads used in practice. Even millifarads are large units for capacitance; components of more than 1,000 μF (which would be 1 mF) are rarely found in real-world electrical systems.

Capacitive reactance varies inversely with the frequency. This means that the function X_C versus f appears as a curve when graphed, and this curve "blows up negatively" as the frequency nears zero. Capacitive reactance also varies inversely with the actual value of capacitance given a fixed frequency. The function of X_C versus C also appears as a curve that "blows up negatively" as the capacitance approaches zero. The negative of X_C is inversely proportional to frequency, as well as to capacitance. Relative graphs of these functions are shown in Fig. 15-9.

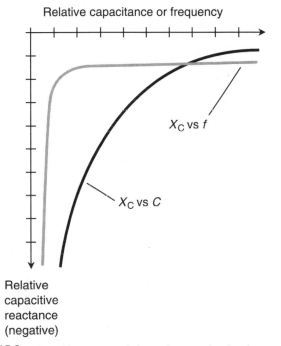

Relative capacitance or frequency

X_C vs f

X_C vs C

Relative
capacitive
reactance
(negative)

Fig. 15-9.   Capacitive reactance is inversely proportional to the negative
of the capacitance, as well as to the negative of the frequency.

PROBLEM 15-7
A capacitor has a value of 0.00100 µF at a frequency of 1.00 MHz. What is the capacitive reactance?

SOLUTION 15-7
Use the formula and plug in the numbers. You can do this directly because the data are specified in microfarads (millionths) and in megahertz (millions):

$$X_C = -1/(6.2832 \times 1.00 \times 0.00100) = -1/(0.0062832) = -159 \text{ ohms}$$

This is rounded to three significant figures because the data are specified only to this many digits.

PROBLEM 15-8

What will be the capacitive reactance of the preceding capacitor if the frequency decreases to zero—that is, if the power source is dc?

SOLUTION 15-8

In this case, if you plug the numbers into the formula, you'll get zero in the denominator. Division by zero is not defined. In reality, however, there is nothing to prevent you from connecting a dc battery to a capacitor! You might say, "The reactance is extremely large negatively and, for all practical purposes, is negative infinity." More appropriately, you should call the capacitor an open circuit for dc.

PROBLEM 15-9

Suppose that a capacitor has a reactance of −100 ohms at a frequency of 10.0 MHz. What is its capacitance?

SOLUTION 15-9

In this problem you need to put the numbers in the formula and solve for the unknown C. Begin with this equation:

$$-100 = -(6.2832 \times 10.0 \times C)^{-1}$$

Dividing through by −100:

$$1 = (628.32 \times 10.0 \times C)^{-1}$$

Multiply each side of this by C:

$$C = (628.32 \times 10.0)^{-1}$$
$$= 6283.2^{-1}$$

This can be solved easily enough. Divide out $C = 1/6283.2$ on your calculator, getting $C = 0.00015915$. Because the frequency is given in megahertz, this capacitance comes out in microfarads, so $C = 0.00015915 \, \mu F$. This must be rounded to $C = 0.000159 \, \mu F$ in this scenario. You also can say $C = 159 \, pF$ (remember that 1 pF = 0.000001 μF).

The arithmetic for dealing with capacitive reactance is a little messier than that for inductive reactance for two reasons. First, you have to work with reciprocals, and therefore, the numbers sometimes get awkward. Second, you have to watch those negative signs. It's easy to leave them out or to put them in when they should not be there. They are important when looking at reactances in the coordinate plane because the minus sign tells you that the reactance is capacitive rather than inductive.

POINTS IN THE *RC* QUARTER-PLANE

Capacitive reactance can be plotted along a half-line or ray just as can inductive reactance. Capacitive and inductive reactance, considered as one, form a real-number line. The point where they join is the zero-reactance point.

In a circuit containing resistance and capacitive reactance, the characteristics are two-dimensional in a way that is analogous to the situation with the *RL* quarter-plane. The resistance ray and the capacitive-reactance ray can be placed end to end at right angles to make the *RC quarter-plane* (Fig. 15-10). Resistance is plotted horizontally, with increasing values toward the right. Capacitive reactance is plotted downward, with increasingly negative values as you go down.

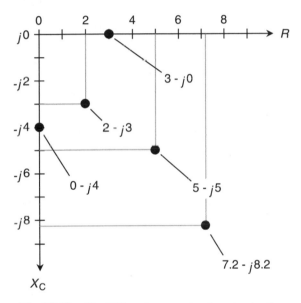

Fig. 15-10.   The *RC* impedance quarter plane showing five points for specific complex-number impedances.

Complex-number impedances that contain resistance and capacitance can be denoted in the form $R + jX_C$; however, X_C is never positive. Because of this, scientists often write $R - jX_C$, dropping the minus sign from X_C and replacing addition with subtraction in the complex-number rendition.

If the resistance is pure, say, $R = 3$ ohms, then the complex-number impedance is $3 - j0$, and this corresponds to the point (3, 0) on the *RC* quarter-plane. You might suspect that $3 - j0$ is the same as $3 + j0$ and that you need not even write the $j0$ part at all. In theory, both these notions are correct. However, writing the $j0$ part indicates that you are open to the possibility that there might be reactance in the circuit and that you're working in two dimensions.

If you have a pure capacitive reactance, say, $X_C = -4$ ohms, then the complex-number impedance is $0 - j4$, and this is at the point $(0, -4)$ on the RC quarter-plane. Again, it's important, for completeness, to write the 0 and not just the $-j4$. The points for $3 - j0$ and $0 - j4$, and three others, are plotted on the RC quarter-plane in Fig. 15-10.

In practical circuits, all capacitors have some *leakage conductance*. If the frequency goes to zero, that is, if the source is dc, a tiny current will flow because no dielectric is a perfect electrical insulator. Some capacitors have almost no leakage conductance, but none are completely free of it. Conversely, all electrical conductors have a little capacitive reactance simply because they occupy physical space. Thus there is no such thing as a mathematically pure conductor of ac either. The points $3 - j0$ and $0 - j4$ are idealized.

Remember that the values for X_C are reactances, not capacitances. Reactance varies with the frequency in an RC circuit. If you raise or lower the frequency, the value of X_C changes. A higher frequency causes X_C to get smaller negatively (closer to zero). A lower frequency causes X_C to get larger negatively (farther from zero or lower down on the RC quarter-plane). If the frequency goes to zero, then the capacitive reactance drops off the bottom of the plane, out of sight. In this case you have two plates or sets of plates having opposite electric charges but no "action."

RLC Impedance

We've seen how inductive and capacitive reactance can be represented along a line perpendicular to resistance. In this section we'll put all three of these quantities—R, X_L, and X_C—together, forming a complete working definition of *impedance*.

THE *RX* HALF-PLANE

Recall the quarter-planes for resistance R and inductive reactance X_L from the preceding sections. This is the same as the upper-right quadrant of the complex-number plane. Similarly, the quarter-plane for resistance R and capacitive reactance X_C is the same as the lower-right quadrant of the complex-number plane. Resistances are represented by nonnegative real numbers. Reactances, whether they are inductive (positive) or capacitive (negative), correspond to imaginary numbers.

There is no such thing, strictly speaking, as negative resistance. That is to say, one cannot have anything better than a perfect conductor. In some cases, a source of dc, such as a battery, can be treated as a negative resistance; in other cases, a device can behave as if its resistance were negative under certain changing conditions. Generally, however, in the RX (resistance-reactance) *half-plane,* the resistance value is nonnegative (Fig. 15-11).

REACTANCE IN GENERAL

Now you should get a better idea of why capacitive reactance X_C is considered negative. In a sense, it is an extension of inductive reactance X_L into the realm of negatives, in a way that generally cannot occur with resistance. Capacitors act like "negative inductors." Interesting things happen when capacitors and inductors are combined.

Reactance can vary from extremely large negative values, through zero, to extremely large positive values. Engineers and physicists always quantify reactances as imaginary numbers. In the mathematical model of impedance, capacitances and inductances manifest themselves perpendicularly to resistance. Thus ac reactance occupies a different and independent dimension from dc resistance. The general symbol for reactance is X; this encompasses both inductive reactance X_L and capacitive reactance X_C.

VECTOR REPRESENTATION OF IMPEDANCE

Any impedance Z can be represented by a complex number $R + jX$, where R can be any nonnegative real number and X can be any real number. Such numbers can be plotted as points in the RX half-plane or as vectors with their end points at the origin $(0 + j0)$. Such vectors are called *impedance vectors.*

Imagine how an impedance vector changes as either R or X or both are varied. If X remains constant, an increase in R causes the vector to get longer. If R remains constant and X_L gets larger, the vector also grows longer. If R stays the same as X_C gets larger (negatively), the vector grows longer again. Think of point representing $R + jX$ moving around in the plane, and imagine where the corresponding points on the resistance and reactance axes lie. These points can be found by drawing straight lines from

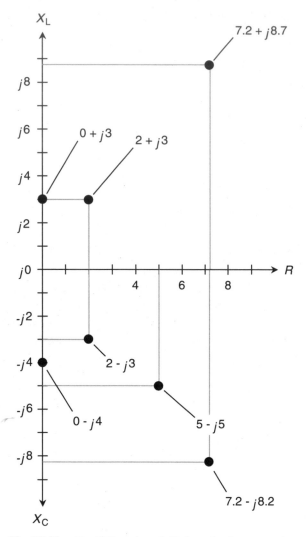

Fig. 15-11.   The *RX* impedance half-plane showing some points
for specific complex-number impedances.

the point $R + jX$ to the *R* and *X* axes so that the lines intersect the axes at
right angles. This is shown in Fig. 15-11 for several different points.

Now think of the points for *R* and *X* moving toward the right and left or up
and down on their axes. Imagine what happens to the point $R + jX$ and the
corresponding vector from $0 + j0$ to $R + jX$ in various scenarios. This is how
impedance changes as the resistance and reactance in a circuit are varied.

ABSOLUTE-VALUE IMPEDANCE

Sometimes you'll read or hear that the "impedance" of some device or component is a certain number of ohms. For example, in audio electronics, there are "8-ohm" speakers and "600-ohm" amplifier inputs. How can manufacturers quote a single number for a quantity that is two-dimensional and needs two numbers to be completely expressed? There are two answers to this.

First, figures like this generally refer to devices that have purely resistive impedances. Thus the "8-ohm" speaker really has a complex-number impedance of $8 + j0$, and the "600-ohm" input circuit is designed to operate with a complex-number impedance at or near $600 + j0$. Second, engineers sometimes talk about the length of the impedance vector, calling this a certain number of "ohms." If you talk about "impedance" in this way, then theoretically you are being ambiguous because you can have an infinite number of different vectors of a given length in the RX half-plane.

The expression "$Z = 8$ ohms," if no specific complex impedance is given, can refer to the complex vectors $8 + j0, 0 + j8, 0 - j8$, or any vector in the RX half-plane whose length is 8 units. This is shown in Fig. 15-12. There can exist an infinite number of different complex impedances with $Z = 8$ ohms in a purely technical sense.

> **PROBLEM 15-10**
> Name seven different complex impedances having an absolute value of $Z = 10$ ohms.
>
> **SOLUTION 15-10**
> It's easy to name three: $0 + j10, 10 + j0$, and $0 - j10$. These are pure inductance, pure resistance, and pure capacitance, respectively.
>   A right triangle can exist having sides in a ratio of 6:8:10 units. This is true because $6^2 + 8^2 = 10^2$. (It's the age-old Pythagorean theorem!) Therefore, you can have $6 + j8, 6 - j8, 8 + j6$, and $8 - j6$. These are all complex-number impedances whose absolute values are 10 ohms.

If you're not specifically told which particular complex-number impedance is meant when a single-number ohmic figure is quoted, it's best to assume that the engineers are talking about *nonreactive impedances*. This means that they are pure resistances and that the imaginary, or reactive, factors are zero. Engineers often will speak of nonreactive impedances as "low-Z" or "high-Z." There is no formal dividing line between the realms of low and high impedance; it depends to some extent on the application. Sometimes a reactance-free impedance is called a *pure resistance* or *purely resistive impedance*.

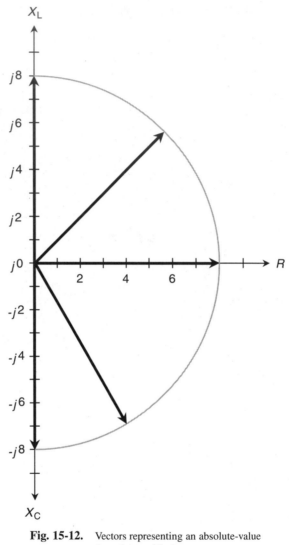

Fig. 15-12.   Vectors representing an absolute-value
impedance of 8 ohms.

Purely resistive impedances are desirable in various electrical and electronic circuits. Entire volumes have been devoted to the subject of impedance in engineering applications. Insofar as basic physics is concerned, we
have gone far enough here. A somewhat more detailed yet still introductory
treatment of this subject can be found in *Teach Yourself Electricity and
Electronics,* published by McGraw-Hill. Beyond that, college textbooks in

electrical, electronics, and telecommunications engineering are recommended.

 **Quiz**

Refer to the text in this chapter if necessary. A good score is eight correct. Answers are in the back of the book.

1. Three 300-pF capacitors are connected in series. What is the reactance of the combination?
   (a) -100 ohms
   (b) -300 ohms
   (c) -900 ohms
   (d) We need more information to calculate it.

2. Three 300-pF capacitors are connected in parallel. What is the capacitance of the combination?
   (a) 100 pF
   (b) 300 pF
   (c) 900 pF
   (d) We need more information to calculate it.

3. The complex-number impedance of a 47-ohm pure resistance is
   (a) $47 + j0$.
   (b) $0 - j47$.
   (c) $0 + j47$.
   (d) $47 + j47$.

4. A component is specified as having a complex-number impedance of $-25 + j30$. From this you can reasonably conclude that
   (a) there is a typographical error in the document.
   (b) the reactance is capacitive.
   (c) the impedance is a pure resistance.
   (d) the device is operating with dc.

5. A solid material with a high dielectric constant between the plates of a capacitor can
   (a) decrease the capacitance compared with an air dielectric.
   (b) increase the capacitance compared with an air dielectric.
   (c) increase the frequency.
   (d) decrease the frequency.

6. If the inductance of a coil is doubled, then X_L at any particular frequency
    (a) becomes twice as large.
    (b) becomes four times as large.
    (c) becomes half as large.
    (d) becomes one-fourth as large.

7. When a dc voltage is applied to an inductor, the reactance, in theory, is
    (a) negative infinity.
    (b) positive infinity.
    (c) zero.
    (d) dependent on the voltage.

8. The complex-number impedance vectors for a pure resistance of 30 ohms and a pure capacitance of 100 μF:
    (a) are the same length.
    (b) are perpendicular to each other.
    (c) point in opposite directions.
    (d) are none of the above.

9. As the frequency of an ac signal through a 33-pF capacitor increases,
    (a) the capacitor offers less and less opposition to the ac.
    (b) the capacitor offers more and more opposition to the ac.
    (c) the opposition to the ac does not change.
    (d) the opposition to the ac might either increase or decrease depending on how fast the frequency changes.

10. A complex-number impedance of $500 + j0$ represents:
    (a) a pure resistance.
    (b) a pure inductive reactance.
    (c) a pure capacitive reactance.
    (d) a short circuit.

CHAPTER 16

Semiconductors

The term *semiconductor* arises from the ability of certain materials to conduct part time. Various mixtures of elements can work as semiconductors. There are two types of semiconductors, called *n type,* in which most of the charge carriers are electrons, and *p type,* in which most of the charge carriers are electron absences called *holes.* In this chapter we will learn a little about semiconductor electronic components.

The Diode

When wafers of *n*-type and *p*-type material are in physical contact, the result is a *p-n junction* with certain properties. Figure 16-1 shows the electronic symbol for a semiconductor *diode.* The *n*-type material is represented by the short, straight line in the symbol and is the *cathode.* The *p*-type material is represented by the arrow and is the *anode.*

In a perfect diode, electrons flow in the direction opposite the arrow but cannot flow in the direction the arrow points. If a battery and resistor are connected in series with a diode, current flows if the cathode is negative relative to the anode (Fig. 16-2*a*) but does not flow if the battery is reversed (see Fig. 16-2*b*). This is an example of something with which you should by now be familiar: the idealized scenario! In the real world, diodes can approach but never attain this state of perfect one-way conduction.

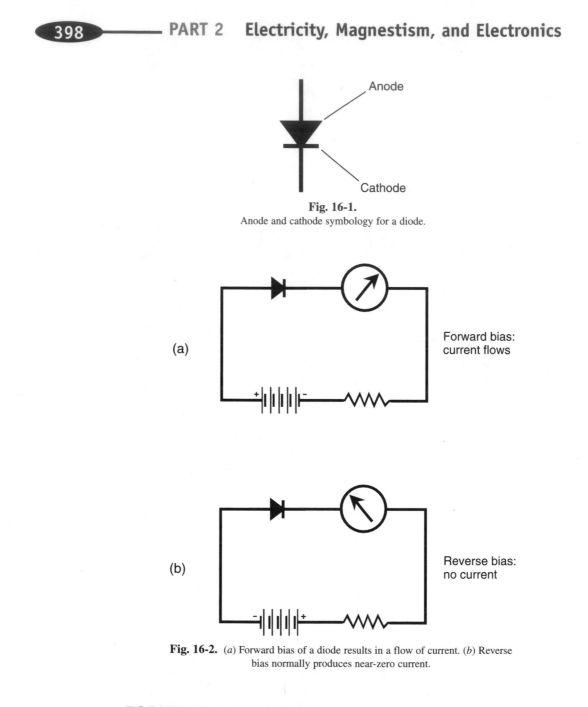

Fig. 16-1.
Anode and cathode symbology for a diode.

(a)

Forward bias:
current flows

(b)

Reverse bias:
no current

Fig. 16-2. (*a*) Forward bias of a diode results in a flow of current. (*b*) Reverse bias normally produces near-zero current.

FORWARD BREAKOVER

It takes a certain minimum voltage for conduction to occur when a *p-n* junction is connected in the manner shown by Fig. 16-2*a*. This minimum voltage is called the *forward breakover voltage* (or simply the *forward breakover*) for the diode. Depending on the type of material from which a diode is manufactured, the forward breakover voltage can range from about 0.3 V to about 1 V. If the voltage across the *p-n* junction is not at least as great as the forward breakover, the diode will not conduct appreciably.

Forward breakover voltages of multiple diodes add together as if the diodes were batteries. When two or more diodes are connected in series with their *p-n* junctions all oriented the same way, the forward breakover voltage of the combination is equal to the sum of the forward breakover voltages of each diode. When two or more diodes are connected in parallel with their *p-n* junctions all oriented the same way, the forward breakover voltage of the combination is the same as that of the diode whose forward breakover voltage is the smallest. The *p-n* junction is unique in this respect. It doesn't conduct perfectly in the forward direction, but it doesn't act quite like a dc resistor when it is conducting.

BIAS

In a *p-n* junction, when the *n*-type material is negative with respect to the *p*-type material, electrons flow easily from *n* to *p*. This is *forward bias*; the diode conducts well. When the polarity is switched so that the *n*-type material is positive with respect to the *p*-type material, it is in a state of *reverse bias,* and the diode conducts poorly.

When a diode is reverse-biased, electrons in the *n*-type material are pulled toward the positive charge, away from the junction. In the *p*-type material, holes are pulled toward the negative charge, also away from the junction. The electrons (in the *n*-type material) and holes (in the *p*-type material) become depleted in the vicinity of the junction. This impedes conduction, and the resulting *depletion region* behaves as a dielectric or electrical insulator.

JUNCTION CAPACITANCE

Under conditions of reverse bias, a *p-n* junction can act as a capacitor. A special type of diode called a *varactor* (a contraction of the words *variable reactor*) is manufactured with this property specifically in mind. The *junction capacitance* of a varactor can be varied by changing the reverse-bias voltage because this voltage affects the width of the depletion region. The greater the reverse voltage, the wider the depletion region gets, and the smaller the capacitance becomes. A good varactor can exhibit a capacitance that fluctuates rapidly, following voltage variations up to high frequencies.

AVALANCHE

If a diode is reverse-biased and the voltage becomes high enough, the *p-n* junction will conduct. This is known as *avalanche effect*. The reverse current, which is near zero at lower voltages, rises dramatically. The *avalanche voltage* varies among different kinds of diodes. Figure 16-3 is a graph of the characteristic current-versus-voltage curve for a typical semiconductor diode showing the *avalanche point*. The avalanche voltage is considerably greater than and is of the opposite polarity from the forward breakover voltage. A *Zener diode* makes use of avalanche effect. Zener diodes are specially manufactured to have precise avalanche voltages. They form the basis for voltage regulation in dc power supplies.

RECTIFICATION

A *rectifier diode* passes current in only one direction under ideal operating conditions. This makes it useful for changing ac to dc.

Generally speaking, when the cathode is negative with respect to the anode, current flows; when the cathode is positive relative to the anode, current does not flow. The constraints on this behavior are the forward breakover and avalanche voltages. During a little less than half the cycle, the diode conducts, and during a little more than half the cycle, it does not conduct. This cuts off slightly more than 50 percent of every ac cycle. Depending on which way the diode is connected in the circuit, either the positive part or the negative part of the ac cycle is blocked.

DETECTION

A diode can recover the audio signal from radiofrequency (rf) ac. This is called *demodulation* or *detection*. If diode detection is to be effective, the diode should have low junction capacitance. Then it works as a rectifier at rf, passing current in one direction but not in the other.

Some rf diodes are microscopic versions of the old *cat whisker,* in which a fine wire was placed in contact with a crystal of a mineral called *galena.* Components of this type are known as *point-contact diodes* and are

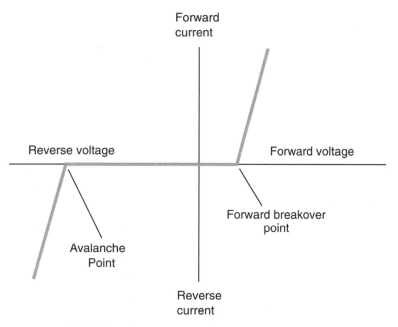

Forward
current

Reverse voltage   Forward voltage

Forward breakover
point

Avalanche
Point

Reverse
current

Fig. 16-3. Characteristic curve for a semiconductor diode.

designed to minimize the junction capacitance. In this way, as the frequency becomes higher and higher (up to a certain maximum), the diodes keep acting like rectifiers rather than starting to behave like capacitors. This makes point-contact diodes good for use at rf.

GUNN DIODES

A *Gunn diode* is made from a compound known as *gallium arsenide* (GaAs). When a voltage is applied to this device, it oscillates because of the *Gunn effect,* named after J. Gunn of International Business Machines (IBM), who first observed the phenomenon in the 1960s. Oscillation occurs as a result of a property called *negative resistance.* This is a misnomer because, as we have learned, there is no such thing as a device that conducts better than perfectly. In this sense, *negative resistance* refers to the fact that over a certain limited portion of the characteristic curve, the current through a Gunn diode decreases as the voltage increases, contrary to what normally takes place in electrical systems.

IMPATT DIODES

The acronym *IMPATT* (pronounced "IM-pat") comes from the words *impact avalanche transit time.* We won't concern ourselves in this book with the exact nature of this effect, except to note that it's similar to negative resistance. An IMPATT diode is a microwave oscillating device like a Gunn diode, but it is manufactured with silicon rather than gallium arsenide.

An IMPATT diode can be used as a low-power *amplifier* for microwave radio signals. As an *oscillator* (a circuit that generates rf ac), an IMPATT diode produces about the same amount of output power, at comparable frequencies, as a Gunn diode.

TUNNEL DIODES

Another type of diode that can oscillate at microwave frequencies is the *tunnel diode,* also known as the *Esaki diode.* It produces a very small amount of rf power.

Tunnel diodes work well as amplifiers in microwave receivers. This is especially true of GaAs devices, which act to increase the amplitudes of weak signals without introducing any unwanted rf *noise,* or signals of their own that cover a large range of frequencies. (An example of noise is the hiss that you hear in a stereo hi-fi amplifier with the gain turned up and no audio input. The less noise, the better.)

LEDs AND IREDs

Depending on the exact mixture of semiconductors used in manufacture, visible light of any color, as well as infrared (IR), can be produced when current is passed through a diode in the forward direction. The most common color for a *light-emitting diode* (LED) is bright red, although LEDs are available in many different colors. An *infrared-emitting diode* (IRED) produces energy at wavelengths slightly longer than those of visible red light. These are called *near-infrared* (NIR) rays.

The intensity of energy emission from an LED or IRED depends to some extent on the forward current. As the current rises, the brightness increases up to a certain point. If the current continues to rise, no further increase in brilliance takes place. The LED or IRED is then said to be in a state of in *saturation.*

INJECTION LASERS

An *injection laser*, also called a *laser diode*, is a special form of LED or IRED with a relatively large and flat *p-n* junction. The injection laser emits coherent electromagnetic (EM) waves, provided the applied current is sufficient. Coherent waves are all lined up and all have the same frequency, compared with the incoherent waves typical of most LEDs and light-producing devices in general.

Figure 16-4 is a simplified diagram of a laser diode. The *substrate* is the material on which the component is built; it is like the foundation of a building. It also serves to carry away excess heat so that the device can carry fairly high current without being destroyed. There are mirrors at opposite ends of the piece of *n*-type material. One of the mirrors (the one labeled in the drawing) is partially reflective. The opposite mirror (not shown) is totally reflective. The coherent rays emerge from the end with the partially reflective mirror.

SILICON PHOTODIODES

A silicon diode housed in a transparent case and constructed in such a way that visible light can strike the barrier between the *p*-type and *n*-type materials

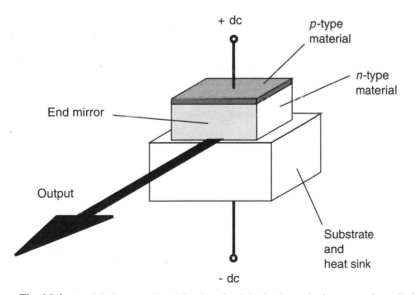

Fig. 16-4. Simplified cross-sectional drawing of an injection laser, also known as a *laser diode*.

forms a *photodiode.* This is essentially the opposite of an LED or IRED. A voltage is applied to the device in the reverse direction, so it ordinarily does not conduct current. When visible light, IR, or ultraviolet (UV) rays strike the *p-n* junction, current flows. The current is proportional to the intensity of the energy, within certain limits. Silicon photodiodes are more sensitive at some wavelengths than at others.

When energy of varying intensity strikes the *p-n* junction of a reverse-biased silicon photodiode, the output current follows the fluctuations. This makes silicon photodiodes useful for receiving modulated-light signals of the kind used in fiberoptic and free-space laser communication systems. This effect diminishes as the frequency increases. At very high frequencies, the diode acts like a capacitor because of its relatively high junction capacitance, and the efficiency of the device as a modulated-light sensor is degraded.

PHOTOVOLTAIC (PV) CELLS

Some types of silicon diodes can generate dc all by themselves if sufficient IR, visible-light, or UV energy strikes their *p-n* junctions. This is known as the *photovoltaic effect,* and it is the principle by which solar cells work.

Photovoltaic (PV) cells have large *p-n* junction surface area (Fig. 16-5). This maximizes the amount of energy that strikes the junction after passing through the thin layer of *p*-type material. A single silicon PV cell produces approximately 0.6 V dc in direct sunlight under no-load conditions (that is, when there is nothing connected to it that will draw current from it). The maximum amount of current that a PV cell can deliver depends on the surface area of the *p-n* junction.

Silicon PV cells are connected in series-parallel combinations to provide solar power for solid-state electronic devices such as portable radios. A large assembly of such cells constitutes a *solar panel.* The dc voltages of the cells add when they are connected in series. A typical *solar battery* supplies 6, 9, or 12 V dc. When two or more identical sets of series-connected PV cells are connected in parallel, the output voltage is not increased, but the solar battery becomes capable of delivering more current. The current-delivering capacity increases in direct proportion to the number of sets of series-connected cells that are connected in parallel.

PROBLEM 16-1
How many silicon PV cells does it take to make a 13.8-V solar battery?

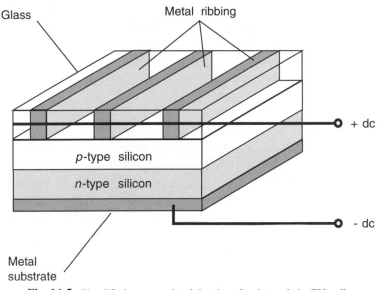

Fig. 16-5. Simplified cross-sectional drawing of a photovoltaic (PV) cell.

SOLUTION 16-1
The PV cells must be connected in series so that the voltages add. Each silicon PV cell produces approximately 0.6 V dc. Therefore, in order to get 13.8 V dc from the solar battery, we must connect 13.8/0.6, or 23, of the PV cells in series.

The Bipolar Transistor

Bipolar transistors have two *p-n* junctions connected together. This can be done in either of two ways: a *p*-type layer between two *n*-type layers or an *n*-type layer between two *p*-type layers.

NPN AND PNP

A simplified drawing of an *npn transistor* and the symbol that is used to represent it in schematic diagrams are shown in Fig. 16-6. The *p*-type, or center, layer is the *base*. The thinner of the *n*-type semiconductors is the *emitter,* and the thicker is the *collector.* Sometimes these are labeled *B, E,* and *C* in schematic diagrams, but the transistor symbol indicates which is which (the arrow is at the emitter). A *pnp transistor* (parts *c* and *d*) has two *p*-type layers, one on either side of a thin *n*-type layer. In the *npn* symbol, the arrow points outward. In the *pnp* symbol, the arrow points inward.

Generally, *pnp* and *npn* transistors can perform identical tasks. The only difference is the polarities of the voltages and the directions of the currents. In most applications, an *npn* device can be replaced with a *pnp* device, or vice versa, and the power-supply polarity reversed, and the circuit will still work if the new device has the appropriate specifications.

There are various kinds of bipolar transistors. Some are used for rf amplifiers and oscillators; others are intended for audiofrequencies (af). Some can handle high power for rf wireless transmission or af hi-fi amplification, and others are made for weak-signal rf reception, microphone preamplifiers, and transducer amplifiers. Some are manufactured for switching, and others are intended for signal processing.

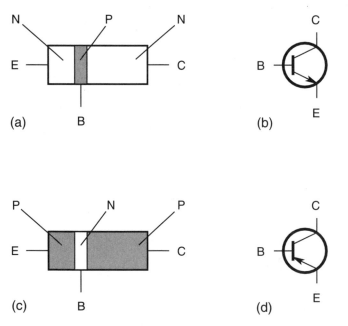

Fig. 16-6. Pictorial diagram of *npn* transistor (*a*), schematic symbol for *npn* transistor (*b*), pictorial diagram of *pnp* transistor (*c*), and schematic symbol for *pnp* transistor (*d*).

NPN BIASING

The normal method of biasing an *npn* transistor is to have the emitter more negative than the collector. In most cases, the emitter is at or near zero potential while the collector is connected to a source of positive dc voltage. This is shown by the connection of the battery in Fig. 16-7. Typical voltages range from 3 V to approximately 50 V.

The base is labeled "control" because the flow of current through the transistor depends on the base bias voltage, denoted E_B or V_B, relative to the emitter-collector bias voltage, denoted E_C or V_C.

ZERO BIAS

When the base is not connected to anything, or when it is at the same potential as the emitter, a bipolar transistor is at *zero bias*. Under this condition, which is called *cutoff*, no appreciable current can flow through a *p-n* junction unless the forward bias is at least equal to the forward breakover voltage. For silicon, the critical voltage is 0.6 V; for germanium, it is 0.3 V.

With zero bias, the emitter-base (*E-B*) current I_B is zero, and the *E-B* junction does not conduct. This prevents current from flowing in the collector unless a signal is injected at the base to change the situation. This signal must have a positive polarity for at least part of its cycle, and its peaks must be sufficient to overcome the forward breakover of the *E-B* junction for at least a portion of the cycle.

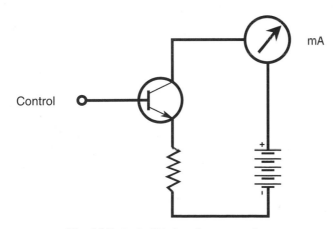

Fig. 16-7. Typical biasing of an *npn* transistor.

REVERSE BIAS

Suppose that another battery is connected to the base of the *npn* transistor at the point marked "control" so that the base is negative with respect to the emitter. The addition of this new battery will cause the *E-B* junction to be in a condition of *reverse bias*. Let's assume that this new battery is not of such a high voltage that avalanche breakdown takes place at the *E-B* junction.

A signal might be injected to overcome the reverse-bias battery and the forward-breakover voltage of the *E-B* junction, but such a signal must have positive voltage peaks high enough to cause conduction of the *E-B* junction for part of the cycle. Otherwise, the transistor will remain cut off for the entire cycle.

FORWARD BIAS

Suppose that the bias at the base of an *npn* transistor is positive relative to the emitter, starting at small levels and gradually increasing. This is *forward bias*. If this bias is less than forward breakover, no current flows. However, when the voltage reaches forward breakover, the *E-B* junction conducts current.

Despite reverse bias at the base-collector (*B-C*) junction, the emitter-collector (*E-C*) current, more often called *collector current* and denoted I_C, flows when the *E-B* junction conducts. A small rise in the positive-polarity signal at the base, attended by a small rise in the base current I_B, causes a large increase in I_C. This is the principle by which a bipolar transistor can amplify signals.

SATURATION

If I_B continues to rise, a point is reached eventually where I_C increases less rapidly. Ultimately, the I_C versus I_B function, or *characteristic curve,* of the transistor levels off. The graph in Fig. 16-8 shows a *family of characteristic curves* for a hypothetical bipolar transistor. The actual current values depend on the particular type of device; values are larger for power transistors and smaller for weak-signal transistors. Where the curves level off, the transistor is in a state of *saturation*. Under these conditions, the tran-

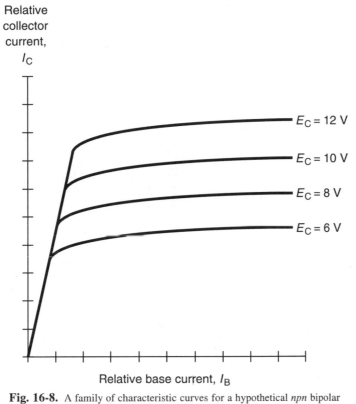

Fig. 16-8. A family of characteristic curves for a hypothetical *npn* bipolar transistor.

sistor loses its ability to efficiently amplify signals. However, the transistor can still work for switching purposes.

PNP BIASING

For a *pnp* transistor, the bias situation is a mirror image of the case for an *npn* device, as shown in Fig. 16-9. The power-supply polarity is reversed. To overcome forward breakover at the *E-B* junction, an applied signal must have sufficient negative polarity.

Either the *pnp* or the *npn* device can serve as a "current valve." Small changes in the base current I_B induce large fluctuations in the collector current I_C when the device is operated in that region of the characteristic

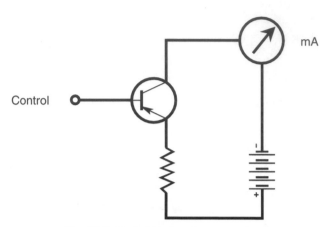

Fig. 16-9. Typical biasing of a *pnp* transistor.

curve where the slope is steep. While the internal atomic activity is different in the *pnp* device as compared with the *npn* device, the performance of the external circuitry is, in most situations, identical for practical purposes.

Current Amplification

Because a small change in I_B results in a large I_C variation when the bias is right, a transistor can operate as a *current amplifier*. The extent of such amplification can be expressed in terms of what happens with either static (steady) or dynamic (varying) input signal current.

STATIC CURRENT AMPLIFICATION

The maximum obtainable current amplification factor of a bipolar transistor is known as the *beta* of the transistor. Depending on the way in which a transistor is manufactured, the beta can range from a factor of a few times up to hundreds of times. One method of expressing the beta of a transistor is as the *static forward current transfer ratio,* symbolized H_{FE}. This is the ratio of the collector current to the base current:

$$H_{FE} = I_C/I_B$$

For example, if a base current I_B of 1 mA produces a collector current I_C of 35 mA, then $H_{FE} = 35/1 = 35$. If $I_B = 0.5$ mA and $I_C = 35$ mA, then $H_{FE} = 35/0.5 = 70$.

DYNAMIC CURRENT AMPLIFICATION

Another way of specifying current amplification is as the ratio of a difference in I_C to a small incremental difference in I_B that produces it. This is the *dynamic current amplification,* also known as *current gain*. It is customary to abbreviate the words *the difference in* by the uppercase Greek letter delta (Δ) in mathematical expressions. Then, according to this definition,

$$\text{Current gain} = \Delta I_C / \Delta I_B$$

The ratio $\Delta I_C / \Delta I_B$ is greatest where the slope of the characteristic curve is steepest. Geometrically, $\Delta I_C / \Delta I_B$ at any given point on the curve is the slope of a line tangent to the curve at that point.

When the *operating point* of a transistor is on the steep part of the characteristic curve, the device has the largest possible current gain, as long as the input signal is small. This value is close to H_{FE}. Because the characteristic curve is a straight line in this region, the transistor can serve as a *linear amplifier* if the input signal is not too strong. This means that the output signal waveform is a faithful reproduction of the input signal waveform, except that the output amplitude is greater than the input amplitude.

As the operating point is shifted into the part of the characteristic curve where the graph is not straight, the current gain decreases, and the amplifier becomes nonlinear. The same thing can happen if the input signal is strong enough to drive the transistor into the nonlinear part of the curve during any portion of the signal cycle.

GAIN VERSUS FREQUENCY

In any particular bipolar transistor, the gain decreases as the signal frequency increases. There are two expressions for gain-versus-frequency behavior.

The *gain bandwidth product,* abbreviated f_T, is the frequency at which the current gain becomes equal to unity (1) with the emitter connected to ground. This means, in effect, that the transistor has no current gain; the output current amplitude is the same as the input current amplitude, even under ideal operating conditions. The *alpha cutoff* is the frequency at which the current gain becomes 0.707 times (that is, 70.7 percent of) its value at exactly 1 kHz (1,000 Hz). Most transistors can work as current amplifiers at frequencies above the alpha cutoff, but no transistor can work as a current amplifier at frequencies higher than its gain bandwidth product.

PROBLEM 16-2
A bipolar transistor has a current gain, under ideal conditions, of 23.5 at an operating frequency of 1,000 Hz. The alpha cutoff is specified as 900 kHz. What is the maximum possible current gain that the device can have at 900 kHz?

SOLUTION 16-2
Multiply 23.5 by 0.707 to obtain 16.6. This is the maximum possible current gain that the transistor can produce at 900 kHz.

PROBLEM 16-3
Suppose that the peak-to-peak (pk-pk) signal input current in the aforementioned transistor is 2.00 mA at a frequency of 1,000 Hz. Further suppose that the operating conditions are ideal and that the transistor is not driven into the nonlinear part of the characteristic curve during any part of the input signal cycle. If the frequency is changed to 900 kHz, what will be the pk-pk signal output current?

SOLUTION 16-3
First, note that the current gain of the transistor is 23.5 at a frequency of 1,000 Hz. This means that the pk-pk output signal current at 1,000 Hz is 2.00 μA $\times$ 23.5 = 47.0 μA. At 900 kHz, the pk-pk output signal current is thus 0.707 $\times$ 47.0 μA = 33.2 μA.

The Field-Effect Transistor

The other major category of transistor, besides the bipolar device, is the *field-effect transistor* (FET). There are two main types of FETs: the *junction FET* (JFET) and the *metal-oxide FET* (MOSFET).

PRINCIPLE OF THE JFET

In a JFET, the current varies because of the effects of an electrical field within the device. Electrons or holes move along a current path called a *channel* from the *source* (S) electrode to the *drain* (D) electrode. This results in a drain current I_D that is normally the same as the source current

I_S. The drain current depends on the voltage at the *gate* (*G*) electrode. As the gate voltage E_G changes, the effective width of the channel varies. Thus fluctuations in E_G cause changes in the current through the channel. Small fluctuations in E_G can cause large variations in the flow of charge carriers through the JFET. This effect makes it possible for the device to act as a *voltage amplifier.*

N-CHANNEL AND P-CHANNEL

A simplified drawing of an *n-channel JFET* and its schematic symbol are shown in Fig. 16-10 *a* and *b*. The *n*-type material forms the path for the current. The majority carriers are electrons. The drain is positive with respect to the source. The gate consists of *p*-type material. Another, larger section of *p*-type material, the substrate, forms a boundary on the side of the channel opposite the gate. The voltage on the gate produces an electrical field that interferes with the flow of charge carriers through the channel. The more negative E_G becomes, the more the electrical field chokes off the current though the channel, and the smaller I_D gets.

A *p-channel JFET* (see Fig. 16-10*c* and *d*) has a channel of *p*-type semiconductor. The majority of charge carriers are holes. The drain is negative with respect to the source. The gate and substrate are *n*-type material. The more positive E_G becomes, the more the electrical field chokes off the current through the channel, and the smaller I_D gets.

In engineering circuit diagrams, the *n*-channel JFET can be recognized by an arrow pointing inward at the gate and the *p*-channel JFET by an arrow pointing outward. The power-supply polarity also shows which type of device is used. A positive drain indicates an *n*-channel JFET, and a negative drain indicates a *p*-channel type.

An *n*-channel JFET almost always can be replaced with a *p*-channel JFET and the power-supply polarity reversed, and the circuit will still work if the new device has the correct specifications.

DEPLETION AND PINCHOFF

A JFET works because the voltage at the gate causes an electrical field that interferes, more or less, with the flow of charge carriers along the channel. As the drain voltage E_D increases, so does the drain current I_D, up to a certain level-off value. This is true as long as the gate voltage E_G is constant

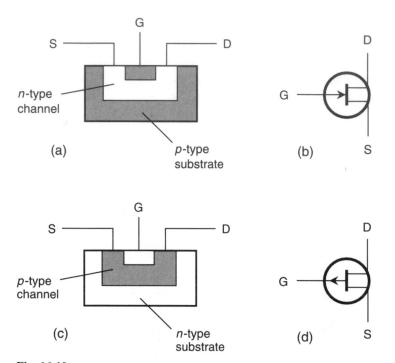

Fig. 16-10. Pictorial diagram of an *n*-channel JFET (*a*), schematic symbol for an *n*-channel JFET (*b*), pictorial diagram of a *p*-channel JFET (*c*), and schematic symbol for a *p*-channel JFET (*d*).

and is not too large. As E_G increases (negatively in an *n* channel or positively in a *p* channel), a *depletion region* develops within the channel. Charge carriers cannot flow in the depletion region, so when there is such a region, they must pass through a narrowed channel. The larger E_G becomes, the wider the depletion region gets, and the more constricted the channel becomes. If E_G is high enough, the depletion region completely obstructs the flow of charge carriers, and the channel cannot conduct current at all. This condition is known as *pinchoff*. It is like pressing down on a garden hose until the water can't flow.

Voltage Amplification

The graph in Fig. 16-11 shows the drain (channel) current, I_D as a function of the gate bias voltage E_G for a hypothetical *n*-channel JFET when no signal is applied to the gate electrode. The drain voltage E_D is assumed to be constant.

When E_G is fairly large and negative, the JFET is pinched off, and no current flows through the channel. As E_G gets less negative, the channel opens up, and current begins flowing. As E_G gets still less negative, the channel gets wider, and the current I_D increases. As E_G approaches the point where the source-gate (*S-G*) junction is at forward breakover, the channel conducts as well as it possibly can. If E_G becomes positive enough so that the *S-G* junction conducts, the JFET no longer works properly. Some of the current in the channel is shunted through the gate. This is like a garden hose springing a leak.

The best amplification for weak signals is obtained when E_G is such that the slope of the curve in Fig. 16-11 is steepest. This is shown roughly by the range marked *X* in the graph. For power amplification, results are often best when the JFET is biased at or beyond pinchoff, in the range marked *Y*.

DRAIN CURRENT VERSUS DRAIN VOLTAGE

Drain current I_D can be plotted as a function of drain voltage E_D for various values of gate bias voltage E_G. The resulting set of curves is called a *family of characteristic curves* for the device. Figure 16-12 shows a family of

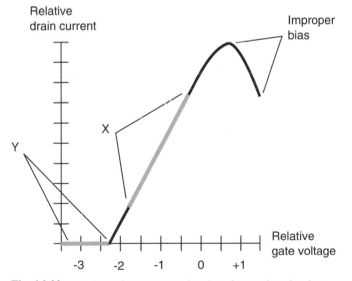

Fig. 16-11. Relative drain current as a function of gate voltage in a hypothetical *n*-channel JFET.

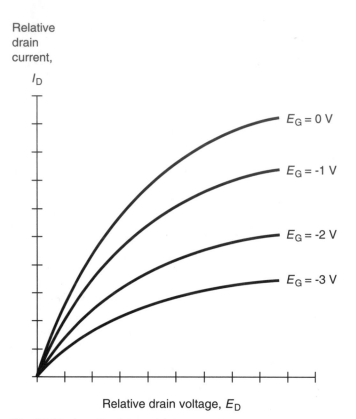

Fig. 16-12. A family of characteristic curves for a hypothetical *n*-channel JFET.

characteristic curves for a hypothetical *n*-channel JFET. Also of importance is the curve of I_D versus E_G, one example of which is shown in Fig. 16-11.

TRANSCONDUCTANCE

Look back for a moment at the discussion of dynamic current amplification for bipolar transistors earlier in this chapter. The JFET analog of this is called *dynamic mutual conductance* or *transconductance*.

Refer to Fig. 16-11. Suppose that E_G is a certain value, with a corresponding I_D that flows as a result. If the gate voltage changes by a small amount ΔE_G, then the drain current also will change by a certain increment ΔI_D. The transconductance is the ratio $\Delta I_D / \Delta E_G$. Geometrically, this translates to the slope of a line tangent to the curve of Fig. 16-11 at some point.

The value of $\Delta I_D/\Delta E_G$ is not the same everywhere along the curve. When the JFET is biased beyond pinchoff, as in the region marked Y in Fig. 16-11, the slope of the curve is zero. There is no drain current, even if the gate voltage changes. Only when the channel conducts some current will there be a change in I_D when there is a change in E_G. The region where the transconductance is the greatest is the region marked X, where the slope of the curve is steepest. This is where the most amplification can be obtained. A small change in E_G produces a large change in I_D, which in turn causes a large variation in a resistive load placed in series with the line connecting the drain to the power supply.

PROBLEM 16-4
Examine Fig. 16-12. Note that the curves in the graph become farther apart as the drain voltage E_D increases (that is, as we move toward the right). Extrapolating on this graph, it is apparent that if E_D exceeds a certain level, the curves become horizontal lines, and they no longer spread out any farther. What can we infer about the ability of this JFET to amplify signals as its E_D increases indefinitely?

SOLUTION 16-4
When a JFET is operated at relatively low drain voltages, a certain pk-pk gate signal voltage (say, from -2 to -1 V) produces a small change in drain current I_D. As E_D increases, the curves represented by gate voltages $E_G = -2$ V and $E_G = -1$ V grow farther apart; this means that the same input signal will result in larger changes in I_D. This translates into more amplification. As E_D continues to increase, the curves represented by $E_G = -2$ V and $E_G = -1$ V level off, and their separation becomes constant. The amplification factor does not increase significantly once E_D exceeds this limiting value. This is illustrated in Fig. 16-13. This same thing will happen for all ac signals with relatively small pk-pk voltages that fall within the ranges indicated by the curves. Of course, there is a limit to all this. If E_D becomes too large, the device will be physically damaged. Most JFETs are designed for operation with E_D values of no more than a few tens of volts.

The MOSFET

The acronym *MOSFET* (pronounced "MOSS-fet") stands for *metal-oxide-semiconductor field-effect transistor*. A simplified cross-sectional drawing of an *n*-channel MOSFET and its schematic symbol are shown in Fig. 16-14*a* and *b*. The *p*-channel device is shown at Fig. 16-14 *c* and *d*.

When the MOSFET was first developed, it was called an *insulated-gate FET* or *IGFET*. This is perhaps more descriptive of the device than the currently accepted name. The gate electrode is actually insulated, by a thin

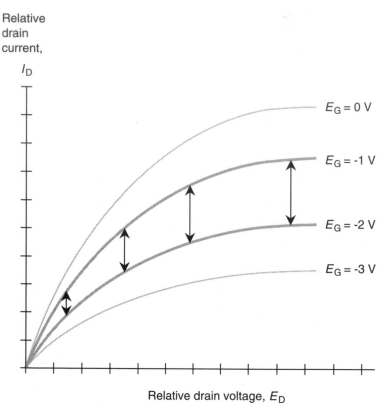

Fig. 16-13. Illustration for Problem 16-4.

layer of dielectric, from the channel. As a result of this, the input resistance (and hence the impedance) is extremely high. The MOSFET draws essentially no current from the input signal source. This is an asset in weak-signal amplifiers.

THE MAIN PROBLEM

The trouble with MOSFETs is that they can be damaged easily by electrostatic discharge. When building or servicing circuits containing MOS devices, technicians must use special equipment to ensure that their hands do not carry electrostatic charges that might ruin the components. If a static

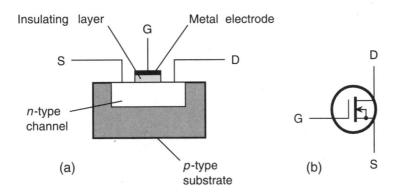

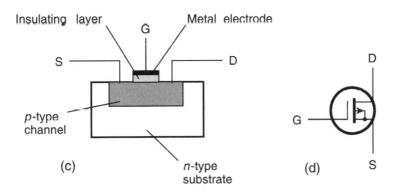

Fig. 16-14. Pictorial diagram of an *n*-channel MOSFET (*a*), schematic symbol for an *n*-channel MOSFET (*b*), pictorial diagram of a *p*-channel MOSFET (*c*), and schematic symbol for a *p*-channel MOSFET (*d*).

discharge occurs through the dielectric of a MOS device, the component will be destroyed permanently. A humid environment does not offer significant protection against this hazard.

FLEXIBILITY

In practical circuits, an *n*-channel JFET sometimes can be replaced with an *n*-channel MOSFET; *p*-channel devices similarly can be interchanged. However, the characteristic curves for MOSFETs are not the same as those for JFETs. The source-gate (*S-G*) junction in a MOSFET is not a *p-n* junction. Forward breakover does not occur under any circumstances. If this junction

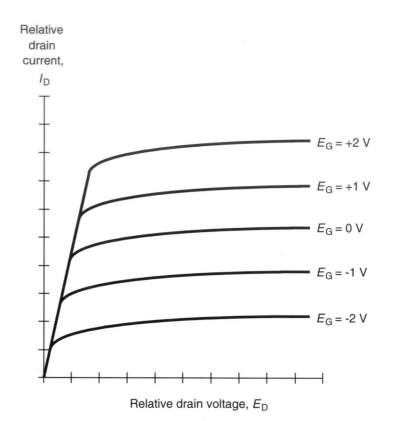

Fig. 16-15. A family of characteristic curves for a hypothetical *n*-channel MOSFET.

conducts, it is because the *S-G* voltage is so great that arcing takes place, damaging the MOSFET permanently. A family of characteristic curves for a hypothetical *n*-channel MOSFET is shown in Fig. 16-15.

DEPLETION VERSUS ENHANCEMENT

In a JFET, the channel conducts with zero bias, that is, when the potential difference between the gate and the source is zero. As the depletion region grows, charge carriers pass through a narrowed channel. This is known as *depletion mode.* A MOSFET can work in the depletion mode, too. The drawings and schematic symbols of Fig. 16-14 show depletion-mode MOSFETs.

Metal-oxide-semiconductor technology allows a second mode of operation. An *enhancement-mode MOSFET* has a pinched-off channel at zero bias. It is necessary to apply a gate bias voltage E_G to create a channel. If $E_G = 0$, the drain current I_D is zero when there is no signal input. The schematic symbols for *n*-channel and *p*-channel enhancement-mode devices are shown in Fig. 16-16. In schematic diagrams, they can be differentiated from depletion-mode devices by looking at the vertical lines inside the circles. Depletion-mode MOSFETs have solid vertical lines; enhancement-mode devices have broken vertical lines.

Integrated Circuits

Most *integrated circuits* (ICs) look like plastic boxes with protruding metal pins. Common configurations are the *single-inline package* (SIP), the *dual-inline package* (DIP), and the *flatpack*. Another package looks like a transistor with too many leads. This is a *metal-can package,* sometimes also called a *T.O. package.* The schematic symbol for an IC is a triangle or rectangle with the component designator written inside.

COMPACTNESS

IC devices and systems are tiny compared with equivalent circuits made from discrete components. More complex circuits can be built and kept down to a reasonable size using ICs as compared with discrete components. Thus, for example, there are notebook computers with capabilities more advanced than the first computers that were built in the middle 1900s and that took up entire rooms.

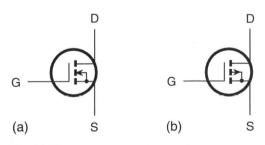

Fig. 16-16. (*a*) The symbol for an *n*-channel enhancement-mode MOSFET. (*b*) The symbol for a *p*-channel enhancement-mode MOSFET.

HIGH SPEED

In an IC, the interconnections among components are physically tiny, making high switching speeds possible. Electric currents travel fast, but not instantaneously. The faster the charge carriers move from one component to another, the more operations can be performed per unit time, and the less time is required for complex tasks.

LOW POWER REQUIREMENT

ICs generally consume less power than equivalent discrete-component circuits. This is important if batteries are used. Because ICs draw so little current, they produce less heat than their discrete-component equivalents. This results in better energy efficiency and minimizes problems that plague equipment that gets hot with use, such as frequency drift and generation of internal noise.

RELIABILITY

Systems using ICs fail less often per component-hour of use than systems that make use of discrete components. This is so mainly because all interconnections are sealed within an IC case, preventing corrosion or the intrusion of dust. The reduced failure rate translates into less downtime.

IC technology lowers service costs because repair procedures are simple when failures occur. Many systems use sockets for ICs, and replacement is simply a matter of finding the faulty IC, unplugging it, and plugging in a new one. Special desoldering equipment is used for servicing circuit boards that have ICs soldered directly to the foil.

MODULAR CONSTRUCTION

Modern IC appliances employ *modular construction.* Individual ICs perform defined functions within a circuit board; the circuit board or card, in turn, fits into a socket and has a specific purpose. Computers programmed with customized software are used by technicians to locate the faulty card in a system. The card can be pulled and replaced, getting the system back to the user in the shortest possible time.

Quiz

Refer to the text in this chapter if necessary. A good score is eight correct. Answers are in the back of the book.

1. A JFET biased beyond pinchoff can work as
   (a) a power amplifier.
   (b) a solar panel.
   (c) an injection laser.
   (d) none of the above.

2. As the depletion region in a JFET becomes wider,
   (a) the channel capacitance increases.
   (b) the channel capacitance decreases.
   (c) the channel conducts less current.
   (d) the channel conducts more current.

3. Fill in the blank in the following sentence to make it true. "The _____ in a bipolar transistor is the frequency at which the current gain becomes equal to unity in a grounded-emitter circuit."
   (a) forward breakover
   (b) junction capacitance
   (c) alpha cutoff
   (d) gain bandwidth product

4. A bipolar transistor has a current gain, under ideal conditions, of 16.0 at an operating frequency of 1,000 Hz. The alpha cutoff is specified as 600 kHz. What is the maximum possible current gain that the device can have at 75 MHz?
   (a) 11.3
   (b) 16.0
   (c) 22.6
   (d) It cannot be determined from this information.

5. At the *p-n* junction in a diode that is conducting in the forward direction,
   (a) the *n*-type semiconductor is always saturated.
   (b) the majority carriers are electrons.
   (c) the *p*-type material is positively charged relative to the *n*-type material.
   (d) the collector is positively charged relative to the base.

6. An enhancement-mode device
   (a) conducts well at zero bias.
   (b) amplifies best when the *p-n* junction is in saturation.
   (c) can be replaced with a depletion-mode device if the polarity is reversed.
   (d) does none of the above.

7. Six silicon diodes, each with a forward breakover voltage of 0.60 V, are connected in parallel, with their *p-n* junctions all oriented the same way. What is the forward breakover voltage of the combination?
   (a) 0.60 V
   (b) 0.10 V
   (c) 3.60 V
   (d) It cannot be determined from this information.

8. The path between the source and drain of a JFET
   (a) always conducts.
   (b) is called the channel.
   (c) is called the depletion region.
   (d) never conducts.

9. A *pnp* transistor can act as a current amplifier when
   (a) small changes in the base current produce large changes in the collector current.
   (b) small changes in the collector voltage produce large changes in the base voltage.
   (c) small changes in the source current produce large changes in the gate current.
   (d) small changes in the gate voltage produce large changes in the drain current.

10. When a *p-n* junction in a bipolar transistor is reverse-biased and the voltage is increased indefinitely, the junction eventually will
   (a) act like an inductor.
   (b) amplify.
   (c) act like a resistor.
   (d) conduct.

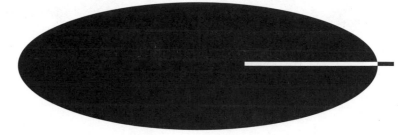

Test: Part Two

Do not refer to the text when taking this test. A good score is at least 37 correct. Answers are in the back of the book. It is best to have a friend check your score the first time so that you won't memorize the answers if you want to take the test again.

1. According to Kirchhoff's voltage law,
   (a) the sum of the voltages going around any branch in a dc circuit is equal to zero.
   (b) the current in a series dc circuit is equal to the voltage divided by the power.
   (c) the power in a parallel dc circuit is equal to the voltage divided by the current.
   (d) the voltage in any dc circuit is equal to the power divided by the resistance.
   (e) the voltage in any dc circuit is equal to the resistance squared divided by the power.

2. Suppose that an air-core loop of wire consisting of one turn carries a certain amount of dc. If the number of turns in the loop is doubled while the current remains the same, the magnetomotive force is
   (a) cut to one quarter its previous value.
   (b) cut in half.
   (c) unchanged.
   (d) doubled.
   (e) quadrupled.

3. A diamagnetic material
   (a) has permeability of zero.
   (b) has permeability of less than 1.
   (c) has permeability equal to 1.
   (d) has permeability of greater than 1.
   (e) can have any permeability.

4. Conductance is specified in units called
   - (a) ohms.
   - (b) farads.
   - (c) henrys.
   - (d) siemens.
   - (e) coulombs.

5. Suppose that a coil carries 100 mA of dc, and then this current is reduced to 10 mA. Assume that all other factors remain the same. The intensity of the magnetic field inside and around the coil
   - (a) does not change.
   - (b) becomes 1 percent as great.
   - (c) becomes 10 percent as great.
   - (d) becomes 10 times as great.
   - (e) changes, but to an extent that cannot be determined unless we have more information.

6. If all other factors are held constant, the capacitance between a pair of parallel, flat, identical metal plates
   - (a) increases as the surface area of the plates increases.
   - (b) does not change as the surface area of the plates increases.
   - (c) decreases as the surface area of the plates increases.
   - (d) depends on the voltage applied to the plates.
   - (e) depends on the current flowing between the plates.

7. A certain ac wave has a positive peak voltage of $+10.0$ V and a negative peak voltage of -5.00 V. What is the peak-to-peak voltage?
   - (a) $+5.00$ V
   - (b) 5.00 V
   - (c) -5.00 V
   - (d) 15.0 V
   - (e) It cannot be calculated from this information.

8. What happens to the inductance of a coil of wire if a ferromagnetic core is placed inside it?
   - (a) The inductance decreases.
   - (b) The inductance stays the same.
   - (c) The inductance increases.
   - (d) The inductance might increase or decrease depending on the frequency.
   - (e) There is no way to say.

9. Which of the following certainly would *not* contain a solenoid?
   - (a) A bell ringer
   - (b) An electromagnet
   - (c) A relay
   - (d) An inductor
   - (e) A capacitor

10. An ionized gas
    (a) is poisonous and radioactive.
    (b) can be a fair conductor of electricity.
    (c) has too few protons in the nuclei of its atoms.
    (d) has too few neutrons in the nuclei of its atoms.
    (e) has too many electrons in the nuclei of its atoms.

11. The aurorae ("northern lights" or "southern lights") are caused indirectly by
    (a) the geomagnetic wind.
    (b) human-made electromagnetic fields.
    (c) the motions of the planets around the Sun.
    (d) bombardment of the Earth's upper atmosphere by meteors.
    (e) solar flares.

12. Suppose that an air-core wire coil has an inductance of 10.0 μH at a frequency
    of 2.00 MHz. What is the inductance of this same coil if the frequency is dou-
    bled to 4.00 MHz?
    (a) 10.0 μH
    (b) 20.0 μH
    (c) 5.00 μH
    (d) 40.0 μH
    (e) 2.50 μH

13. A certain radio wave has a frequency of 0.045 GHz. This would more likely be
    expressed as
    (a) 45 Hz.
    (b) 450 Hz.
    (c) 45 kHz.
    (d) 450 kHz.
    (e) none of the above.

14. As the frequency of an ac voltage across a capacitor increases, a point will be
    reached eventually where the capacitor behaves, with respect to the ac, like
    (a) a resistor.
    (b) an inductor.
    (c) an open circuit.
    (d) a short circuit.
    (e) a diode.

15. Complex impedance consists of
    (a) resistance and reactance.
    (b) conductance and resistance.
    (c) inductance and capacitance.
    (d) conductance and inductance.
    (e) conductance and capacitance.

16. A resistor has a value of 220 ohms and carries 100 mA of direct current. The voltage across this resistor is
    (a) 22 kV.
    (b) 2.2 V.
    (c) 22.0 V.
    (d) 0.22 V.
    (e) impossible to calculate from this information.

17. A photovoltaic cell
    (a) is a rechargeable electrochemical cell.
    (b) is used as a variable capacitor.
    (c) generates dc when visible light strikes its *p-n* junction.
    (d) lights up when reverse-biased.
    (e) is useful as a voltage regulator.

18. Magnetic flux density can be expressed in terms of
    (a) lines per pole.
    (b) lines per centimeter.
    (c) lines per meter.
    (d) lines per meter squared.
    (e) lines per meter cubed.

19. In a bipolar transistor, what happens to the maximum obtainable amplification as the operating frequency becomes higher and higher?
    (a) It increases.
    (b) It increases up to a certain value and then levels off.
    (c) It does not change.
    (d) It decreases.
    (e) It decreases to zero and then becomes negative.

20. Resistances in parallel
    (a) add together directly.
    (b) add together like capacitances in series.
    (c) add together like inductances in series.
    (d) all draw the same amount of current no matter what their individual ohmic values.
    (e) all dissipate the same amount of power no matter what their individual ohmic values.

21. Which of the following is a general difference between a *pnp* bipolar transistor circuit and an *npn* bipolar transistor circuit?
    (a) The operating frequencies are different.
    (b) The current-handling capabilities are different.
    (c) The devices exhibit opposite types of reactance.
    (d) The devices exhibit different impedances.
    (e) The power-supply polarities are opposite.

22. Suppose that you read in a technical paper that two ac signals have identical frequencies and identical waveforms and that their negative peak voltages are the same as their positive peak voltages. You are also told that the signals are in phase coincidence. From this you can conclude that
    (a) the peak-to-peak voltage of the composite signal is twice the peak-to-peak voltage of either signal taken alone.
    (b) the peak-to-peak voltage of the composite signal is half the peak-to-peak voltage of either signal taken alone.
    (c) the peak-to-peak voltage of the composite signal is 1.414 times the peak-to-peak voltage of either signal taken alone.
    (d) the peak-to-peak voltage of the composite signal is 2.828 times the peak-to-peak voltage of either signal taken alone.
    (e) the paper contains a mistake.

23. The width of the channel in a JFET depends on
    (a) the gate voltage.
    (b) the base current.
    (c) the alpha cutoff.
    (d) the beta.
    (e) the collector current.

24. The instantaneous amplitude of an ac wave is
    (a) the amplitude as measured or defined at some specific moment in time.
    (b) the amplitude averaged over any exact whole number of wave cycles.
    (c) equal to approximately 0.707 times the peak amplitude.
    (d) equal to approximately 1.414 times the peak amplitude.
    (e) constant with the passage of time.

25. In an electromagnet,
    (a) the current in the coil temporarily magnetizes the core.
    (b) the electrical field alternates constantly.
    (c) the field strength is inversely proportional to the current.
    (d) the magnetic lines of flux are all straight.
    (e) there is only one pole.

26. What happens when the emitter-base (*E-B*) junction of a field-effect transistor (FET) is reverse-biased?
    (a) A large current flows.
    (b) The FET is optimally efficient.
    (c) The reactance is zero.
    (d) The device can be used as an amplifier but not as a switch.
    (e) This is a meaningless question! No FET has an *E-B* junction.

27. Suppose that you read in a technical paper that two particular sine waves have different frequencies but are in phase coincidence. You conclude that
    (a) the phase angle is 0°.
    (b) the phase angle is 180°.
    (c) the phase angle is +90°.
    (d) the phase angle is −90°.
    (e) the paper contains a mistake.

28. Five capacitors are connected in parallel. They all have capacitances of 100 pF. The total capacitance is
    (a) 20 pF.
    (b) 100 pF.
    (c) 500 pF.
    (d) dependent on the frequency.
    (e) dependent on the voltage.

29. A microphone is specified as having an "impedance of 500 ohms." The engineers mean that the microphone is designed to operate best with a circuit whose complex-number impedance is
    (a) $0 + j500$.
    (b) $0 - j500$.
    (c) $300 + j400$.
    (d) $300 - j400$.
    (e) $500 + j0$.

30. Four resistances are connected in series. Three of them have values of 100 ohms; the fourth resistance is unknown. A battery of 6.00 V is connected across the series combination, producing a current of 10.0 mA through the circuit. What is the value of the unknown resistor?
    (a) 300 ohms
    (b) 600 ohms
    (c) 900 ohms
    (d) 1,200 ohms
    (e) It cannot be calculated from this information.

31. Magnetic tape most likely would be found in
    (a) a transistorized amplifier.
    (b) a high-capacity computer data storage system.
    (c) a high-fidelity compact-disc player.
    (d) a computer hard drive.
    (e) none of the above; magnetic tape is never used anymore.

32. Which of the following is *not* an asset of integrated circuits?
    (a) Low power requirements
    (b) Excellent reliability
    (c) Substantial physical bulk
    (d) Ease of maintenance
    (e) Modular construction

33. The angular frequency of a certain ac wave is 450 rad/s. What is the frequency in kilohertz?
    (a) 0.716 kHz
    (b) 0.450 kHz
    (c) 0.0716 kHz
    (d) 71.6 kHz
    (e) 2,830 kHz

34. In a varactor,
    (a) the resistance varies with the current.
    (b) the *p-n* junction does not conduct when forward-biased.
    (c) the capacitance varies with the applied reverse voltage.
    (d) the inductance depends on the reverse current.
    (e) the avalanche voltage is about 0.3 V.

35. When two waves having identical frequencies are in phase opposition, they are displaced by approximately
    (a) 0.785 radians.
    (b) 1.57 radians.
    (c) 3.14 radians.
    (d) 6.28 radians.
    (e) an amount that cannot be determined unless we are given more information.

36. A certain component has 50 ohms of resistance and −70 ohms of reactance. The complex-number impedance is
    (a) $50 + j70$.
    (b) $50 - j70$.
    (c) $-50 + j70$.
    (d) $-50 - j70$.
    (e) impossible to determine without more information.

37. Three light bulbs are connected in parallel across a 12.0-V battery. The first bulb consumes 5.00 W of power, the second bulb consumes 15.0 W, and the third bulb consumes 20.0 W. What is the current through the third bulb?
    (a) 139 mA
    (b) 600 mA
    (c) 1.67 A
    (d) 7.20 A
    (e) It cannot be calculated from this information.

38. In an *n*-channel JFET, as the gate voltage becomes more and more negative, a point is reached eventually at which
    (a) the channel conducts current as well as it possibly can.
    (b) the channel does not conduct current.
    (c) the amplification factor levels off.
    (d) the device attains a state of saturation.
    (e) the beta of the device becomes equal to 1.

39. In an ideal reverse-biased diode in which the applied dc voltage is less than the avalanche voltage,
    (a) the current through the *p-n* junction is zero.
    (b) the current through the *p-n* junction is high.
    (c) electrons flow from the *p*-type material to the *n*-type material.
    (d) electrons flow from the *n*-type material to the *p*-type material.
    (e) electrons and holes flow in the same direction.

40. If you are told that wave *X* leads wave *Y* by 225°, you would do better by saying that
    (a) wave *X* lags wave *Y* by 135°.
    (b) wave *X* lags wave *Y* by 45°.
    (c) wave *X* leads wave *Y* by 45°.
    (d) wave *X* and wave *Y* are in phase opposition.
    (e) wave *X* and wave *Y* are in phase coincidence.

41. Suppose that three inductors are connected in series. Suppose also that there is no mutual inductance among them and that they each exhibit a reactance of $0 + j300$ at a frequency of 500 kHz. Also imagine that all three inductors are coils made of perfectly conducting wire; that is, they all have zero resistance. What is the complex-number impedance of the series combination at 500 kHz?
    (a) $0 + j100$
    (b) $0 - j100$
    (c) $0 + j900$
    (d) $0 - j900$
    (e) It is impossible to say without more information.

42. Imagine that a dc circuit carries *I* amperes, has an emf source of *E* volts, and has a resistance of *R* ohms in which *P* watts of power are dissipated. Which of the following formulas is incorrect?
    (a) $P = I^2 R$
    (b) $E = IR$
    (c) $R = E/I$
    (d) $P = E^2/R$
    (e) $I = ER$

43. The rms ac input signal current to a bipolar-transistor amplifier is 10.0 μA, and the alpha cutoff of the device is 100 kHz. What is the rms ac output signal current?
    (a) 10.0 mA
    (b) 10.0 μA
    (c) 0.100 μA
    (d) 0.00100 μA
    (e) It cannot be calculated without more information.

44. A simple dc circuit consists of a 12-V battery and a lamp whose resistance is 144 Ω when the battery is connected to it and the lamp is glowing. How much power is dissipated by the lamp?
    (a) 83 mW
    (b) 1.0 W
    (c) 12 W
    (d) 1.7 kW
    (e) 21 kW

45. In a rectangular ac wave,
    (a) the amplitude transitions occur instantaneously.
    (b) the amplitude becomes more positive at a steady rate and becomes more negative instantaneously.
    (c) the amplitude becomes more negative at a steady rate and becomes more positive instantaneously.
    (d) the amplitude changes both positively and negatively at a steady rate.
    (e) the waveform looks like the mathematical sine function.

46. A purely resistive impedance is also said to be
    (a) nonconductive.
    (b) noncapacitive.
    (c) noninductive.
    (d) nonreactive.
    (e) purely imaginary.

47. The average amplitude of a pure ac sine wave is
    (a) the same as the peak amplitude.
    (b) approximately 0.707 times the peak amplitude.
    (c) approximately 0.707 times the instantaneous amplitude.
    (d) approximately 0.707 times the peak-to-peak amplitude.
    (e) zero.

48. An energy transducer that converts mechanical motion into ac is
    (a) an electric motor.
    (b) a solenoid.
    (c) an electromagnet.
    (d) an electric generator.
    (e) an inductor.

49. The vertical component of the geomagnetic field at any particular location is known as the
    (a) flux density.
    (b) flux angle.
    (c) inclination.
    (d) declination.
    (e) right ascension.

50. Conventional current
    (a) flows from minus to plus.
    (b) flows from plus to minus.
    (c) flows only in perfect insulators.
    (d) cannot flow in an ionized gas.
    (e) is measured in amperes per second.

Waves, Particles, Space, and Time

Wave Phenomena

The universe is awash with ripples. Waves can occur in any mode and medium we care to imagine. Waves propagate in gases, liquids, and solids. Waves undulate throughout the space-time continuum and through what appears to be the absence of any medium at all. Consider

- The air during a musical concert
- The surface of a pond after a pebble falls in
- The surface of a lake on a windy day
- The surface of the ocean off Maverick's Beach in California
- The tops of the stalks in a wheat field
- The surface of a soap bubble when you blow on it
- High clouds near the jet stream
- Earth's surface during a major quake
- Earth's interior after a major quake
- The human brain at any time
- A guitar string after it has been plucked
- A utility power line carrying alternating current
- A radio or television transmitting antenna
- An optical fiber carrying a laser beam
- The electromagnetic field inside a microwave oven
- Space time near a coalescing binary neutron star

Some wave phenomena are easier to comprehend than others. If a wave is easy to see in the mind's eye, it is not necessarily common in nature. If it's difficult to envision, it is not necessarily rare.

Intangible Waves

The easiest waves to think about are the ones we can see and feel. Water waves are the best example. If you sit on a beach where swells are rolling in, you can feel their force and rhythm. Prehistoric humans no doubt spent hours staring at the waves on lakes and oceans, wondering where they came from, why they were sometimes big and sometimes small, why they were sometimes smooth and other times choppy, why they sometimes came from the west and other times came from the north, and why they sometimes moved with the wind and other times moved against the wind.

Imagine a child, 100,000 years ago, dropping a pebble into a pond or watching a fish jump and noticing that the ripples emanating from the disturbance were just like ocean swells, except smaller. This must have been a revelation, but it was nothing compared to the discoveries scientists would later make with the help of instruments, mathematics, and a fascination with the intangible.

ELECTROMAGNETIC WAVES

Think about the electromagnetic (EM) fields generated by wireless broadcasting. In the journal of cosmic history, these waves have existed in our corner of the universe for only a short while. Our galaxy is ancient, but television programs have been airing for less than 2×10^{-8} of its life.

Television waves don't arise directly out of nature. They are manufactured by specialized equipment invented by a particular species of living thing on the third planet in orbit around a medium-small star. Perhaps such waves are generated by other species who live on other planets in orbit around other stars. If so, we haven't heard any of their signals yet.

GRAVITATIONAL WAVES

Some scientists think *gravitational waves* exist and that the fabric of space is swarming with them, just as the sea is filled with swells and chop. There's a theory to the effect that the evolution of the known universe is a single cycle in an oscillating system, a wave with a period unimaginably long.

How many people walk around thinking they are submicroscopic specks on a particle in an expanding and collapsing bubble in the laboratory of the heavens? Not many, but even to them, gravitational waves, assuming that they exist, are impossible to see with our eyes and difficult to envision even with the keenest mind's eye because gravitational waves are four-dimensional.

PERFECT WAVES

Surfers dream of riding perfect waves; engineers strive to synthesize them. For a surfer, a perfect wave might be "tubular," "glassy," and part of an "overhead set" in a swell at one of the beaches on the north shore of Oahu in February. In the mind of a communications engineer, a perfect wave is a sinusoid, also called a *sine wave*. You may recall this type of wave from Chapter 13.

Even to a person who has never heard of the sine function, a sinusoidal waveform is easy to remember. An acoustic sinusoid makes an unforgettable noise. It focuses sound at a single frequency. A visual sinusoid has an unforgettable appearance. It concentrates light at a single wavelength. A perfect set of ocean waves can give a surfer an unforgettable thrill, and it, too, concentrates a lot of energy at a single wavelength.

Back in Chapter 13 the sine wave was represented by a child swinging an object around and around. As viewed from high above, the path of the object is a circle. As its trajectory is observed edge-on, the object seems to move toward the left, speed up, slow down, reverse, move toward the right, speed up, slow down, reverse, move toward the left again, speed up, slow down, and reverse. The actual motion of the object is circular and constant. Suppose that it takes place at a rate of one revolution per second. Then the object travels through 180° of arc every half second, 90° of arc every quarter second, 45° of arc every $1/8$ second, and 1° of arc every $1/360$ second. A scientist or engineer will say that the object has an angular speed of 360°/s.

GRAPHING A SINE WAVE

Suppose that you draw a precise graph of the swinging object's position with respect to time as seen from edge-on. Time is plotted horizontally; the past is toward the left, and the future is toward the right. One complete revolution of the object appears on the graph as a sine wave. Degree values can be assigned along this wave corresponding to the degrees around the circle (Fig. 17-1).

Constant rotational motion, such as that of the object on the string, takes place all over the universe. The child whirling the object can't make the object abruptly slow down and speed up, or instantly stop and change direction,

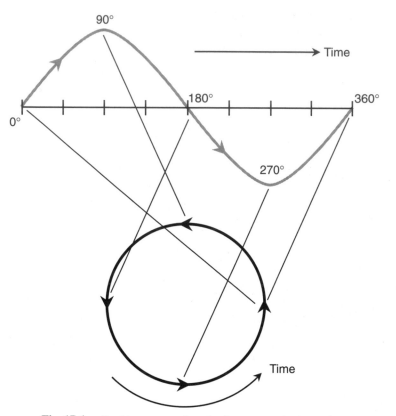

Fig. 17-1. Graphic representation of a sine wave as circular motion.

or revolve in steps like a ratchet wheel. However, once that mass is moving, it doesn't take much energy to keep it going. Uniform circular motion is a theoretical ideal. There's no better way to whirl an object around and around. A sinusoid is a theoretical ideal, too. There's no better way to make a wave.

Fundamental Properties

All waves, no matter what the medium or mode, have three different but interdependent properties. The *wavelength* is the distance between identical points on two adjacent waves. It is measured in meters. The *frequency* is the number of wave cycles that occur or that pass a given point per unit time. It is specified in cycles per second, or hertz. The *propagation speed* is the rate at which the disturbance travels through the medium. It is clocked in meters per second. These three properties are related: *Speed equals wavelength times frequency.* Consistent units must be used for this relation to have meaning.

PERIOD, FREQUENCY, WAVELENGTH, AND PROPAGATION SPEED

Sometimes it's easier to talk about a wave's *period* rather than its frequency. The period T (in seconds) of a sine wave is the reciprocal of the frequency f (in hertz). Mathematically, the following formulas hold:

$$f = 1/T = T^{-1}$$
$$T = 1/f = f^{-1}$$

If a wave has a frequency of 1 Hz, its period is 1 s. If the frequency is one cycle per minute ($1/60$ Hz), the period is 60 s. If the frequency is one cycle per hour ($1/3600$ Hz), the period is 3,600 s, or 60 min.

The period of a wave is related to the wavelength λ (in meters) and the propagation speed c (in meters per second) as follows: *Wavelength equals speed times period.* Mathematically:

$$\lambda = cT$$

This gives rise to other formulas:

$$\lambda = c/f$$
$$c = f\lambda$$
$$c = \lambda/T$$

PROBLEM 17-1
If the child whirling the object on the string slows down the speed so that the object whirls at the rate of one revolution every 2 seconds instead of one revolution per second, what happens to the wavelength of the graph in Fig. 17-1, assuming that time is plotted on the same scale horizontally?

SOLUTION 17-1
Consider the formula $\lambda = c/f$ above. Cutting the frequency in half doubles the wavelength. If each horizontal division represents a constant amount of time, then if the object revolves half as fast, the wavelength becomes twice as long.

FREQUENCY UNITS

Audible acoustic waves repeat at intervals that are only a small fraction of a second. The lowest sound frequency a human being can hear is about 20 cycles per second, or 20 hertz (20 Hz). The highest frequency an acoustic wave can have and still be heard by a person with keen ears is a thousand times higher: 20,000 Hz.

Radio waves travel in a different medium than sound waves. Their lowest frequency is a few thousand hertz, and their highest frequencies range into the trillions of hertz. Infrared (IR) and visible-light waves occur at frequencies much higher than radio waves. Ultraviolet (UV) waves, x-rays, and gamma (γ) rays range into quadrillions and quintillions of hertz, vibrating more than 1 trillion times faster than middle C on the musical scale.

To denote high frequencies, scientists and engineers use frequency units of kilohertz (kHz), megahertz (MHz), gigahertz (GHz), and terahertz (THz). Each unit is a thousand times higher than the previous one in this succession. That is, 1 kHz = 1,000 Hz, 1 MHz = 1,000 kHz, 1 GHz = 1,000 MHz, and 1 THz = 1,000 GHz.

MORE ABOUT SPEED

The fastest wave speed ever measured is 299,792 kilometers (186,282 miles) per second in a vacuum. This is commonly rounded off to 300,000 km/s or 3.00×10^8 m/s. This is the proverbial *speed of light,* the absolute maximum speed with which anything can travel. (Lately, some experiments have suggested that certain effects travel faster than this, but over long distances, 3.00×10^8 m/s is the speed limit as far as we know.) Lesser disturbances than light, in media humbler than intergalactic space, travel slower. Even light itself moves along more slowly than this in media other than a vacuum.

In air at sea level, sound waves poke along at about 335 m/s, or about 700 miles per hour (mi/h). This is called *Mach 1.* When you speak to someone across a room, your voice travels at Mach 1. Sound in air propagates at Mach 1 no matter what the frequency and no matter how strong (loud or soft) it is. The exact figure varies a little, depending on the altitude, the temperature, and the relative humidity, but 335 m/s is a good number to remember.

Electromagnetic waves do not always push the absolute cosmic speed limit. In glass or under water, light waves propagate at significantly less than 3.00×10^8 m/s. Radio waves slow down when they pass through

Earth's ionosphere. These variations in speed affect the wavelength, even if the frequency remains constant.

PROBLEM 17-2

One nanometer (1 nm) is 10^{-9} m. Suppose that a light beam has a wavelength of 500 nm in free space and then enters a new medium where the speed of light is only 2.00×10^8 m/s. The frequency does not change. What happens to the wavelength?

SOLUTION 17-2

Note the preceding formula that defines wavelength in terms of speed and frequency:

$$\lambda = c/f$$

The speed is cut to 200/300 of its initial value. Therefore, the wavelength is also cut to 200/300 of its initial value. The wavelength in the new medium is 500 nm $\times$ 200/300, or 333 nm.

AMPLITUDE

In addition to the frequency or period, the wavelength, and the propagation speed, waves have another property: *amplitude.* This is the strength or height of the wave, the relative distance between peaks and troughs. When all other factors are constant, the more energy a wave contains, the greater is its amplitude.

The energy in a light wave is directly proportional to the amplitude, directly proportional to the frequency, and inversely proportional to the wavelength. The same is true for gamma rays, x-rays, UV, IR, and radio waves. However, for waves on the surface of a liquid, these exact mathematical relations do not apply. Amplitude is sometimes, but not always, a precise indicator of the energy in a wave.

SEICHE AND HARMONICS

Any child who lives in a house with a bathtub knows about *seiche* (pronounced "saysh"). Any enclosed or semienclosed body of water can be made to slosh back and forth at a rate that depends on the size and shape of the container. In a bathtub, this sloshing can be set up with a period of 1 or 2 seconds. Give the water a little push, and then another, and then another. Keep this up at a certain regular repetitive rate, and soon there is water all over the bathroom. The same thing can happen in a swimming pool during an earthquake, although the period is longer. When waves moving in opposite directions collide, the peaks and troughs are exaggerated (Fig. 17-2).

Harmonics are familiar to anyone who plays a musical instrument such as a clarinet, flute, trumpet, or trombone. If you can blast out a note with certain keys pressed or with the slide at a given position, then if you tighten your lips enough, you can sound a note one octave higher. The higher note is the *second harmonic* of the first note. The chamber of the instrument contains twice as many wave peaks and valleys at the higher note as compared with the lower note. If you're a virtuoso, you might get the instrument to toot at three times the original, or fundamental, frequency. This is the *third harmonic*. Mathematically, there is no limit to how far this can go (Fig. 17-3). When the frequency of one wave is a harmonic of the frequency of another wave, the two waves are said to be *harmonically related*.

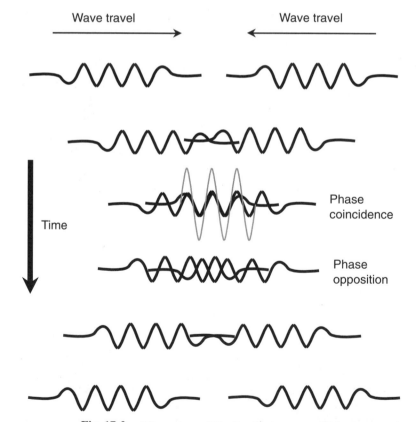

Fig. 17-2.   When waves collide, the effects are magnified.

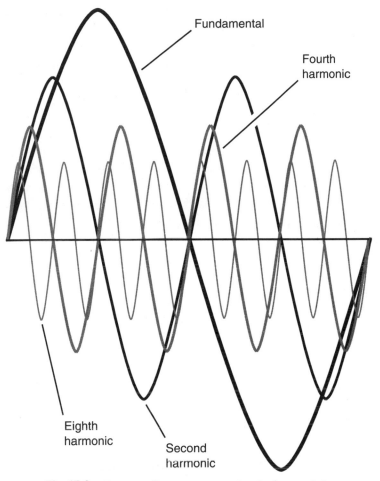

Fundamental

Fourth harmonic

Eighth harmonic

Second harmonic

Fig. 17-3.   Resonant effects occur at wavelengths that are whole-number fractions of the wavelength of the fundamental.

RESONANCE

You can demonstrate harmonics if you have a piece of rope about 10 m long. Anchor one end of the rope to an immovable object such as a fence post or a hook in a wall. Be sure that the rope is tied securely so that it won't shake loose. Hold the other end, and back off until the rope is tight. Then start pumping, slowly at first and then gradually faster. At a certain pumping speed, the rope will get into the rhythm and will seem to move up and down with a will of its own. This is a condition of *resonance*. Get it going this way for awhile. Then double the rate of pumping. If you keep at it, you'll get a full wave cycle to appear along the rope. The wave will reverse itself in phase each time you pump, and its curvature will attain a familiar shape: the sinusoid. Keep on pumping at this rate for awhile. Then, if you can, double the pumping speed once more. This experiment requires some conditioning and coordination, but eventually you'll get two complete wave cycles to appear along the length of the rope. You're at the second harmonic of the previous oscillation, and resonance occurs again.

If you're strong and fast enough, and if you have enough endurance, you might double the frequency again, getting four complete standing waves to appear along the rope (the fourth harmonic). If you're a professional athlete, maybe you can double it yet another time, getting eight standing waves (the eighth harmonic). Theoretically, there is no limit to how many cycles can appear between the shaker and the anchor. In the real world, of course, the diameter and elasticity of the rope impose a limit.

When you pump a rope, the wave impulses have *longitudinal motion*; they travel lengthwise along the rope. The individual molecules in the rope undergo *transverse motion*; they move from side to side (or up and down). The waves along the rope resemble swells on the surface of the ocean.

Stop shaking the rope and let it come to rest. Then give it a quick, hard, single pump. A lone wave shoots from your hand toward the far end of the rope and then reflects from the anchor and travels back toward you. As the pulse travels, its amplitude decays. Hold your hand steady as the pulse returns. The pulse energy is partially absorbed by your arm and partially reflected from your hand, heading down toward the far end again. After several reflections, the wave dies down. Some of its energy has been dissipated in the rope. Some has been imparted to the object to which the far end of the rope is anchored. Some has been absorbed by your body. Even the air has taken up a little of the original wave energy.

STANDING WAVES

Start shaking the rope rhythmically once again. Set up waves along its length just as you did before. Send sine waves down the rope. At certain shaking frequencies, the impulses reflect back and forth between your hand and the anchor so that their effects add together: Each point on the rope experiences a force upward, then downward, and then up again, then down again. The reflected impulses reinforce; the sideways motion of the rope is exaggerated. *Standing waves* appear.

Standing waves get their name from the fact that they do not, in themselves, travel anywhere. But they can acquire tremendous power. Some points along the rope move up and down a lot, some move up and down a little, and others stand completely still, only rotating slightly as the rest of the rope wags. The points where the rope moves up and down the furthest are called *loops*; the points where the rope doesn't move are known as *nodes*. There are always two loops and two nodes in a complete standing-wave cycle. They are all equally spaced from one another.

> **PROBLEM 17-3**
> How far apart is a standing-wave loop from an adjacent node in terms of degrees of phase?

> **SOLUTION 17-3**
> As you should remember from Chapter 13, there are 360° of phase in a complete cycle. From the preceding description, there are two loops and two nodes in a complete cycle, all equally spaced from each other; this means that they are all one-quarter cycle, or 90°, apart. Any given loop is 90° from the node on either side; any given node is 90° from the loop on either side.

IRREGULAR WAVES

Not all waves are sine waves. Some nonsinusoidal waves are simple but are seen rarely in nature. Some of these waves have abrupt transitions; unlike the smooth sinusoid, they jump or jerk back and forth. If you've used a laboratory oscilloscope, you're familiar with waves like this. The simplest nonsinusoids are the *square wave,* the *ramp wave,* the *sawtooth wave,* and the *triangular wave.* You learned about these in Chapter 13. These can be generated with an electronic music synthesizer, and they have a certain mathematical perfection, but you'll never see them on the sea. Irregular waves come in myriad shapes, like fingerprints or snowflakes. The sea is filled with these. In the world of waves, simplicity is scarce, and chaos is common.

Most musical instruments produce irregular waves, like the chop on the surface of a lake. These are complex combinations of sine waves. Any waveform can be broken down into sinusoid components, although the mathematics that define this can become complicated. Cycles superimpose themselves on longer cycles, which in turn superimpose themselves on still longer cycles, *ad infinitum.* Even square, ramp, sawtooth, and triangular waves, with their straight edges and sharp corners, are composites of smooth sinusoids that exist in precise proportions. Waves of this sort are easier on the ear than sine waves. They are also easier to generate. Try setting a music synthesizer or signal generator to produce square, ramp, sawtooth, triangular, and irregular waves, and listen to the differences in the way they sound. They all have the same pitch, but the *timbre,* or tone, of the sound is different.

Wave Interaction

Two or more waves can combine to produce interesting effects and, in some cases, remarkable patterns. Amplitudes can be exaggerated, waveforms altered, and entirely new waves generated. Common wave-interaction phenomena include *interference, diffraction,* and *heterodyning.*

INTERFERENCE

Imagine that you are a surfer. You spend the winter on the north shore of Oahu. In the maritime sub-Arctic, storms parade across the Pacific, spinning off from a parent vortex near the Kamchatka peninsula. You watch the satellite images of these storms on the Internet. The Kamchatka low is strong and stable, breeding systems that swoop southeastward and vent their fury in North America. Swells are propelled across the entire Pacific from these storms; there is nothing between the storm tracks and the shores of Hawaii. The swells arrive at places like Pipeline and break over coral and sand, reaching heights that often exceed 5 m (16 ft). Trade winds blow from east to west, producing smaller swells across the big ones. Gusts of wind and local squalls add chop. On a good day—the kind you, as a surfer, live for—the storm swell is strong, and the wind is light. You can ride the big breakers without being bumped around by the small stuff. On a bad day, waves pile onto waves in a haphazard way. The main swell is as big and well defined as on a good day, but the interference makes surfing difficult.

When two major marine storms are separated by a great distance, each producing significant swells, things get interesting. Such conditions are more likely to attract scientists than surfers. This type of situation can occur during the winter on the north shore of Oahu, but it is more often found in the tropics during hurricane season. Tropical storms produce some of the largest surf in the world. When a hurricane prowls the sea, swells radiate in expanding circles from the storm's central vortex. If two storms of similar size and intensity are separated by a vast distance, complex swell patterns span millions of square kilometers. Between the storms, swells alternately cancel and reinforce each other, producing wild seas.

Interference patterns created by multiple wave sources appear at all scales, from swells at sea to sound waves in a concert hall, from radio broadcast towers to holographic apparatus. The slightest change in the relative positions or wavelengths of two sources can make a profound difference in the way the composite pattern emerges. Examples are shown in Figs. 17-4 and 17-5.

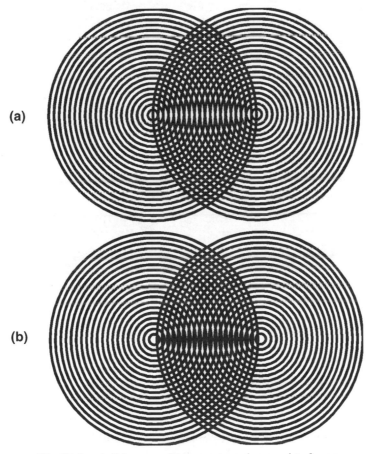

(a)

(b)

Fig. 17-4.   A slight source displacement can change an interference pattern dramatically. Notice the difference between the pattern formed by the intersecting lines in part *a*, compared with the pattern in part *b*.

(a)

(b)

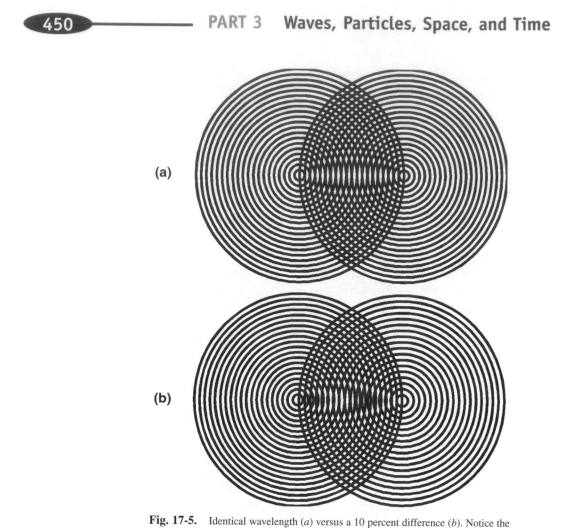

Fig. 17-5. Identical wavelength (*a*) versus a 10 percent difference (*b*). Notice the difference in the interference pattern caused by the wavelength change.

Suppose that two tropical storms are sweeping around the Atlantic basin, steered by currents in the upper atmosphere. The interference pattern produced by their swells evolves from moment to moment. Multiple crests and troughs conspire along a front hundreds of kilometers long: a *rogue wave*. Such a monster wave can capsize ocean liners and freighters. Veteran sailors tell stories about walls of water that break in the open sea, seeming to defy the laws of hydrodynamics.

Wave interference on the high seas, while potentially frightening in its proportions, is not easy for scientists to observe. Patterns are sometimes seen from aircraft, and sophisticated radar can reveal subtleties of the surface, but oceans do not lend themselves to controlled experiments. Nor can you go out in a boat and sail into storm-swell interference patterns and expect to return with meaningful data, although you might end up with stories to tell your grandchildren if you survive. However, there are ways that even a child can conduct memorable experiments with wave convergence and interference.

Soap bubbles, with their rainbow-colored surfaces, are tailor made for this. Visible-light waves add and cancel across the visible spectrum, reflecting from the inner and outer surfaces of the soap film and teasing the eye with red, green, violet, then red again.

Adults can play with wave interference, too. Any building with a large rotunda provides a perfect venue. According to legend, long ago in the halls of Congress, a few elected officials were able to eavesdrop on certain supposedly private whisperings because the vast dome overhead reflected and focused the sound waves from one politician's mouth to another's ears.

DIFFRACTION

Waves can gang up, fight each other, and travel in illogical directions at unreasonable speeds. They also can turn corners. Do you remember playing in the side yard when you were a child, hoping that you would not hear your mother call you from the front door? When the time came, the voice reached your ears anyway. How could sound find its way around the house? Your mother was out of your sight, and you were out of hers. Why should sound go places visible light could not? Doesn't sound travel in straight lines, like light? Was your mother's voice reflecting from other houses in the neighborhood? To find out, you conducted an experiment with a friend near an old, abandoned barn in the middle of nowhere, and her voice found its way around the structure even though there was nothing nearby from which the sound could reflect.

Sound turns corners, especially sharp corners, because of waves' ability to *diffract*. When a wave disturbance encounters a "sharp" obstruction, the obstruction behaves as a new source of energy at the same wavelength (Fig. 17-6). The phenomenon can occur repeatedly. Even if you hid in the garage, you could still hear that voice. In a side yard three houses down the street, you could hear it. You noticed that the sounds from musical instruments, car engines, lawn mowers, and all sorts of other noisemakers can diffract too. As anyone who lives in a city knows that no one can hide from noise. Diffraction is one of the reasons sound is so pervasive.

Long waves diffract more readily than shorter waves. As an edge or corner becomes sharper relative to the wavelength of sound, diffraction occurs

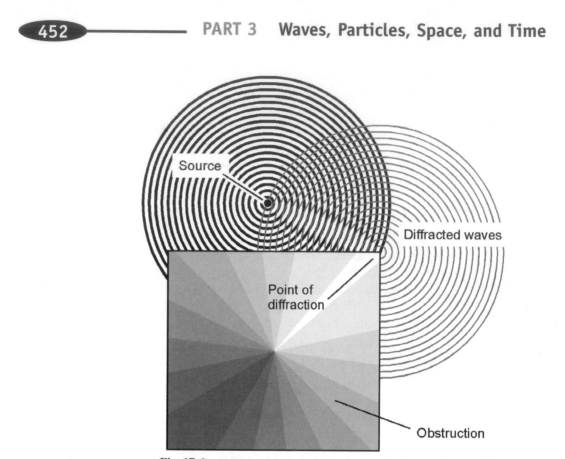

Fig. 17-6. Diffraction makes it possible for waves to "go around corners."

more efficiently. As the frequency decreases, the wavelength increases so that all edges and corners become, in effect, sharper. This effect is not unique to sound waves. It happens with water waves, as any surfer knows. It happens with radio waves; this is why you can hear the broadcast stations on your car radio, especially on the AM broadcast band, where the EM waves are hundreds of meters long, even when there are buildings or hills between you and the transmitter. It happens with visible light waves too, although the effect is more subtle and can be observed only under certain conditions. All waves diffract around sharp corners. One of the tests by which scientists ascertain the wave nature of a disturbance is to see whether or not the effect can be observed from around a corner.

When an obstruction is tiny in comparison with the wavelength of a disturbance, the waves diffract so well that they pass the object as if it's not there. A flagpole has no effect on low-frequency sound waves. The pilings of a pier likewise are ignored by ocean surf.

HETERODYNING

No matter what the mode, and regardless of the medium, waves mix to produce other waves. When this happens with sound, the effect is called *beating*; when it happens with radio signals, it is called *heterodyning*, or *mixing*. Two sound waves that are close to each other in pitch will beat to form a new wave at a much lower frequency and another wave at a higher frequency. If you have access to a music synthesizer or a laboratory signal generator (or failing those, a couple of loud horns), you can do an experiment to demonstrate this. When two notes in the treble clef are played loudly at the same time, you'll hear a low-frequency hum. This is the lower-pitched of the two beat notes. The higher-pitched note is harder to notice. Figure 17-7 shows examples of wave beating in which the low-frequency notes are visually evident. In part *a,* the waves, shown by sets of vertical lines, differ in frequency by 10 percent (f and $1.1f$); in part *b,* by 20 percent (f and $1.2f$), and in part *c,* by 30 percent (f and $1.3f$).

Beat and heterodyne waves always occur at a frequencies equal to the sum and difference between the frequencies of the waves that produce them. If you play two notes together, one at 1,000 Hz and another at 1,100 Hz, a beat note appears at 100 Hz. When notes at 1,000 Hz and 1,200 Hz are combined, a 200-Hz beat wave is the result. If you adjust a music synthesizer to play notes at 1,000 Hz and 1,300 Hz, there is a 300-Hz beat note. If you hold one note steady and continuously vary the frequency of the other, the beat notes rise and fall in pitch. If you can secure the use of a good synthesizer, you ought to try this experiment. The beat notes will seem to come from a direction you cannot identify. It is a bizarre sensation, not to be missed by the truly devoted audiophile.

Radio-frequency heterodyning was discovered by engineers in the early 1900s. Under certain conditions, two wireless signals combine to produce a new signal at the difference frequency. It's easy to design a circuit to produce this effect. In fact, the phenomenon, not always desired, is difficult to prevent.

Given two waves having different frequencies f and g (in hertz), where $g > f$, they beat or heterodyne together to produce new waves at frequencies x and y (also in hertz) as follows:

$$x = g - f$$
$$y = g + f$$

These formulas also apply for frequencies in kilohertz, megahertz, gigahertz, and terahertz—provided, of course, that you stick with the same units through out any given calculation.

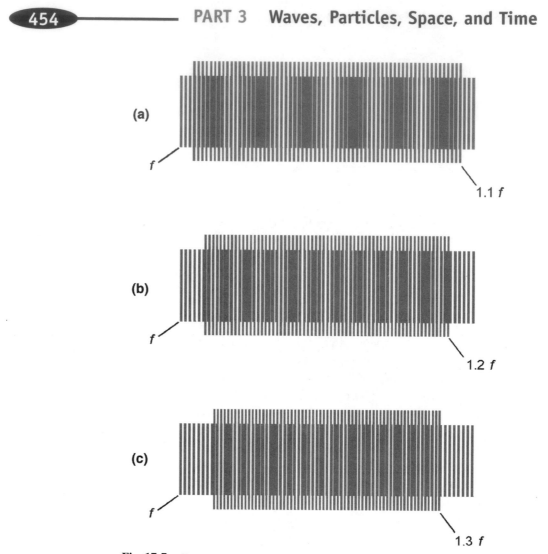

Fig. 17-7. Two waves can beat to form a new wave at the difference frequency. (*a*) The two waves differ in frequency by 10 percent. (*b*) By 20 percent. (*c*) By 30 percent.

PROBLEM 17-4

Suppose that you have two waves, one at 500 Hz and another at 2.500 kHz. What are the beat frequencies?

SOLUTION 17-4

Convert the frequencies to hertz. Then let $f = 500$ Hz and $g = 2,500$ Hz. This produces beat frequencies x and y as follows:

$$x = g - f = 2,500 - 500 = 2,000 \text{ Hz}$$

$$y = g + f = 2,500 + 500 = 3,000 \text{ Hz}$$

If you want to get particular about significant digits here, you can consider these frequencies to be 2.00 kHz and 3.00 kHz, respectively.

Wave Mysteries

The more we delve into the mysteries of wave phenomena, the less we seem to know about them. Studying waves is bound to produce more questions than answers.

LENGTHWISE VERSUS SIDEWAYS

When waves travel through matter, the molecules oscillate to and fro, up and down, or back and forth. The nature of the particle movement differs from the nature of the wave as it travels. The atoms or molecules rarely move more than a few meters—sometimes less than a centimeter—but the wave can travel thousands of kilometers. Sometimes the particles vibrate in line with the direction of wave travel; this is a *compression wave,* also called a *longitudinal wave.* In other instances, the particles move at right angles to the direction of propagation; this is a *transverse wave.* The distinction is illustrated in Fig. 17-8.

What is it that wags or wiggles or compresses or stretches when a wave travels through a particular medium? It depends on the medium and on the nature of the wave disturbance. Sound waves in air are longitudinal, but radio waves are transverse. The waves on the surface of the ocean are transverse, but when a big wave arrives on a beach, plenty of longitudinal motion is involved as well.

FORCE FIELDS

When waves travel through a vacuum, they manifest as force fields (in the case of EM waves) or as undulations in space-time (gravitational waves). It took scientists a long time to accept the fact that waves can propagate without any apparent medium to carry them.

Both EM and gravitational waves are transverse disturbances. The electrical and magnetic force fields in radio waves, IR, visible light, UV, x-rays, or gamma rays pulsate at right angles to each other and at right angles to

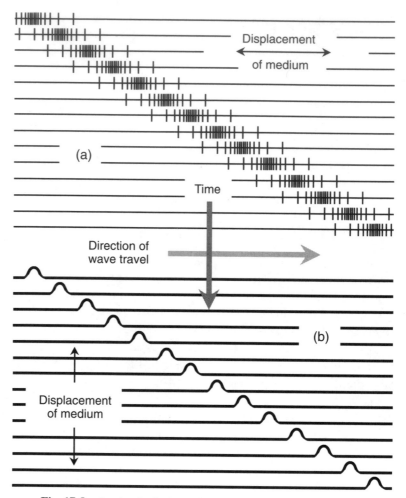

Fig. 17-8. In a longitudinal wave (*a*), the particles vibrate parallel to the direction of wave travel. In a transverse wave (*b*), the particles vibrate laterally.

the direction of propagation. This occurs in three dimensions, so it can be envisioned in the "mind's eye." Gravitational waves are more esoteric; they cause space-time to oscillate in four dimensions. If anyone claims to be able to envision four-dimensional wave oscillation directly, you are justified in suspecting that they're either joking or crazy. Nevertheless, they can be defined quite easily with mathematics.

CORPUSCLES OF LIGHT

The theory of EM-wave propagation is a relatively recent addition to the storehouse of physics knowledge. Isaac Newton, the seventeenth-century English physicist and mathematician known for his theory of gravitation and his role in the invention of calculus, believed that visible light consists of submicroscopic particles. To the casual observer, visible light travels in straight lines through air or free space. Shadows are cast in such a way as to suggest that there are no exceptions to this rule, at least in a vacuum. Today scientists know that light behaves, in some ways, like a barrage of bullets. Particles of EM energy, called *photons,* have momentum, and they exert measurable pressure on objects they strike. The energy in a beam of light can be broken down into packets of a certain minimum size but no smaller.

However, one need not search hard to find complications with Newton's *corpuscular theory* of the nature of light. At a boundary between air and water, photons do inexplicable things. Ask any child who has ever stuck a fishing pole into a lake or who has looked into the deep end of a swimming pool and seen 4 m of water look like 1 m. Photons change direction abruptly when they pass at a sharp angle from water to air or vice versa (Fig. 17-9), but there is no apparent force to give them a sideways push. When light passes through a glass prism, things get stranger still; not only are light beams bent by the glass, but the extent to which they are bent depends on the color!

ALTERNATIVES TO THE CORPUSCULAR THEORY

Some of Newton's colleagues thought he painted an overly simplistic picture of the nature of light, so they set out to find alternative models. Christian Huygens, a Dutch physicist fond of optics, was one of the first to suggest that visible light is a wave disturbance, like the ripples on a pond or the vibrations of a violin string. Today, even laypeople speak of *light waves* as if the two words go together. To scientists in the 1600s, however, the connection was not obvious.

Huygens kept at his research and showed that light waves interfere with each other in the same way as waves on water and in the same way as waves from musical instruments. This explains the ripples or concentric rings seen around images in high-powered optical instruments.

The bending of light rays at the surface of a lake or pool is consistent with the wave theory. When a light beam strikes the surface, the beam is

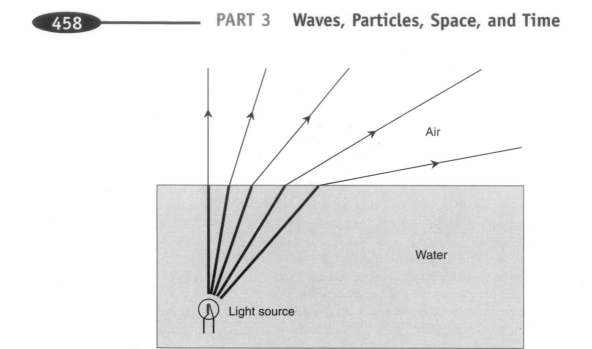

Fig. 17-9.   If light rays consist of particles, what pushes
them sideways at the water surface?

bent at an angle (Fig. 17-10). The extent of the bending depends on the angle at which the wavefronts strike the surface. Wavefronts parallel to the boundary are not bent. Wavefronts striking the water surface at a large enough angle from within the water do not pass through the surface boundary but are reflected instead.

WHAT "DOES THE WAVING"?

Huygens's colleagues had trouble accepting his wave theory, even when they saw the demonstrations with their own eyes. They had some bothersome questions to ask. What "does the waving" in a light-wave disturbance? When light passes through the atmosphere, does the air vibrate? If so, why can't we hear it? If not, why not? When light waves enter water, does the water undulate? If so, why doesn't it make ripples on the surface? If not, why not? How about light passing through an evacuated chamber? If visible light is caused by physical oscillations, why does a glass jar with all the air pumped out appear transparent rather than opaque? After all, there's nothing in an evacuated jar to "do the waving"—is there?

In the 1800s, an attempt was made to answer these questions by postulating the existence of a *luminiferous ether,* a medium that was imagined to

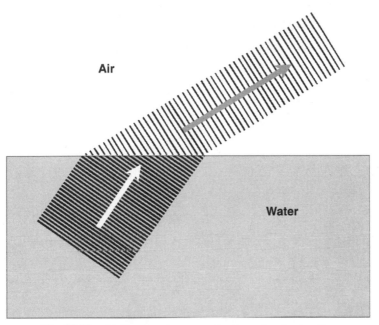

Fig. 17-10.   Light waves change speed and wavelength when they strike a boundary between media having different indices of refraction.

permeate all of space. In the early 1900s, a free-thinking European theorist named Albert Einstein decided that the ether theory was nonsense. You will learn about the consequences of Einstein's rejection of the ether theory in Chapter 20.

What "does the waving" in an EM disturbance? This question still baffles scientists. Magnetic and electrical fields existing at right angles to each other and oscillating at extremely high rates act in synergy to propagate through all kinds of media. The intensity of the fields "do the waving," but these fields are not material things. They are presences, or effects, that cause certain things to happen to matter and energy even though the fields themselves are intangible.

Particle or Wave?

The question, "Is an EM field a barrage of particles or a wave disturbance?" has never been fully and rigorously answered. There is a relation, however, between photon energy, frequency, and wavelength.

ENERGY, FREQUENCY, AND WAVELENGTH

The energy contained in a single photon of EM energy can be found in terms of the frequency by this formula:

$$e = hf$$

where e is the energy (in joules) contained in a photon, f is the frequency of the EM wave disturbance (in hertz), and h is a constant known as *Planck's constant,* approximately equal to 6.6262×10^{-34}.

If the wavelength λ (in meters) is known, and c is the propagation speed of the EM disturbance (in meters per second), then

$$e = hc/\lambda$$

This can be rearranged to determine the wavelength of a photon in terms of the energy it contains:

$$\lambda = hc/e$$

For EM rays in free space, the product hc is approximately equal to 1.9865×10^{-25} because c is approximately equal to 2.99792×10^{8} m/s.

PROBLEM 17-5
What is the energy contained in a photon of visible light whose wavelength is 550 nm in free space?

SOLUTION 17-5
First, convert 550 nm to meters; 550 nm = 550×10^{-9} m = 5.50×10^{-7} m. Then use the formula for energy in terms of wavelength:

$$e = hc/\lambda$$
$$= (1.9865 \times 10^{-25})/(5.50 \times 10^{-7})$$
$$= 3.61 \times 10^{-19} \text{ J}$$

PROBLEM 17-6
What is the wavelength of an EM disturbance consisting of photons that all have 1.000×10^{-25} J of energy in free space?

SOLUTION 17-6
Use the formula for wavelength in terms of energy:

$$\lambda = hc/e$$
$$= (1.9865 \times 10^{-25})/(1.000 \times 10^{-25})$$
$$= 1.9865 \text{ m}$$

This turns out to be a signal in the very-high-frequency (VHF) radio range. You can calculate the exact frequency if you like.

DOUBLE-SLIT EXPERIMENTS

If a beam of light gets dim enough, its photons emanate from the source at intervals that can be measured in seconds, minutes, hours, days, or years. If a beam of light gets brilliant enough, its photons rain down at the rate of trillions per second. These particles can be detected and their energy content determined just as if they were tiny bullets traveling at 2.99792×10^8 m/s in free space. However, the particle theory of EM radiation does not explain *refraction* of the sort that occurs at the surface of a body of water. The corpuscular theory also fails to explain beating and interference effects that are observed with visible light and high-speed subatomic particles. The classic *double-slit experiment* has been used as a demonstration of the wavelike nature of visible light. The following is a somewhat oversimplified description of this experiment.

An English physicist named Thomas Young devised an experiment in the hope of resolving the particle/wave question. He shone a beam of light having a certain color and a nearly perfect point source at a barrier with two narrow slits cut in it. Beyond the barrier was a photographic film. The light would, Young supposed, pass through the two slits and land on the film, producing a pattern. If light is comprised of corpuscles, then the pattern on the film ought to be two bright vertical lines. If light is a train of waves, an interference pattern ought to appear in the form of alternating bright and dark bands. When the experiment was carried out, the verdict was clear: Light is a wave disturbance. Interference bands showed up (Fig. 17-11), indicating that the beam was diffracted as it passed through the slits. The crests and troughs from the two diffracted rays alternately added together and canceled out as they arrived at various points on the film. This would happen with a wave disturbance but not with stream of corpuscles—or at least not with any sort of particle ever imagined up to that time.

However, it had been shown by other experiments that light has a particle nature. What about the pressure that visible light exerts? What about the discovery that its energy can be broken down into certain minimum packets? Is light both a wave and a particle? Or is it something else, something that is actually neither but with characteristics of both?

Suppose that photons are hurled one by one at a barrier with two slits and are allowed to land on a sensitive surface? Experiments of this sort

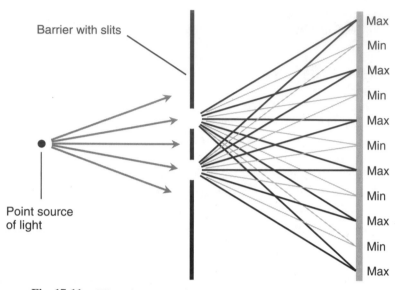

Fig. 17-11. When photons pass through a pair of slits in a barrier, a wave-like interference pattern results.

have been done, and interference bands appear on the surface no matter how weak the beam. Even if the source is made so dim that only one particle hits the surface every minute, the pattern of light and dark bands appears after a period of time long enough to expose the film (Fig. 17-12). This pattern changes depending on the distance between the two slits, but it is the same pattern at all energy intensities.

What happens in experiments like this? Do photons "know" where to land on the film based on the wavelength of the light they represent? How can a single photon passing through one slit "ascertain" the separation between the slits, thus "knowing" where it "can" and "cannot" land on the film? Is it possible that a photons split in two and pass through both slits at the same time? Does some effect take place backwards in time so that the photons from the light source "know" about the sort of barrier they are going to have to pass through?

Researchers have a saying: "One theorist can keep a thousand experimentalists busy." The double-slit experiments show that the converse of this is also true. In the quest for knowledge, turnabout is fair play.

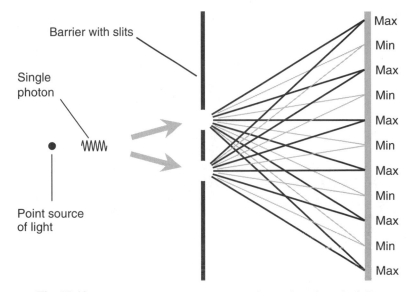

Fig. 17-12.   How can "wave particles" pass one by one through a pair of slits
and still produce an interference pattern?

Refer to the text in this chapter if necessary. A good score is eight correct. Answers
arc in thc back of the book.

1. Two radio waves, one having a frequency of 400.0 kHz and the other having a
    frequency of 1.250 MHz, are heterodyned together. Which of the following is
    a beat frequency?
    (a) 320.0 kHz
    (b) 500.0 kHz
    (c) 3.125 MHz
    (d) 0.850 MHz

2. Which of the following types of waves are longitudinal?
    (a) Swells on the surface of the ocean
    (b) EM waves in a vacuum
    (c) Sound waves in the air
    (d) Gravitational waves in interstellar space

3. What is the wavelength of a disturbance whose frequency is 500 Hz?
   (a) 2.00 mm
   (b) 20.0 mm
   (c) 200 mm
   (d) It cannot be calculated from this information.

4. In which of the following situations would you expect diffraction to occur to the greatest extent?
   (a) Visible-light waves around a utility pole
   (b) High-pitched sound waves around a cylindrical grain silo
   (c) Low-pitched sound waves around the corner of a building
   (d) Diffraction would occur to an equal extent in all of these scenarios.

5. The corpuscular theory of light is good for explaining
   (a) the pressure a ray of light exerts when it strikes an obstruction.
   (b) the interference patterns produced in double-slit experiments.
   (c) refraction of light at the surface of a body of water.
   (d) the way a prism splits light into the colors of the rainbow.

6. What is the period of an EM wave in free space whose wavelength is 300 m?
   (a) 1.00×10^{-6} s
   (b) 3.33×10^{-3} s
   (c) 1.00×10^{6} s
   (d) It cannot be calculated from this information.

7. An acoustic sine wave
   (a) spans the entire range of audible frequencies.
   (b) concentrates sound at a single frequency.
   (c) is high-pitched.
   (d) is low-pitched.

8. What is the wavelength of an EM disturbance consisting of photons that all have 2.754×10^{-20} J of energy in free space?
   (a) 7.213×10^{-6} m
   (b) 5.471×10^{-6} m
   (c) 7.213×10^{-45} m
   (d) 5.471×10^{-45} m

9. What is the energy contained in each photon of an acoustic disturbance whose frequency is 700 Hz?
   (a) 4.64×10^{-31} J
   (b) 9.47×10^{-37} J
   (c) 0.00143 J
   (d) The question makes no sense; acoustic disturbances have no photons.

10. Sound waves propagate at approximately 335 m/s in the Earth's atmosphere at
    sea level. What is the wavelength of an acoustic disturbance whose frequency
    is 440 Hz?
    (a) 1.31 m
    (b) 147 km
    (c) 76.1 cm
    (d) It cannot be calculated from this information.

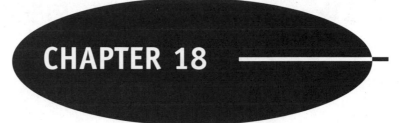

CHAPTER 18

Forms of Radiation

Isaac Newton believed that visible light is composed of tiny particles, or *corpuscles*. Today we recognize these particles as photons. However, light is more complex than can be represented by the particle theory. It has wave-like characteristics too. The same is true of all forms of radiant energy.

EM Fields

The wave nature of radiant energy is the result of interaction of electricity and magnetism. Charged particles, such as electrons and protons, are surrounded by *electrical fields*. Magnetic poles or moving charged particles produce *magnetic fields*. When the fields are strong enough, they extend a considerable distance. When electrical and magnetic fields vary in intensity, we have an *electromagnetic (EM) field*.

STATIC FIELDS

You have observed the attraction between opposite poles of magnets and the repulsion between like poles. Similar effects occur with electrically charged objects. These forces seem to operate only over short distances under laboratory conditions, but this is so because such fields weaken rapidly, as the distance between poles increases to less than the smallest intensity we can detect. In theory, the fields extend into space indefinitely.

A constant electric current in a wire produces a magnetic field around the wire. The *lines of magnetic flux* are perpendicular to the direction of the current. The existence of a constant voltage difference between two nearby objects produces an electrical field; the *lines of electrical flux* are parallel to the gradient of the charge differential. When the intensity of a current or voltage changes with time, things get more interesting.

FLUCTUATING FIELDS

A fluctuating current in a wire or a variable *charge gradient* between two nearby objects gives rise to a magnetic field and an electrical field in combination. These fields leapfrog through space so that the EM field can travel long distances with less attenuation than either an electrical field or a magnetic field alone. The electrical and magnetic fields in such a situation are perpendicular to each other everywhere in space. The direction of travel of the resulting EM field is perpendicular to both the electrical and magnetic lines of flux, as shown in Fig. 18-1.

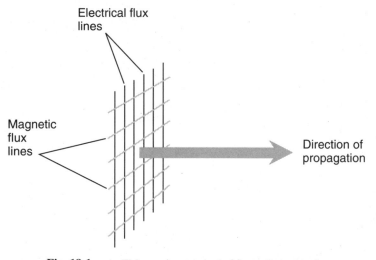

Fig. 18-1.   An EM wave is comprised of fluctuating, mutually perpendicular electrical and magnetic lines of flux. The field travels perpendicular to both sets of flux lines.

In order for an EM field to exist, the electrons in a wire or other conductor not only must be set in motion, but they also must be accelerated. That is, their velocity must be made to change. The most common method of creating this sort of situation is the introduction of an alternating current (ac) in an electrical conductor. It also can result from the bending of charged-particle beams by electrical or magnetic fields.

FREQUENCY AND WAVELENGTH

EM waves travel through space at the speed of light, which is approximately 2.99792×10^8 m/s (1.86262×10^5 mi/s). This is often rounded up to 3.00×10^8 m/s, expressed to three significant figures. The wavelength of an EM field in free space gets shorter as the frequency becomes higher. At 1 kHz, the wavelength is about 300 km. At 1 MHz, the wavelength is about 300 m. At 1 GHz, the wavelength is about 300 mm. At 1 THz, an EM signal has a wavelength of 0.3 mm—so small that you would need a magnifying glass to see it.

The frequency of an EM wave can get much higher than 1 THz; some of the most energetic known rays have wavelengths of 0.00001 *Ångström* (10^{-5} Å). The Ångström is equivalent to 10^{-10} m and is used by some scientists to denote extremely short EM wavelengths. A microscope of great magnifying power would be needed to see an object with a length of 1 Å. Another unit, increasingly preferred by scientists these days, is the *nanometer* (nm), where 1 nm $= 10^{-9}$ m $= 10$ Å.

The formula for wavelength λ, in meters, as a function of the frequency f, in hertz, for an EM field in free space is

$$\lambda = 2.99792 \times 10^8/f$$

This same formula can be used for λ in millimeters and f in kilohertz, for λ in micrometers and f in megahertz, and for λ in nanometers and f in gigahertz. Remember your prefix multipliers: 1 millimeter (1 mm) is 10^{-3} m, 1 micrometer (1 μm) is 10^{-6} m, and 1 nanometer (1 nm) is 10^{-9} m.

The formula for frequency f, in hertz, as a function of the wavelength λ, in meters, for an EM field in free space is given by transposing f and λ in the preceding formula:

$$f = 2.99792 \times 10^8/\lambda$$

As in the preceding case, this formula will work for f in kilohertz and λ in millimeters, for f in megahertz and λ in micrometers, and for f in gigahertz and λ in nanometers.

MANY FORMS

The discovery of EM fields led ultimately to the variety of wireless communications systems we know today. *Radio waves* are not the only form of EM radiation. As the frequency increases above that of conventional radio, we encounter new forms. First comes *microwaves.* Then comes infrared (IR) or "heat rays." After that comes visible light, ultraviolet (UV) radiation, x-rays, and gamma (γ) rays.

In the opposite, and less commonly imagined, sense, EM fields can exist at frequencies far below those of radio signals. In theory, an EM wave can go through one complete cycle every hour, day, year, thousand years, or million years. Some astronomers suspect that stars and galaxies generate EM fields with periods of years, centuries, or millennia.

THE EM WAVELENGTH SCALE

To illustrate the range of EM wavelengths, we use a logarithmic scale. The logarithmic scale is needed because the range is so great that a linear scale is impractical. The left-hand portion of Fig. 18-2 is such a logarithmic scale that shows wavelengths from 10^8 m down to 10^{-12} m. Each division, in the direction of shorter wavelength, represents a 100-fold decrease, or two orders of magnitude. Utility ac is near the top of this scale; the wavelength of 60-Hz ac in free space is quite long. The gamma rays are denoted approximately at the bottom; their EM wavelengths are tiny. It is apparent here that visible light takes up only a tiny sliver of the *EM spectrum.* In the right-hand scale, visible wavelengths are denoted in nanometers (nm).

HOW LITTLE WE SEE!

To get some idea of what a small EM "window" is represented by the visible-light wavelengths, try looking through a red- or blue-colored piece of glass or cellophane. Such a color filter greatly restricts the view you get of the world because only a narrow range of visible wavelengths can pass through it. Different colors cannot be ascertained through the filter. For example, when a scene is viewed through a red filter, everything is a shade of red or nearly red. Blue appears the same as black, bright red appears the same as white, and maroon red appears the same as gray. Other colors look red with varying degrees of saturation, but there is little or no variation in the hue. If our eyes had built-in red color filters, we would be pretty much color-blind.

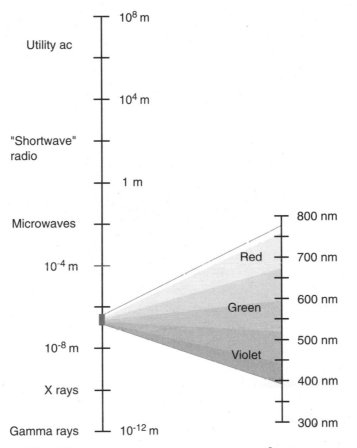

Fig. 18-2. The EM spectrum from wavelengths of 10^8 m down to 10^{-12} m, and an exploded view of the visible-light spectrum within.

When considered with respect to the entire EM spectrum, all optical instruments suffer from the same sort of handicap we would have if the lenses in our eyeballs were tinted red. The range of wavelengths we can detect with our eyes is approximately 770 nm at the longest and 390 nm at the shortest. Energy at the longest visible wavelengths appears red to our eyes, and energy at the shortest visible wavelengths appears violet. The intervening wavelengths show up as orange, yellow, green, blue, and indigo.

PROBLEM 18-1
What is the frequency of a red laser beam whose waves measure 7400 Å?

SOLUTION 18-1

Use the formula for frequency in terms of wavelength. Note that 7400 Å = 7400 $\times$ 10^{-10} m = 7.400 $\times$ 10^{-7} m. Then the frequency in hertz is found as follows:

$$f = 2.99792 \times 10^8 / \lambda$$
$$= 2.99792 \times 10^8 / (7.400 \times 10^{-7})$$
$$= 4.051 \times 10^{14} \text{ Hz}$$

This is 405.1 THz. To give you an idea of how high this frequency is, compare it with a typical frequency-modulated (FM) broadcast signal at 100 MHz. The frequency of the red light beam is more than 4 million times as high.

PROBLEM 18-2

What is the wavelength of the EM field produced in free space by the ac in a common utility line? Take the frequency as 60.0000 Hz (accurate to six significant figures).

SOLUTION 18-2

Use the formula for wavelength in terms of frequency:

$$\lambda = 2.99792 \times 10^8 / f$$
$$= 2.99792 \times 10^8 / 60.0000$$
$$= 4.99653 \times 10^6 \text{ m}$$

This is about 5,000 km, or half the distance from the Earth's equator to the north geographic pole as measured over the surface of the globe.

ELF Fields

Many electrical and electronic devices produce EM fields. Some of these fields have wavelengths much longer than standard broadcast and communications radio signals. The fields have *extremely low frequencies* (ELFs); this is how the term *ELF fields* has arisen.

WHAT ELF IS (AND ISN'T)

The *ELF spectrum* begins, technically, at the lowest possible frequencies (less than 1 Hz) and extends upward to approximately 3 kHz. This corresponds to wavelengths longer than 100 km. The most common ELF field in the modern world has a frequency of 60 Hz. These ELF waves are emitted by all live utility wires in the United States and many other countries. (In some countries it is 50 Hz.) In the Great Lakes area of the United States, the military has an ELF installation that is used to communicate with submarines. The ELF waves travel underground and underwater more efficiently than radio waves at higher frequencies.

The term *extremely-low-frequency radiation* and the media attention it has received have led some people to unreasonably fear this form of EM energy. An ELF field is not like a barrage of x-rays or gamma rays, which can cause sickness and death if received in large doses. Neither does ELF energy resemble UV radiation, which has been linked to skin cancer, or intense IR radiation, which can cause burns. An ELF field will not make anything radioactive. Some scientists suspect, nevertheless, that long-term exposure to high levels of ELF energy is linked to an abnormally high incidence of certain health problems. This is a hotly debated topic and, like any such issue, has become politicized.

COMMON SOURCES

One ELF source that has received much publicity is the common cathode-ray tube (CRT) monitor of the sort used in desktop personal computers. (Actually, CRT monitors produce EM energy at higher frequencies, not only at ELF.) Other parts of a computer are not responsible for much EM energy. Laptop and notebook computers produce essentially none.

In the CRT, the characters and images are created as electron beams strike a phosphor coating on the inside of the glass. The electrons change direction constantly as they sweep from left to right and from top to bottom on the screen. The sweeping is caused by deflecting coils that steer the beam across the screen. The coils generate magnetic fields that interact with the negatively charged electrons, forcing them to change direction. The fields thus fluctuate at low frequencies. Because of the positions of the coils and the shapes of the fields surrounding them, there is more EM energy "radiated" from the sides of a CRT monitor cabinet than from the front. If there's any health hazard with ELF, therefore, it is greatest for someone sitting off to the side of a monitor and least for someone watching the screen from directly in front.

PROTECTION

The best "shielding" from ELF energy is physical distance. This is especially true for people sitting next to (rather than in front of) a desktop computer monitor. The ELF field dies off rapidly with distance from the monitor cabinet. Computer workstations in an office environment should be at least 1.5 m (about 5 ft) apart. You should keep at least 0.5 m (about 18 in) away from the front of your own monitor. A monitor can be shut off when it's not in use.

Special monitors designed to minimize ELF fields are available. They are rather expensive, but they can offer peace of mind for people concerned about possible long-term health effects from exposure to ELF fields.

You'll sometimes see devices marketed with claims to eliminate or greatly reduce ELF fields. Some such schemes are effective; others are not. Electrostatic screens that you can place in front of the monitor glass to keep it from attracting dust will not stop ELF fields. Neither will glare filters.

The ELF issue has attracted the attention of fear mongers and quacks, as well as the interest of legitimate scientists. It is best to avoid blowing it out of proportion and not to succumb to unsubstantiated media hype. If you are concerned about ELF radiation in your home or work environment, consult someone whose word you can trust, such as a computer hardware engineer or a wireless communications engineer.

Rf Waves

An EM disturbance is called a *radio-frequency (rf) wave* if its wavelength falls within the range of 100 km to 1 mm. This is a frequency range of 3 kHz to 3000 GHz.

FORMAL RF BAND DESIGNATORS

The rf spectrum is split into eight *bands,* each representing one order of magnitude in terms of frequency and wavelength. These bands are called *very low, low, medium, high, very high, ultrahigh, superhigh,* and *extremely high frequencies.* They are abbreviated, respectively, as VLF, LF, MF, HF, VHF, UHF, SHF, and EHF. These are depicted in Table 18-1 in terms of the frequency and the free-space wavelength.

These bands have alternative names. Energy at VLF and LF is sometimes called *longwave radio* or *long waves.* Energy in the HF range is sometimes called *shortwave radio* or *short waves* (even though the waves aren't short compared with most EM waves in wireless communications used today). Superhigh-frequency and extremely-high-frequency rf waves are sometimes called *microwaves.*

Table 18-1   The Bands in the Radio-frequency (rf) Spectrum. Each band spans one mathematical order of magnitude in terms of frequency and wavelength.

Designator	Frequency	Wavelength
Very low frequency (VLF)	3–30 kHz	100–10 km
Low frequency (LF)	30–300 kHz	10–1 km
Medium frequency (MF)	300 kHz–3 MHz	1 km–100 m
High frequency (HF)	3–30 MHz	100–10 m
Very high frequency (VHF)	30–300 MHz	10–1 m
Ultrahigh frequency (UHF)	300 MHz–3 GHz	1 m–100 mm
Superhigh frequency (SHF)	3–30 GHz	100–10 mm
Extremely high frequency (EHF)	30–300 GHz	10–1 mm

Radio-frequency waves propagate through the Earth's atmosphere and through space in various ways, depending on the wavelength. Some waves are affected by the ionosphere; this is especially true at VLF, LF, MF, and HF. The troposphere can bend, reflect, or scatter waves at VHF, UHF, SHF, and EHF.

EARTH'S IONOSPHERE

The atmosphere of our planet becomes less dense with increasing altitude. Because of this, the energy received from the Sun is much greater at high altitudes than it is at the surface. High-speed subatomic particles, UV rays, and x-rays cause ionization of the rarefied gases in the upper atmosphere. Ionized regions occur at specific altitudes and comprise the *ionosphere*. The ionosphere causes absorption and refraction of radio waves. This makes long-distance communication or reception possible at some radiofrequencies.

Ionization in the upper atmosphere occurs in four fuzzy layers. The lowest region is called the *D layer*. It exists at an altitude of about 50 km (30 mi) and ordinarily is present only on the daylight side of the planet. This layer does not contribute to long-distance radio communications, and sometimes impedes them. The *E layer*, about 80 km (50 mi) above the surface, also exists mainly during the day, although nighttime ionization is sometimes observed. The E layer can facilitate medium-range radio communications at certain frequencies. The uppermost layers are called the *F1 layer* and the *F2 layer*. The F1 layer, normally present only on the daylight side of the Earth, forms at about 200 km (125 mi) altitude. The F2 layer, which exists more or less around the clock, is about 300 km (180 mi) above the surface. On the dark side of the Earth, when the F1 layer disappears, the F2 layer is sometimes called simply the *F layer.*

Figure 18-3 illustrates the relative altitudes of the ionospheric D, E, F1, and F2 layers above the Earth's surface. All these layers have some effect on the way radio waves travel at very low, low, medium, and high frequencies. Sometimes, ionospheric effects can even be observed into the VHF portion of the radio spectrum. These layers not only make long-distance wireless communications possible between points on the Earth's surface; they also prevent radio waves at frequencies below approximately 5 MHz from reaching the surface from outer space.

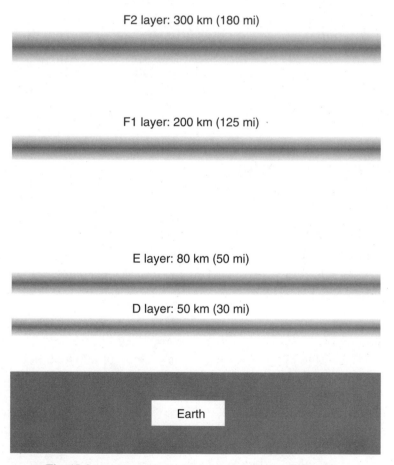

F2 layer: 300 km (180 mi)

F1 layer: 200 km (125 mi)

E layer: 80 km (50 mi)

D layer: 50 km (30 mi)

Earth

Fig. 18-3. Layers of the Earth's ionosphere. These ionized regions affect the behavior of EM waves at some radio frequencies.

SOLAR ACTIVITY

The number of sunspots is not constant but changes from year to year. The variation is periodic and dramatic. This fluctuation of sunspot numbers is called the *sunspot cycle*. It has a period of approximately 11 years. The rise in the number of sunspots is generally more rapid than the decline, and the maximum and minimum sunspot counts vary from cycle to cycle.

The sunspot cycle affects propagation conditions at frequencies up to about 70 MHz for F1- and F2-layer propagation and 150 to 200 MHz for E-layer propagation. When there are not many sunspots, the *maximum usable frequency* (MUF) is comparatively low because the ionization of the upper atmosphere is not dense. At or near the time of a sunspot peak, the MUF is higher because the upper atmosphere is more ionized.

A *solar flare* is a violent storm on the surface of the Sun. Solar flares cause an increase in the level of radio noise that comes from the Sun and cause the Sun to emit an increased quantity of high-speed subatomic particles. These particles travel through space and arrive at the Earth a few hours after the first appearance of the flare. Because the particles are electrically charged, they are accelerated by the Earth's magnetic field. Sometimes a *geomagnetic storm* results. Then we see the "northern lights" or "southern lights" (*aurora borealis* or *aurora australis,* often called simply the *aurora*) at high latitudes during the night and experience a sudden deterioration of ionospheric radio-propagation conditions. At some frequencies, communications can be cut off within seconds. Even wire communications circuits are sometimes affected.

Solar flares can occur at any time, but they seem to take place most often near the peak of the 11-year sunspot cycle. Scientists do not know exactly what causes solar flares, but the events seem to be correlated with the relative number of sunspots.

GROUND-WAVE PROPAGATION

In radio communication, the *ground wave* consists of three distinct components: the *direct wave* (also called the *line-of-sight wave*), the *reflected wave,* and the *surface wave.* The direct wave travels in a straight line. It plays a significant role only when the transmitting and receiving antennas are connected by a straight geometric line entirely above the Earth's surface. At most radiofrequencies, EM fields pass through objects such as trees and frame houses with little attenuation. Concrete-and-steel structures cause some loss in the direct wave at higher frequencies. Earth barriers such as hills and mountains block the direct wave.

A radio signal can be reflected from the Earth or from certain structures such as concrete-and-steel buildings. The reflected wave combines with the direct wave (if any) at the receiving antenna. Sometimes the two are exactly out of phase, in which case the received signal is weak even if the transmitter and receiver lie along a direct line of sight. This effect occurs mostly at frequencies above 30 MHz (wavelengths less than 10 m).

The surface wave travels in contact with the Earth, and the Earth forms part of the circuit. This happens only with vertically polarized EM fields (those in which the electrical flux lines are vertical) at frequencies below about 15 MHz. Above 15 MHz, there is essentially no surface wave. At frequencies from about 9 kHz up to 300 kHz, the surface wave propagates for hundreds or even thousands of kilometers. Sometimes the surface wave is called the *ground wave,* but technically this is a misnomer.

SPORADIC-E PROPAGATION

At certain radio-frequencies, the ionospheric E layer occasionally returns signals to the Earth. This effect is intermittent, and conditions can change rapidly. For this reason, it is known as *sporadic-E propagation.* It is most likely to occur at frequencies between approximately 20 and 150 MHz. Occasionally it is observed at frequencies as high as 200 MHz. The propagation range is on the order of several hundred kilometers, but occasionally communication is observed over distances of 1,000 to 2,000 km (600 to 1,200 mi).

The standard FM broadcast band is sometimes affected by sporadic-E propagation. The same is true of the lowest television (TV) broadcast channels, especially channels 2 and 3. Sporadic-E propagation is sometimes mistaken for effects that take place in the lower atmosphere independently of the ionosphere.

AURORAL AND METEOR-SCATTER PROPAGATION

In the presence of unusual solar activity, the aurora often reflect radio waves at some frequencies. This is called *auroral propagation.* The aurorae occur in the ionosphere at altitudes of 25 km (40 mi) to 400 km (250 mi) above the surface. Theoretically, auroral propagation is possible, when the aurorae are active, between any two points on the Earth's surface from which the same part of the aurora lies on a line of sight. Auroral propagation seldom occurs when either the transmitter or the receiver is at a latitude less than 35 degrees north or south of the equator. Auroral propagation can take place at frequencies well above 30 MHz and often is accompanied by deterioration in ionospheric propagation via the E and F layers.

When a meteor from space enters the upper part of the atmosphere, an ionized trail is produced because of the heat of friction. Such an ionized region reflects EM energy at certain wavelengths. This phenomenon, known as *meteor-scatter propagation,* can result in over-the-horizon radio communication or reception.

A meteor produces a trail that persists for a few tenths of a second up to several seconds depending on the size of the meteor, its speed, and the angle at which it enters the atmosphere. This amount of time is not sufficient for the transmission of very much information, but during a *meteor shower,* ionization can be almost continuous. Meteor-scatter propagation has been observed at frequencies considerably above 30 MHz and occurs over distances ranging from just beyond the horizon up to about 2,000 km (1,200 mi) depending on the altitude of the ionized trail and on the relative positions of the trail, the transmitting station, and the receiving station.

TROPOSPHERIC BENDING

The lowest 13 to 20 km (8 to 12 mi) of the Earth's atmosphere comprise the *troposphere.* This region has an effect on radio-wave propagation at certain frequencies. At wavelengths shorter than about 15 m (frequencies above 20 MHz), refraction and reflection can take place within and between air masses of different density. The air also produces some scattering of EM energy at wavelengths shorter than about 3 m (frequencies above 100 MHz). All these effects generally are known as *tropospheric propagation,* which can result in communication over distances of hundreds of kilometers.

A common type of tropospheric propagation takes place when radio waves are refracted in the lower atmosphere. This is most dramatic near weather fronts, where warm, relatively light air lies above cool, more dense air. The cooler air has a higher index of refraction than the warm air, causing EM fields to be bent downward at a considerable distance from the transmitter. This is *tropospheric bending.* It is often responsible for anomalies in reception of FM and TV broadcast signals.

TROPOSCATTER

At frequencies above about 100 MHz, the atmosphere has a scattering effect on radio waves. The scattering allows over-the-horizon communication at VHF, UHF, and microwave frequencies. This is called *tropospheric scatter,* or *troposcatter.* Dust and clouds in the air increase the scattering effect, but some troposcatter occurs regardless of the weather. Troposcatter takes place mostly at low altitudes where the air is the most dense. Some effects occur at altitudes up to about 16 km (10 mi). Troposcatter can provide reliable communication over distances of several hundred kilometers when the appropriate equipment is used.

Figure 18-4 shows tropospheric scatter and bending. The transmitting station is at the lower left. There is a temperature inversion in this example; it exaggerates the bending. If the boundary between the cool air near the surface and the warm air above is well defined enough, reflection can occur in addition to the bending. If the inversion covers a large geographic area, signals can bounce repeatedly between the inversion boundary and the surface, providing exceptional long-range communication, especially if the surface is salt water.

DUCT EFFECT

The *duct effect* is a form of tropospheric propagation that takes place at approximately the same frequencies as bending and scattering. Also called

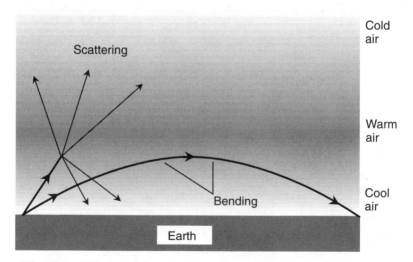

Fig. 18-4.   The troposphere can bend and scatter radio waves at some frequencies.

ducting, this form of propagation is most common very close to the surface, sometimes at altitudes of less than 300 m.

A *duct* forms when a layer of cool air becomes sandwiched between two layers of warmer air. This is common along and near weather fronts in the temperate latitudes. It also takes place frequently above water surfaces during the daylight hours and over land surfaces at night. Radio waves can be trapped inside the region of cooler air, in much the same way that light waves are trapped inside an optical fiber. Ducting often allows over-the-horizon communication of exceptional quality over distances of hundreds of kilometers at VHF and UHF.

PROBLEM 18-3
Suppose that you are using a handheld radio transceiver to talk with someone across town. You stand on a hill and can see the house where the other person is located, and the two of you are well within the quoted communications range for the radios. Yet the signal is extremely weak. You move over a few meters, and the signal gets strong. What might cause this?

SOLUTION 18-3
The direct wave and the reflected wave from the other radio's antenna happen to arrive out of phase at your antenna, so they almost cancel each other out. Moving a few meters remedies this, and the signal becomes strong.

Beyond the Radio Spectrum

The shortest rf waves measure approximately 1 mm; this corresponds to a frequency of 300 GHz. As the wavelength becomes shorter than this, we encounter the IR, visible, UV, x-ray, and gamma-ray spectra in that order.

INFRARED

The longest IR waves are approximately 1 mm in length; the reddest visible light has a wavelength of a little less than 0.001 mm. This is a span of a thousandfold, or three mathematical orders of magnitude. In terms of frequency, the IR spectrum lies below the visible red spectrum, and it is from this fact that it gets its name (*infra-* means "below"). Our bodies sense IR radiation as warmth or heat. The IR rays are not literally heat, but they produce heat when they strike an absorptive object such as the human body.

The Sun is a brilliant source of IR; it emits just about as much IR as visible light. Other sources of IR include incandescent light bulbs, fire, and electrical heating elements. If you have an electric stove and switch on one of the burners to low, you can feel the IR radiation from it even though the element appears black to the eye.

Infrared radiation can be detected by special films that can be used in most ordinary cameras. Some high-end photographic cameras have focus numbers for IR as well as for visible light printed on their lens controls. Glass transmits IR at the shorter wavelengths (*near IR*) but blocks IR at the longer wavelengths (*far IR*). When you take an IR photograph in visible-light darkness, warm objects show up clearly. This is the principle by which some night-vision apparatus works. Infrared-detecting equipment has been used recently in wartime to detect the presence and movement of personnel.

The fact that glass transmits near IR but blocks far IR is responsible for the ability of glass greenhouses to maintain interior temperatures much higher than that of the external environment. It is also responsible for the extreme heating of automobile interiors on sunny days when the windows are closed. This effect can be used to advantage in energy-efficient homes and office buildings. Large windows with southern exposures can be equipped with blinds that are opened on sunny winter days and closed in cloudy weather and at night.

IR radiation at low and moderate levels is not dangerous and in fact has been used therapeutically to help relieve the discomfort of joint injuries and muscle strains. At high intensity, however, IR radiation can cause burns. In massive structural or forest fires, this radiation can scorch the clothing off a person and then literally cook the body alive. The most extreme earthly IR radiation is produced by the detonation of a nuclear bomb or by an asteroid impact. The IR burst from a 20-megaton weapon (equivalent to 2×10^7 tons of conventional explosive) can kill every exposed living organism within a radius of several kilometers.

In some portions of the IR spectrum, the atmosphere of our planet is opaque. In the near IR between about 770 nm (the visible red) and 2,000 nm, our atmosphere is reasonably clear. Water vapor causes attenuation in the IR between the wavelengths of approximately 4,500 and 8,000 nm. Carbon dioxide (CO_2) gas interferes with the transmission of IR at wavelengths ranging from about 14,000 to 16,000 nm. Rain, snow, fog, and dust interfere with the propagation of IR. The presence of CO_2 in the atmosphere keeps the surface warmer than it would be if there were less CO_2. Most scientists agree that increasing CO_2 in the atmosphere will produce a significant rise in the average surface temperature. This *greenhouse effect* gets its name from the fact that the CO_2 in the Earth's atmosphere treats IR in much the same way as the glass in a greenhouse.

VISIBLE LIGHT

The visible portion of the EM spectrum lies within the wavelength range of 770 to 390 nm. Emissions at the longest wavelengths appear red; as the wavelength decreases, we see orange, yellow, green, blue, indigo, and violet in that order.

Visible light is transmitted fairly well through the atmosphere at all wavelengths. Scattering increases toward the blue, indigo, and violet end of the spectrum. This is why the sky appears blue during the daytime. Long-wavelength light is scattered the least; this is why the Sun often appears red or orange when it is on the horizon. Red is the preferred color for terrestrial line-of-sight laser communications systems for this same reason. Rain, snow, fog, smoke, and dust interfere with the transmission of visible light through the air. We'll take a detailed look at the characteristics and behavior of visible light in the next chapter.

ULTRAVIOLET

As the wavelength of an EM disturbance becomes shorter than the smallest we can see, the energy contained in each individual photon increases. The UV range of wavelengths starts at about 390 nm and extends down to about 1 nm. At a wavelength of approximately 290 nm, the atmosphere becomes highly absorptive, and at wavelengths shorter than this, the air is essentially opaque. This protects the environment against damaging ultraviolet radiation from the Sun. *Ozone* (molecules consisting of three oxygen atoms) in the upper atmosphere is primarily responsible for this effect. Ozone pollution, prevalent in large cities during the summer months, further attenuates UV.

Ordinary glass is virtually opaque at UV, so cameras with glass lenses cannot be used to take photographs in this part of the spectrum. Instead, a pinhole-type device is used, and this severely limits the amount of energy that passes into the detector. While a camera lens has a diameter of several millimeters or centimeters, a pinhole is less than a millimeter across. Another type of device that can be used to sense UV and to measure its intensity at various wavelengths is the *spectrophotometer.* A diffraction grating is used to disperse EM energy into its constituent wavelengths from IR through the visible and into the UV range. By moving the sensing device back and forth, any desired wavelength can be singled out for analysis.

The principle of operation of the spectrophotometer is shown Fig. 18-5. At the extremely short-wavelength end of the UV spectrum (*hard UV*), radiation counters are sometimes used, similar to the apparatus employed for the detection of x-rays and gamma rays. For photographic purposes, ordinary camera film will work at the longer UV wavelengths (*soft UV*). A special film, rather like x-ray film, is necessary to make hard-UV photos.

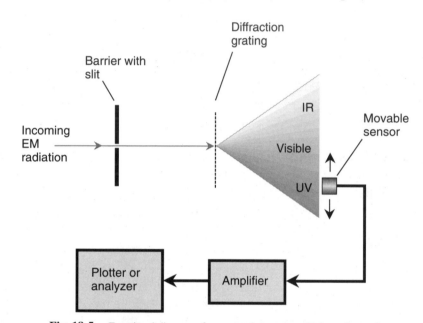

Fig. 18-5. Functional diagram of a spectrophotometer, which can be used to sense and measure EM radiation at IR, visible, and UV wavelengths.

UV rays possess an interesting property that can be observed using a so-called black light. Most hobby shops sell lamps of this sort. They are cylindrical in shape, and superficially, they can be mistaken for small fluorescent tubes. (The incandescent black light bulbs sold in department stores are not especially good sources of UV.) When subjected to UV, certain substances glow brightly in the visible range. This is known as *fluorescence.* Art stores sell acrylic paints that are specially tailored to glow in various colors when UV strikes them. The effect in a darkened room can be striking. The phosphor coatings on CRTs fluoresce under UV too. So will certain living organisms, such as scorpions. If you live in the desert, go outside some night with a black light and switch it on. If there are scorpions around, you'll find out.

Most of the radiation from the Sun occurs in the IR and visible parts of the EM spectrum. If the Sun were a much hotter star, producing more energy in the UV range, life on any earthlike planet in its system would have developed in a different way, if at all. Excessive exposure to UV, even in the relatively small amounts that reach the Earth's surface on bright days, can, over time, cause skin cancer and eye cataracts. There is evidence to suggest that excessive exposure to UV suppresses the activity of the immune system, rendering people and animals more susceptible to infectious diseases. Some scientists believe that the *ozone hole* in the upper atmosphere, prevalent in the southern hemisphere, is growing because of increased production and emission of certain chemicals by humankind. If this is the case, and if the problem worsens, we should expect that it will affect the evolution of life on this planet.

X-RAYS

The x-ray spectrum consists of EM energy at wavelengths from approximately 1 nm down to 0.01 nm. (Various sources disagree somewhat on the dividing line between the hard-UV and x-ray regions.) Proportionately, the x-ray spectrum is large compared with the visible range.

X-rays were discovered accidentally in 1895 by a physicist named Wilhelm Roentgen during experiments involving electric currents in gases at low pressure. If the current was sufficiently intense, the high-speed electrons produced mysterious radiation when they struck the anode (positively charged electrode) in the tube. The rays were called x-rays because of their behavior, never before witnessed. The rays were able to penetrate barriers opaque to visible light and UV. A phosphor-coated object happened to be in the vicinity of the tube containing the gas, and Roentgen noticed that the phosphor glowed. Subsequent experiments showed that the rays possessed so much penetrating power that they passed through the skin and muscles in the human hand, casting shadows of the bones on a phosphor-coated surface. Photographic film could be exposed in the same way.

Modern x-ray tubes operate by accelerating electrons to high speed and forcing them to strike a heavy metal anode (usually made of tungsten). A simplified functional diagram of an x-ray tube of the sort used by dentists to locate cavities in your teeth is shown in Fig. 18-6.

As the wavelengths of x-rays become shorter and shorter, it becomes increasingly difficult to direct and focus them. This is so because of the penetrating power of the short-wavelength rays. A piece of paper with a tiny hole works very well for UV photography; in the x-ray spectrum, the radiation passes right through the paper. Even aluminum foil is relatively

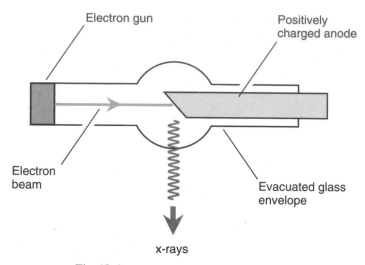

Fig. 18-6. Functional diagram of an x-ray tube.

transparent to x-rays. However, if x-rays land on a reflecting surface at a nearly grazing angle, and if the reflecting surface is made of suitable material, some degree of focusing can be realized. The shorter the wavelength of the x-rays, the smaller the angle of incidence, measured relative to the surface (not the normal), must be if reflection is to take place. At the shortest x-ray wavelengths, the angle must be smaller than 1° of arc. This grazing reflection effect is shown in Fig. 18-7a. A rough illustration of how a high-resolution x-ray observing device achieves its focusing is shown in part b. The focusing mirror is tapered in the shape of an elongated paraboloid. As parallel x-rays enter the aperture of the reflector, they strike its inner surface at a grazing angle. The x-rays are brought to a focal point, where a radiation counter or detector can be placed.

X-rays cause ionization of living tissue. This effect is cumulative and can result in damage to cells over a period of years. This is why x-ray technicians in doctors' and dentists' offices work behind a barrier lined with lead. Otherwise, these personnel would be subjected to dangerous cumulative doses of x radiation. It only takes a few millimeters of lead to block virtually all x-rays. Less dense metals and other solids also can block x-rays, but these must be thicker. The important factor is the amount of mass through which the radiation must pass. Sheer physical displacement also can reduce the intensity of x radiation, which diminishes according to the square of the distance. However, it isn't practical for most doctors or dentists to work in offices large enough to make this a viable alternative.

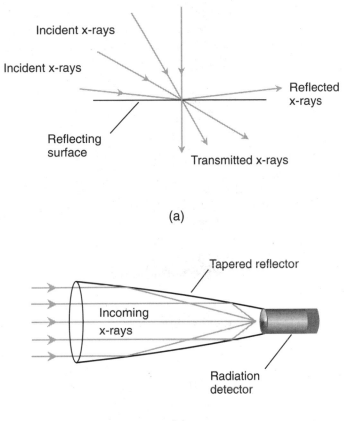

(a)

(b)

Fig. 18-7. (*a*) x-rays are reflected from a surface only when they strike at a grazing angle. (*b*) A functional diagram of an x-ray focusing and observing device.

GAMMA (γ) RAYS

As the wavelength of EM rays becomes shorter and shorter, their penetrating power increases until focusing is impossible. The cutoff point where the x-ray region ends and the gamma-ray region begins is approximately 0.01 nm (10^{-11} m). Gamma rays can, in theory, get shorter than this without limit. The gamma classification represents the most energetic of all EM fields. Short-wavelength gamma rays can penetrate several centimeters of solid lead or more than a meter of concrete. They are even more damaging to living tissue than x-rays. Gamma rays come from radioactive materials, both natural (such as radon) and human-made (such as plutonium).

Radiation counters are the primary means of detecting and observing sources of gamma rays. Gamma rays can dislodge particles from the nuclei of atoms they strike. These subatomic particles can be detected by a counter. One type of radiation counter consists of a thin wire strung within a sealed cylindrical metal tube filled with alcohol vapor and argon gas. When a high-speed subatomic particle enters the tube, the gas is ionized for a moment. A voltage is applied between the wire and the outer cylinder so that a pulse of electric current occurs whenever the gas is ionized. Such a pulse produces a click in the output of an amplifier connected to the device.

A simplified diagram of a radiation counter is shown in Fig. 18-8. A glass window with a metal sliding door is cut in the cylinder. The door can be opened to let in particles of low energy and closed to allow only the most energetic particles to get inside. High-speed subatomic particles, which are tiny yet massive for their size, have no trouble penetrating the window glass if they are moving fast enough. When the door is closed, gamma rays can penetrate it with ease.

COSMIC PARTICLES

If you sit in a room with no radioactive materials present and switch on a radiation counter with the window of the tube closed, you'll notice an occasional click from the device. Some of the particles come from the Earth; there are radioactive elements in the ground almost everywhere (usually in

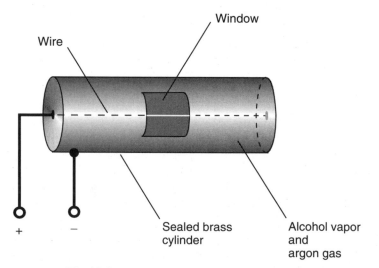

Fig. 18-8.   Simplified diagram of a radiation counter.

small quantities). Some of the radiation comes indirectly from space. Cosmic particles strike atoms in the atmosphere, and these atoms in turn eject other subatomic particles that arrive at the counter tube.

In the early 1900s, physicists noticed radiation apparently coming from space. They found that the strange background radiation increased in intensity when observations were made at high altitude; the radiation level decreased when observations were taken from underground or underwater. This space radiation has been called *secondary cosmic radiation* or *secondary cosmic particles*. The actual particles from space, called *primary cosmic particles,* usually do not penetrate far into the atmosphere before they collide with and break up the nuclei of atoms. To observe primary cosmic particles, it is necessary to ascend to great heights, and as with the UV and x-ray investigations, this was not possible until the advent of the space rocket.

While the radiation in the EM spectrum—the radio waves, IR, visible light, UV, x-rays, and gamma rays—consists of photons traveling at the speed of light, primary cosmic particles are comprised of matter traveling at almost, but not quite, the speed of light. At such high speeds, the protons, neutrons, and other heavy particles gain mass because of relativistic effects, and this renders them almost immune to the effects of the Earth's magnetosphere. Such particles arriving in the upper atmosphere come to us in nearly perfect straight-line paths despite the magnetic field of our planet. By carefully observing the trails of the particles in a device called a *cloud chamber* aboard a low-orbiting space ship, it is possible to ascertain the direction from which they have come. Over time, cosmic-particle maps of the heavens can be generated and compared with maps at various EM wavelengths.

PROBLEM 18-4
What is the energy contained in each photon of a barrage of gamma rays whose wavelength is 0.00100 nm?

SOLUTION 18-4
Recall the formula from Chapter 17 for energy in terms of wavelength λ the speed of EM propagation in free space c, and Planck's constant h:

$$e = hc/\lambda$$

For EM rays in free space, the product hc is approximately equal to 1.9865 $\times 10^{-25}$. (You can go back and check the text in Chapter 17 if you've forgotten how this is derived.) The wavelength 0.00100 nm is equivalent to 1.00 $\times 10^{-12}$ m. Therefore, the energy e, in joules, contained in each photon of the gamma ray is

$$e = (1.9865 \times 10^{-25})/(1.00 \times 10^{-12})$$

$$= 1.99 \times 10^{-13} \text{ J}$$

Radioactivity

The nuclei of most familiar substances are stable. They retain their identities and remain unchanged indefinitely. However, some atomic nuclei change with time; they are unstable. As unstable atomic nuclei disintegrate, they emit high-energy photons and various subatomic particles. *Radioactivity* is a general term that refers to any of these types of radiation arising from the disintegration of unstable atoms.

FORMS

Radioactivity, also called *ionizing radiation* because it can strip electrons from atoms, occurs in various forms. The most common are gamma rays (which we've already discussed), *alpha* (α) particles, *beta* (β) *particles,* and *neutrinos.* There are also some less common forms, such as high-speed protons and antiprotons, neutrons and antineutrons, and the nuclei of atoms heavier than helium.

Alpha particles are helium-4 (^{4}He) nuclei traveling at high speeds. A ^{4}He nucleus consists of two protons and two neutrons. An alpha particle has a positive electric charge because there are no negatively charged electrons surrounding it. As such, all alpha particles are ions. They have significant mass, so if they attain high enough speeds, they can acquire considerable kinetic energy. An alpha particle traveling at a sizable fraction of the speed of light (known as *relativistic speed*) attains an increased mass because of relativistic effects; this gives it additional kinetic energy. You'll learn about relativistic mass increase and other effects in Chapter 20. Most alpha particles can be blocked by modest barriers.

Beta particles are high-speed electrons or positrons. (Remember that a positron is the antimatter counterpart of the electron.) Any beta particle consisting of an electron, also called a *negatron* because it has a negative electric charge, is denoted β^-, and any β particle consisting of a positron, which carries a positive charge, is denoted β^+. All beta particles have nonzero *rest mass* (their mass when not moving at relativistic speed). Their kinetic energies are increased by relativistic effects if they move at near-light speeds.

Neutrinos are an entirely different sort of particle. They have no electric charge and no rest mass. They have tremendous penetrating power. The Earth is constantly being bombarded by neutrinos from space. These neutrinos have their origins in the cores of the Sun and distant stars. Most neutrinos pass through the entire planet unaffected. Sophisticated equipment is required to detect them. Neutrino detectors are placed far underground to block all other forms of radiation so that scientists can be sure the equipment is really detecting neutrinos and not stray particles of some other sort. The neutrino has a counterpart, known as the *antineutrino.*

NATURAL SOURCES

In nature, radioactivity is produced by certain isotopes of elements with atomic numbers up to and including 92 (uranium). These are known as *radioactive isotopes.* An isotope of carbon, known as *carbon-14* (^{14}C), has eight neutrons. Atoms of ^{14}C are unstable; over time, they decay into carbon-12 (^{12}C) atoms, which have six neutrons. Other examples of an unstable atoms include hydrogen-3 (^{3}II), also known as *tritium,* which has a nucleus consisting of one proton and two neutrons; beryllium-7 (^{7}Be), with a nucleus containing four protons and three neutrons; and ^{10}Be, with a nucleus containing four protons and six neutrons.

In some instances, the most common isotope of a naturally occurring element also happens to be radioactive. Examples are radon, radium, and uranium. The barrage of cosmic particles from deep space can be considered a form of radioactivity, but these particles sometimes can create radioactive isotopes when they strike stable atoms in the Earth's upper atmosphere.

HUMAN-MADE SOURCES

Radioactivity is produced by a variety of human activities. The most well known in the early years of atomic research was the *fission bomb.* Its modern descendant is the vastly more powerful *hydrogen fusion bomb.* Such a weapon, when detonated, produces an immediate, intense burst of ionizing radiation. The high-speed subatomic particles produced in the initial blast, especially if the explosion occurs at or near the ground, cause large amounts of material to become radioactive. The resulting radioactive dust, called *fallout,* precipitates back to Earth over a period of time. Some fallout, especially from the largest nuclear bombs, can rise high into the troposphere and enter the jet streams, where it is carried around the globe.

Nuclear *fission reactors* contain radioactive elements. The heat from the decay of these elements is used to generate electrical power. Some by-products of fission are radioactive, and because they cannot be reused to generate more power, they represent *radioactive waste.* Disposal of this waste is a problem because it takes many years, even centuries, to decay. If a *fusion reactor* is ever developed and put into use, it will be a vast improvement over the fission reactor because controlled hydrogen fusion produces no radioactive waste.

Radioactive isotopes can be produced by bombarding the atoms of certain elements with high-speed subatomic particles or energetic gamma rays. Charged particles are accelerated to relativistic speeds by *particle accelerators,* also known informally as *atom smashers.* A *linear particle accelerator* is a long, evacuated tube that employs a high voltage to accelerate particles such as protons, alpha particles, and electrons to speeds so great that they can alter or split certain atomic nuclei that they strike. A *cyclotron* is a large ring-shaped chamber that uses alternating magnetic fields to accelerate the particles to relativistic speeds.

DECAY AND HALF-LIFE

Radioactive substances gradually lose "potency" as time passes. Unstable nuclei degenerate one by one. Sometimes an unstable nucleus decays into a stable one in a single event. In other cases, an unstable nucleus changes into another unstable nucleus, which later degenerates into a stable one. Suppose that you have an extremely large number of radioactive nuclei, and you measure the length of time required for each one to degenerate and then average all the results. The average decay time is called the *mean life* and is symbolized by the lowercase Greek letter tau (τ).

Some radioactive materials give off more than one form of emission. For any given ionizing radiation form (alpha particles, beta particles, gamma rays, or other), there is a separate *decay curve,* or function of intensity versus time. A radioactive decay curve always has a characteristic shape: It starts out at a certain value and tapers down toward zero. Some decay curves decrease rapidly, and others decrease slowly, but the characteristic shape is always the same and can be defined in terms of a time span known as the *half-life,* symbolized $t_{1/2}$.

Suppose that the intensity of radiation of a particular sort is measured at time t_0. After a period of time $t_{1/2}$ has passed, the intensity of that form of radiation decreases to half the level it was at t_0. After the half-life passes again (total elapsed time $2t_{1/2}$), the intensity goes down to one-quarter of its original value. After yet another half-life passes (total elapsed time $3t_{1/2}$), the intensity goes down to one-eighth its original value. In general, after n half-lives pass from the initial time t_0 (total elapsed time $nt_{1/2}$), the intensity goes down to $1/(2^n)$, or 0.5^n, times its original value. If the original intensity is x_0 units and the final intensity is x_f units, then

$$x_f = 0.5^n x_0$$

The general form of a radioactive decay curve is shown in Fig. 18-9. The half-life $t_{1/2}$ can vary tremendously depending on the particular radioactive substance involved. Sometimes $t_{1/2}$ is a tiny fraction of 1 second; in other cases it is millions of years. For each type of radiation emitted by a material, there is a separate value of $t_{1/2}$ and therefore a separate decay curve.

Another way to define radioactive decay is in terms of a number called the *decay constant,* symbolized by the lowercase Greek letter lambda (λ). The decay constant is equal to the natural logarithm of 2 (approximately 0.69315) divided by the half-life in seconds. This is expressed as follows:

$$\lambda = 0.69315/t_{1/2}$$

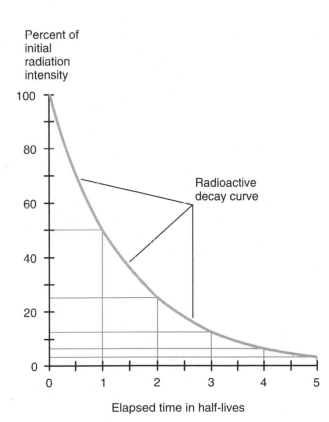

Fig. 18-9.   General form of a radioactive decay curve.

The symbol for the radioactive decay constant happens to be the same as the symbol for EM wavelength. Don't confuse them; they are entirely different and independent quantities. Also, when determining the decay constant, be sure that $t_{1/2}$ is expressed in seconds. This will ensure that the decay constant is expressed in the proper units (s^{-1}). If you start with $t_{1/2}$ expressed in units other than seconds, you'll get a decay constant that is the wrong number because it is expressed inappropriately.

The decay constant is the reciprocal of the mean life in seconds. Therefore, we can state these equations:

$$\lambda = 1/\tau \quad \text{and} \quad \tau = 1/\lambda$$

From these equations we can see that the mean life τ is related to the half-life $t_{1/2}$ as follows:

$$\tau = t_{1/2}/0.69315$$

$$= 1.4427t_{1/2}$$

and

$$t_{1/2} = 0.69315\tau$$

UNITS AND EFFECTS

There are several different units employed to define overall radiation exposure. The unit of radiation in the International System of units is the *becquerel* (Bq), representing one nuclear transition per second ($1\ s^{-1}$). Exposure to radiation is measured according to the amount necessary to produce a coulomb of electric charge, in the form of ions, in a kilogram of pure dry air. The SI unit for this quantity is the *coulomb per kilogram* (C/kg). An older unit, known as the *roentgen* (R), is equivalent to 2.58×10^{-4} C/kg.

When matter such as human tissue is exposed to radiation, the standard unit of *dose equivalent* is the *sievert* (Sv), equivalent to 1 joule per kilogram (1 J/kg). Sometimes you'll hear about the *rem* (an acronym for roentgen equivalent man); 1 rem = 0.01 Sv.

All these units make it confusing to talk about radiation quantity. To make things worse, some of the older, technically obsolete units such as the roentgen and rem have stuck around, especially in laypeople's conversations about radiation, whereas the standard units have been slow to gain acceptance. Have you read that "more than 100 roentgens of exposure to ionizing radiation within a few hours will make a person sick" or that "people are typically exposed to a few rems during a lifetime"? Statements like these were common in civil-defense documents in the 1960s after the Cuban missile crisis, when fears of worldwide nuclear war led to the installation of air-raid sirens and fallout shelters all over the United States.

When people are exposed to excessive amounts of radiation in a short time, physical symptoms such as nausea, skin burns, fatigue, and dehydration commonly occur. In extreme cases, internal ulceration and bleeding lead to death. When people get too much radiation gradually over a period of years, cancer rates increase, and genetic mutations also occur, giving rise to increased incidence of birth defects.

PRACTICAL USES

Radioactivity has numerous constructive applications in science, industry, and medicine. The most well known is the nuclear fission reactor, which was popular during the middle to late 1900s for generating electricity on a large scale. This type of power plant has fallen into disfavor because of the dangerous waste products it produces.

Radioactive isotopes are used in medicine to aid doctors in diagnosing illness, locating tumors inside the body, measuring rates of metabolism, and examining the structure of internal organs. Controlled doses of radiation are sometimes used in an attempt to destroy cancerous growths. In industry, radiation can be used to measure the dimensions of thin sheets of metal or plastic, to destroy bacteria and viruses that might contaminate food and other matter consumed or handled by people, and to x-ray airline baggage. Other applications include the irradiation of food, freight, and mail to protect the public against the danger of biological attack.

Geologists and biologists use *radioactive dating* to estimate the ages of fossil samples and archeological artifacts. The element most commonly used for this purpose is carbon. When a sample is created or a specimen is alive, there is believed to exist a certain proportion of ^{14}C atoms among the total carbon atoms. These gradually decay into ^{12}C atoms. By measuring the radiation intensity and determining the proportion of ^{14}C in archeological samples, anthropologists can get an idea of how long ago the world's great civilizations came into being, thrived, and declined. Climatologists used the technique to discover that the Earth has gone through cycles of generalized global warming and cooling.

Carbon dating has revealed that the dinosaurs suddenly and almost completely disappeared in a short span of time around 65 million years ago. By a process of elimination, it was determined that a large meteorite or comet splashed down in the Gulf of Mexico at that time. The Earth's climate cooled off for years because of debris injected into the atmosphere following the impact that blocked much of the solar IR that normally reaches the surface. Further research has shown that there have been several major *bolide impacts* in the distant past, each of which has radically altered the evolutionary course of life on Earth. Based on this knowledge, most scientists agree that it is only a matter of time before another such event takes place. When—not if—it does, the consequences for humanity will be of biblical proportions.

PROBLEM 18-5

Suppose that the half-life of a certain radioactive substance is 100 days. You measure the radiation intensity and find it to be x_0 units. What will the intensity x_{365} be after 365 days?

SOLUTION 18-5

To determine this, use the equation presented earlier:

$$x_{365} = 0.5^n x_0$$

where *n* is the number of half-lives elapsed. In this case, $n = 365/100 = 3.65$. Therefore:

$$x_{365} = 0.5^{3.65} x_0$$

To determine the value of $0.5^{3.65}$, use a calculator with an x^y function. This yields the following result to three significant digits:

$$x_{365} = 0.0797 x_0$$

PROBLEM 18-6

What is the decay constant of the substance described in Problem 18-5?

SOLUTION 18-6

Use the preceding formula for decay constant λ in terms of half-life $t_{1/2}$. In this case, $t_{1/2} = 100$ days. This must be converted to seconds to get the proper result for the decay constant. There are $24 \times 60 \times 60 = 8.64 \times 10^4$ s in one day. Thus $t_{1/2} = 8.64 \times 10^6$ s, and

$$\lambda = 0.69315/(8.64 \times 10^6)$$

$$= 8.02 \times 10^{-8} \text{ s}^{-1}$$

PROBLEM 18-7

What is the mean life of the substance described in Problem 18-5? Express the answer in seconds and in days.

SOLUTION 18-7

The mean life τ is the reciprocal of the decay constant. To obtain τ in seconds, divide the numbers in the preceding equation with the numerator and denominator interchanged:

$$\tau = (8.64 \times 10^6)/0.69315$$

$$= 1.25 \times 10^7$$

This is expressed in seconds. To express it in days, divide by 8.64×10^4. This gives an answer of approximately 145 days.

Quiz

Refer to the text in this chapter if necessary. A good score is eight correct. Answers are in the back of the book.

1. A radio transmitter is said to operate in the 15-m band. What is the frequency corresponding to a wavelength of 15 m?
   (a) 20 Hz
   (b) 20 kHz
   (c) 20 MHz
   (d) 20 GHz

2. A thick lead barrier is almost perfectly transparent to
   (a) UV radiation.
   (b) beta particles.
   (c) alpha particles.
   (d) neutrinos.

3. The most energetic EM fields (in terms of the energy in each photon) take the form of
   (a) alpha particles.
   (b) ELF radiation.
   (c) rf.
   (d) gamma rays.

4. In order for an EM field to be produced, charge carriers must be
   (a) in motion.
   (b) accelerating.
   (c) perpendicular to electrical lines of flux.
   (d) perpendicular to magnetic lines of flux.

5. A certain radioactive sample loses exactly seven-eighths of its radiation intensity after 240 years. What is the half-life of this material?
   (a) 30 years
   (b) 80 years
   (c) 160 years
   (d) 210 years

6. Medium-frequency radio waves originating in space cannot reach the Earth's surface because
   (a) the EM noise from the Sun drowns them out.
   (b) the solar wind deflects them from the Earth.
   (c) the Sun's magnetic field traps them in solar orbit.
   (d) they cannot penetrate the Earth's ionosphere.

7. When radio waves travel long distances because they are trapped between layers of air having different temperatures, the effect is known as
   (a) ground wave.
   (b) surface wave.
   (c) ducting.
   (d) sporadic-E.

8. Greenhouse effect in the Earth's environment takes place because
   (a) the atmosphere is transparent to IR at some wavelengths and opaque to IR at other wavelengths.
   (b) the atmosphere does not contain enough oxygen to block IR from the Sun.
   (c) the ozone hole is growing, letting in more and more IR.
   (d) the polar ice caps are melting.

9. A major advantage of a hydrogen fusion reactor over a fission reactor is the fact that
   (a) a hydrogen fusion reactor is cooler than a fission reactor.
   (b) a hydrogen fusion reactor is easier to build than a fission reactor.
   (c) a fusion reactor produces no radioactive waste.
   (d) a fusion reactor can be used as a bomb in case of emergency.

10. A good way to protect yourself against ELF
    (a) is to construct thick barriers made of concrete or lead.
    (b) is to put the source behind a glare filter.
    (c) is to put some physical distance between yourself and the source.
    (d) does not exist.

CHAPTER 19

Optics

Until a few hundred years ago, the only instrument available for observation of visible phenomena was the human eye. This changed as experimenters developed telescopes, microscopes, and other devices.

Behavior of Light

Visible light always take the shortest path between two points, and it always travels at the same speed. These two rules hold as long as the light stays in a vacuum. However, if the medium through which light passes is significantly different from a vacuum, and especially if the medium changes as the light ray travels, these axioms do not apply. If a ray of light passes from air into glass or from glass into air, for example, the path of the ray is bent. A light ray also changes direction when reflected from a mirror.

LIGHT RAYS

A thin shaft of light, such as that which passes from the Sun through a pinhole in a piece of cardboard, can be called a *ray* or *beam* of light. In a more technical sense, a *ray* is the path that an individual photon (light particle) follows through space, air, glass, water, or other medium.

Visible light has properties both wavelike and particle-like. This duality has long been a topic of interest among physicists. In some situations, the

PART 3 **Waves, Particles, Space, and Time**

particle model or *corpuscular model* explains light behavior very well, and the wave model falls short. In other scenarios, the opposite is true. No one has actually seen a ray of light; all we can see are the effects produced when a ray of light strikes something. However, there are certain things we can say about the way in which rays of light behave. These things are predictable, both qualitatively and quantitatively.

REFLECTION

Prehistoric people surely knew about *reflection*. It would not take an intelligent creature very long to figure out that the "phantom in the pond" was actually a visual image of himself or herself. Any smooth, shiny surface reflects some of the light that strikes it. If the surface is perfectly flat, perfectly shiny, and reflects all the light that strikes it, then any ray that encounters the surface is reflected away at the same angle at which it hits. You have heard the expression, "The *angle of incidence* equals the *angle of reflection*." This principle, known as the *law of reflection,* is illustrated in Fig. 19-1.

In optics, the angle of incidence and the angle of reflection are both measured relative to a *normal line* (also called an *orthogonal* or *perpendicular*). In Fig. 19-1, these angles are denoted q and can range from $0°$, where the light ray strikes at a right angle with respect to the surface, to almost $90°$, a grazing angle relative to the surface.

If the reflective surface is not perfectly flat, then the law of reflection still applies for each ray of light striking the surface at a specific point. In such a case, the reflection is considered with respect to a flat plane passing through the point and tangent to the surface at that point. When many parallel rays of light strike a curved or irregular reflective surface at many different points, each ray obeys the law of reflection, but the reflected rays do not all emerge parallel. In some cases they converge; in other cases they diverge. In still other cases the rays are scattered haphazardly.

REFRACTION

Early humans doubtless noticed *refraction* as well as reflection; a clear pond looks shallower than it actually is because of this effect. Refraction is associated with the fact that different media transmit light at different speeds. This does not violate the fundamental principle of relativity theory. The speed of light is absolute in a vacuum, where it travels at about 299,792 km/s (186,282 mi/s), but light travels more slowly than this in other media.

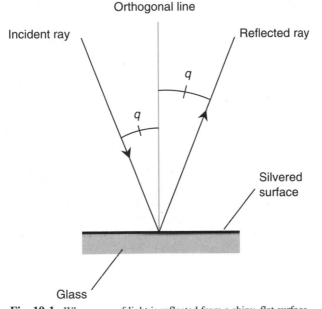

Orthogonal line

Incident ray   Reflected ray

q

q

Silvered
surface

Glass

Fig. 19-1. When a ray of light is reflected from a shiny, flat surface,
the angle of incidence is equal to the angle of reflection. Here both
angles are denoted q.

In air, the difference in the speed of light is slight compared with its speed
in a vacuum, although it can be significant enough to produce refractive
effects at near-grazing angles between air masses having different densities.
In water, glass, quartz, diamond, and other transparent media, light travels
quite a lot more slowly than it does in a vacuum. The *refractive index,* also
called the *index of refraction,* of a medium is the ratio of the speed of light in
a vacuum to the speed of light in that medium. If c is the speed of light in a
vacuum and c_m is the speed of light in medium M, then the index of refraction for medium M, call it r_m, can be calculated simply:

$$r_m = c/c_m$$

Always use the same units when expressing c and c_m. According to this definition, the index of refraction of any transparent material is always greater
than or equal to 1.

The greater the index of refraction for a transparent substance, the more
light is bent when it passes the boundary between that substance and
air. Different types of glass have different refractive indices. Quartz
refracts more than glass, and diamond refracts more than quartz. The high

refractive index of diamond is responsible for the multicolored shine of diamond stones.

PROBLEM 19-1
A certain clear substance has an index of refraction of 1.50 for yellow light. What is the speed at which yellow light travels in this medium?

SOLUTION 19-1
Use the preceding formula and "plug in" the refractive index and the speed of light in a vacuum. Let's express the speeds in kilometers per second and round off c to 3.00×10^5 km/s. Then the speed of the yellow light in the clear substance c_m can be found as follows:

$$1.50 = 3.00 \times 10^5 / c_m$$

$$1.50 c_m = 3.00 \times 10^5$$

$$c_m = 3.00 \times 10^5 / 1.50 = 2.00 \times 10^5$$

LIGHT RAYS AT A BOUNDARY

A qualitative example of refraction is shown in Fig. 19-2, where the refractive index of the first (lower) medium is higher than that of the second (upper) medium. A ray striking the boundary at a right angle (that is, angle of incidence equal to 0°) passes through without changing direction. However, a ray that hits at some other angle is bent; the greater the angle of incidence, the sharper is the turn the beam takes. When the angle of incidence reaches a certain *critical angle,* then the light ray is not refracted at the boundary but instead is reflected back into the first medium. This is known as *total internal reflection.*

A ray originating in the second (upper) medium and striking the boundary at a grazing angle is bent downward. This causes distortion of landscape images when viewed from underwater. You have seen this effect if you are a scuba diver. The sky, trees, hills, buildings, people, and everything else can be seen within a circle of light that distorts the scene like a wide-angle lens.

If the refracting boundary is not flat, the principle shown by Fig. 19-2 still applies for each ray of light striking the boundary at any specific point. The refraction is considered with respect to a flat plane passing through the point and tangent to the boundary at that point. When many parallel rays of light strike a curved or irregular refractive boundary at many different points, each ray obeys the same principle individually.

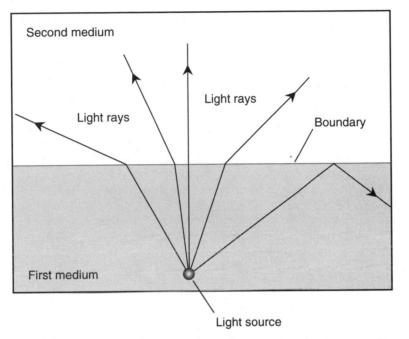

Fig. 19-2. Rays of light are bent more or less as they cross a boundary between media having different properties.

SNELL'S LAW

When a ray of light encounters a boundary between two substances having different indices (or indexes) of refraction, the extent to which the ray is bent can be determined according to an equation called *Snell's law*.

Look at Fig. 19-3. Suppose that B is a flat boundary between two media M_r and M_s, whose indices of refraction are r and s, respectively. Imagine a ray of light crossing the boundary as shown. The ray is bent at the boundary whenever the ray does not strike at a right angle, assuming that the indices of refraction r and s are different.

Suppose that $r < s$; that is, the light passes from a medium having a relatively lower refractive index to a medium having a relatively higher refractive index. Let N be a line passing through some point P on B such that N is normal to B at P. Suppose that R is a ray of light traveling through M_r that strikes B at P. Let x be the angle that R subtends relative to N at P. Let S be the ray of light that emerges from P into M_s. Let y be the angle that S subtends relative to N at P. Then line N, ray R, and ray S all lie in the same

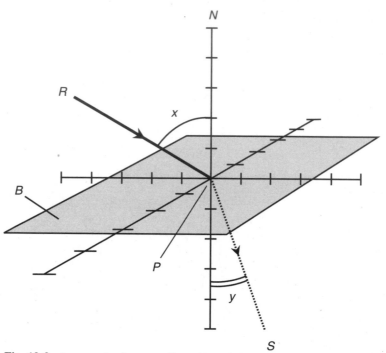

Fig. 19-3. A ray passing from a medium with a relatively lower refractive index to a
medium with a relatively higher refractive index.

plane, and $y < x$. The two angles x and y are equal if, but only if, ray R
strikes the boundary at an angle of incidence of $0°$. The following equation
holds for angles x and y in this situation:

$$\sin y / \sin x = r/s$$

This equation also can be expressed like this:

$$s \sin y = r \sin x$$

Now look at Fig. 19-4. Again, let B be a flat boundary between two media M_r
and M_s whose absolute indices of refraction are r and s, respectively. In this
case, imagine that $r > s$; that is, the ray passes from a medium having a relatively higher refractive index to a medium having a relatively lower refractive index. Let N, B, P, R, S, x, and y be defined as in the preceding example.
As before, $x = y$ if, but only if, ray R strikes B at an angle of incidence of $0°$.
Then line N, ray R, and ray S all lie in the same plane, and $x < y$. Snell's law
holds in this case, just as in the situation described previously:

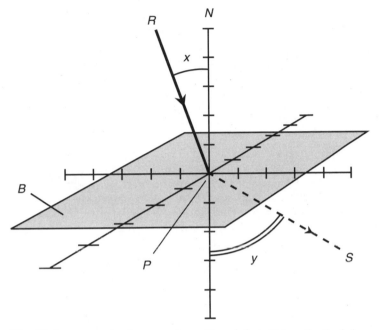

Fig. 19-4. A ray passing from a medium with a relatively higher refractive index
to a medium with a relatively lower refractive index.

$$\sin y/\sin x = r/s \quad \text{and} \quad s \sin y = r \sin x$$

DETERMINING THE CRITICAL ANGLE

Refer again to Fig. 19-4. The light passes from a medium having a relatively higher index of refraction r into a medium having a relatively lower index s. Therefore, $r > s$. As angle x increases, angle y approaches 90°, and ray S gets closer to the boundary plane B. When x, the angle of incidence, gets large enough (somewhere between 0° and 90°), angle y reaches 90°, and ray S lies exactly in plane B. If angle x increases even more, ray R undergoes total internal reflection at the boundary plane B. Then the boundary acts like a mirror.

The critical angle is the largest angle of incidence that ray R can subtend relative to the normal N without being reflected internally. Let's call this angle x_c. The measure of the critical angle is the arcsine of the ratio of the indices of refraction:

$$x_c = \sin^{-1}(s/r)$$

PROBLEM 19-2

Suppose that a laser is placed beneath the surface of a freshwater pond. The refractive index of fresh water is approximately 1.33, whereas that of air is 1.00. Imagine that the surface is perfectly smooth. If the laser is directed upward so that it strikes the surface at an angle of 30.0° relative to the normal (perpendicular), at what angle, also relative to the normal, will the beam emerge from the surface into the air?

SOLUTION 19-2

Envision the situation in Fig. 19-4 "upside down." Then M_r is the water and M_s is the air. The indices of refraction are $r = 1.33$ and $s = 1.00$. The measure of angle x is 30.0°. The unknown is the measure of angle y. Use the equation for Snell's law, plug in the numbers, and solve for y. You'll need a calculator. Here's how it goes:

$$\sin y/\sin x = r/s$$
$$\sin y/(\sin 30.0°) = 1.33/1.00$$
$$\sin y/0.500 = 1.33$$
$$\sin y = 1.33 \times 0.500 = 0.665$$
$$y = \sin^{-1} 0.665 = 41.7°$$

PROBLEM 19-3

What is the critical angle for light rays shining upward from beneath a fresh-water pond?

SOLUTION 19-3

Use the formula for critical angle, and envision the scenario of Problem 19-2, where the laser angle of incidence x can be varied. Plug in the numbers to the equation for critical angle x_c:

$$x_c = \sin^{-1}(s/r)$$
$$x_c = \sin^{-1}(1.00/1.33)$$
$$x_c = \sin^{-1} 0.752$$
$$x_c = 48.8°$$

Remember that the angles in all these situations are defined with respect to the normal to the surface, not with respect to the plane of the surface.

DISPERSION

The index of refraction for a particular substance depends on the wavelength of the light passing through it. Glass slows down light the most at the shortest wavelengths (blue and violet) and the least at the longest wavelengths (red and orange). This variation of the refractive index with wavelength is known as *dispersion.* It is the principle by which a prism works (Fig. 19-5). The more the light is slowed down by the glass, the more its path is deflected when it passes through the prism. This is why prisms cast rainbows when white light is shone through them.

Dispersion is important in optics for two reasons. First, a prism can be used to make a *spectrometer,* which is a device for examining the intensity of visible light at specific wavelengths. (Fine gratings are also used for this.) Second, dispersion degrades the quality of white-light images viewed through simple lenses.

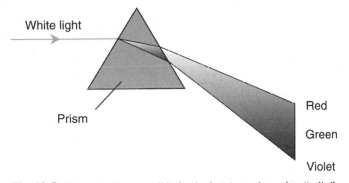

Fig. 19-5. Dispersion is responsible for the fact that a glass prism "splits" white light into its constituent colors.

Lenses and Mirrors

The ways in which visible light is reflected and refracted can be used to advantage. This was first discovered when experimenters found that specially shaped pieces of glass can make objects look larger or smaller. The refractive properties of glass have been used for centuries to help correct deficiencies in vision that occur as a person gets older. Lenses work because they refract light more or less depending on where and at what angle the light strikes their surfaces. Curved mirrors have much the same effect when they reflect light.

THE CONVEX LENS

You can buy a *convex lens* in almost any novelty store or department store. In a good hobby store you should be able to find a "magnifying glass" up to 10 cm or even 15 cm in diameter. The term *convex* arises from the fact that one or both faces of the glass bulge outward at the center. A convex lens is sometimes called a *converging lens.* It brings parallel light rays to a sharp *focus* or *focal point,* as shown in Fig. 19-6*a,* when those rays are parallel to the axis of the lens. It also can *collimate* (make parallel) the light from a point source, as shown in Fig. 19-6*b.*

The properties of a convex lens depend on the diameter and the difference in thickness between the edges and the center. The larger the diameter, the greater is the light-gathering power. The greater the difference in thickness between the center and the edges, the shorter is the distance between the lens and the point at which it brings parallel light rays to a focus. The effective area of the lens, measured in a plane perpendicular to the axis, is known as the *light-gathering area.* The distance between the center of the lens and the focal point (as shown in Fig. 19-6*a* or *b*) is called the *focal length.* If you look through a convex lens at a close-up object such as a coin, the features are magnified; they appear larger than they look with the unaided eye. The light rays from an object at a great distance from a convex lens converge to form a *real image* at the focal point.

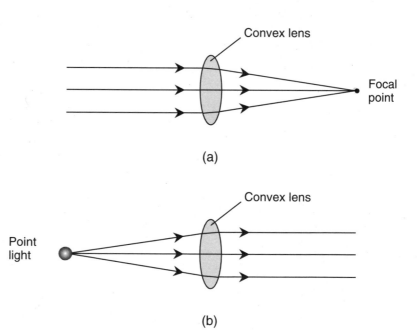

(a)

(b)

Fig. 19-6. (*a*) A convex lens focuses parallel light rays to a point. (*b*) The same lens collimates light from a point source at the focus.

The surfaces of most convex lenses are spherical. This means that if you could find a large ball having just the right diameter, the curve of the lens face would fit neatly inside the ball. Some convex lenses have the same radius of curvature on each face; others have different radii of curvature on their two faces. Some converging lenses have one flat face; these are called *planoconvex lenses.*

THE CONCAVE LENS

You will have some trouble finding a *concave lens* in a department store, but you should be able to order one from a specialty catalog or a Web site. The term *concave* refers to the fact that one or both faces of the glass bulge inward at the center. This type of lens is also called a *diverging lens.* It spreads parallel light rays outward (Fig. 19-7a). It can collimate converging rays if the convergence angle is just right (see Fig. 19-7b).

As with convex lenses, the properties of a concave lens depend on the diameter and the extent to which the surface(s) depart from flat. The greater the difference in thickness between the edges and the center of the lens, the

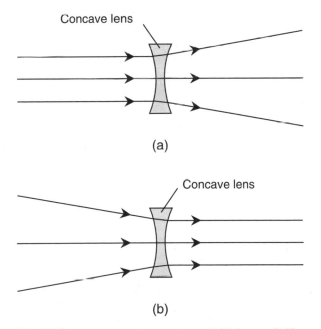

Fig. 19-7. (*a*) A concave lens spreads parallel light rays. (*b*) The same lens collimates converging light rays.

more the lens will cause parallel rays of light to diverge. If you look through a concave lens at a close-in object such as a coin, the features are reduced; they appear smaller than they look with the unaided eye.

The surfaces of concave lenses, like those of their convex counterparts, are generally spherical. Some concave lenses have the same radius of curvature on each face; others have different radii of curvature on their two faces. Some diverging lenses have one flat face; these are called *planoconcave lenses*.

THE CONVEX MIRROR

A *convex mirror* reflects light rays in such a way that the effect is similar to that of a concave lens. Incident rays, when parallel, are spread out (Fig. 19-8a) after they are reflected from the surface. Converging incident rays, if the angle of convergence is just right, are collimated by a convex mirror (see Fig. 19-8b). When you look at the reflection of a scene in a convex

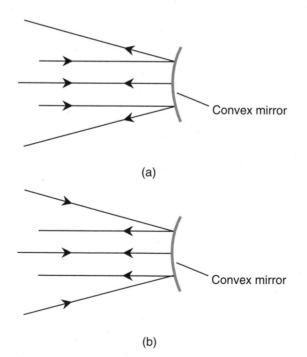

(a)

(b)

Fig. 19-8. (*a*) A convex mirror spreads parallel incident light rays. (*b*) The same mirror collimates converging incident light rays.

mirror, the objects all appear reduced. The field of vision is enlarged, a fact that is used to advantage in some automotive rear-view mirrors.

The extent to which a convex mirror spreads light rays depends on the radius of curvature. The smaller the radius of curvature, the greater is the extent to which parallel incident rays diverge after reflection.

THE CONCAVE MIRROR

A *concave mirror* reflects light rays in a manner similar to the way a convex lens refracts them. When incident rays are parallel to each other and to the axis of the mirror, they are reflected so that they converge at a focal point (Fig. 19-9a). When a point source of light is placed at the focal point, the concave mirror reflects the rays so that they emerge parallel (see Fig. 19-9b).

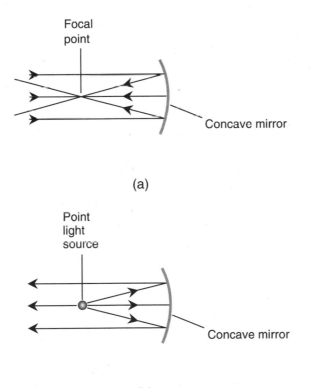

Fig. 19-9. (*a*) A concave mirror focuses parallel light rays to a point. (*b*) The same mirror collimates light from a point source at the focus.

The properties of a concave mirror depend on the size of the reflecting surface, as well as on the radius of curvature. The larger the light-gathering area, the greater is the light-gathering power. The smaller the radius of curvature, the shorter is the focal length. If you look at your reflection in a convex mirror, you will see the same effect that you would observe if you placed a convex lens up against a flat mirror.

Concave mirrors can have spherical surfaces, but the finest mirrors have surfaces that follow the contour of an idealized three-dimensional figure called a *paraboloid*. A paraboloid results from the rotation of a *parabola,* such as that having the equation $y = x^2$ in rectangular coordinates, around its axis. When the radius of curvature is large compared with the size of the reflecting surface, the difference between a *spherical mirror* and a *paraboloidal mirror* (more commonly called a *parabolic mirror*) is not noticeable to the casual observer. However, it makes a big difference when the mirror is used in a telescope.

PROBLEM 19-4
Suppose that a simple convex lens is made from the same material as a prism that casts a rainbow when white light is shone through it. How does the focal length of this lens for red light compare with its focal length for blue light?

SOLUTION 19-4
The glass has a higher refractive index for blue light than for red light. Therefore, the glass bends blue light more, resulting in a shorter focal length for blue than for red.

Refracting Telescopes

The first telescopes were developed in the 1600s. They employed lenses. Any telescope that enlarges distant images with lenses alone is called a *refracting telescope.*

GALILEAN REFRACTOR

Galileo Galilei, the astronomer who became famous during the 1600s for noticing craters on the Moon and natural satellites orbiting Jupiter, devised a telescope consisting of a convex-lens *objective* and a concave-lens *eyepiece*. His first telescope magnified the apparent diameters of distant objects by a factor of only a few times. Some of his later telescopes magnified up to 30 times. The *galilean refractor* (Fig. 19-10*a*) produces an *erect image,* that is, a right-side-up view of things. In addition to appearing right-side up, images are also true in the left-to-right sense. The *magnification factor,* defined as the number of times the angular diameters of distant objects are increased, depends on the focal length of the objective, as well as on the distance between the objective and the eyepiece.

Galilean refractors are available today mainly as novelties for terrestrial viewing. Galileo's original refractors had objective lenses only 2 or 3 cm (about 1 in) across; the same is true of most galilean telescopes today. Some of these telescopes have sliding concentric tubes that provide variable magnification. When the inner tube is pushed all the way into the outer one, the magnification factor is the lowest; when the inner tube is pulled all the way out, the magnification is highest. The image remains fairly clear over the entire magnification-adjustment range. These instruments are sometimes called *spyglasses.*

KEPLERIAN REFRACTOR

Johannes Kepler, whose audience was more friendly than Galileo's when it came to his theories concerning the universe, refined Galileo's telescope design. Kepler's refracting telescope employed a convex-lens objective with a long focal length and a smaller convex-lens eyepiece with a short focal length. Unlike the galilean telescope, the *keplerian refractor* (see Fig. 19-10*b*) produces an *inverted image*; it is upside-down and backwards. The distance between the objective and the eyepiece must be exactly equal to the sum of the focal lengths of the two lenses in order for the image to be clear. The magnification factor depends on the ratio of the focal length of the objective to the focal length of the eyepiece.

The keplerian telescope is preferred over the galilean type primarily because Kepler's design provides a larger *apparent field of view.* Galilean telescopes in general have apparent fields of view so narrow that looking through them is an uncomfortable experience. The magnification factor of a keplerian telescope can be changed by using eyepieces with longer or shorter focal lengths. The shorter the focal length of the eyepiece, the greater is the magnification factor, informally known as *power,* assuming that the focal length of the objective lens remains constant.

The largest refracting telescope in the world is located at the Yerkes Observatory in Wisconsin. Its objective lens has a diameter of 40 in (slightly more than 1 m). Keplerian refractors are used by thousands of amateur astronomers worldwide.

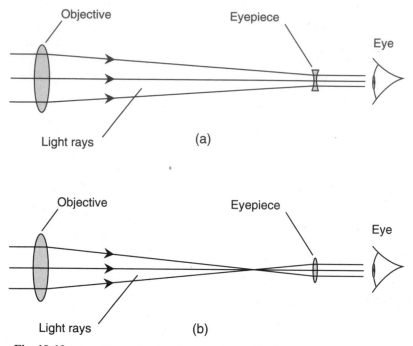

Fig. 19-10. The galilean refractor (*a*) uses a convex objective and a concave eyepiece. The keplerian refractor (*b*) has a convex objective and a convex eyepiece.

LIMITATIONS OF REFRACTORS

There are certain problems inherent to telescopes that use objective lenses. These are known as *spherical aberration, chromatic aberration,* and *lens sag.*

Spherical aberration results from the fact that spherical convex lenses don't bring parallel light rays to a perfect focus. A refracting telescope with a spherical objective focuses a ray passing through its edge a little differently than a ray passing closer to the center. The actual focus of the objective is not a point but a very short line along the lens axis. This effect causes slight blurring of images of objects that have relatively large angular diameters, such as nebulae and galaxies. The problem can be corrected by grinding the objective lens so that it has a paraboloidal rather than a spherical surface.

Chromatic aberration occurs because the glass in a simple lens refracts the shortest wavelengths of light slightly more than the longest wavelengths. The focal length of any given convex lens is shorter for violet light than for blue light, shorter for blue than for yellow, and shorter for yellow than for red. This produces rainbow-colored halos around star images and along sharply defined edges of objects with large angular diameters. Chromatic aberration can be almost, but not completely, corrected by the use of *compound lenses*. These lenses have two or more sections made of different types of glass; the sections are glued together with a special transparent adhesive. Such objectives are called *achromatic lenses* and are standard issue on refracting telescopes these days.

Lens sag occurs in the largest refracting telescopes. When an objective is made larger than approximately 1 m in diameter, it becomes so massive that its own weight distorts its shape. Glass is not perfectly rigid, as you have noticed if you have seen the reflection of the landscape in a large window on a windy day. There is no way to get rid of this problem except to take the telescope out of Earth's gravitational field.

Reflecting Telescopes

The troubles that plague refracting telescopes, particularly lens sag, can be largely overcome by using mirrors instead of lenses as objectives. A *first-surface mirror,* with the silvering on the outside so that the light never passes through glass, can be ground so that it brings light to a focus that does not vary with wavelength. Mirrors can be supported from behind, so it is possible to make them in diameters several times larger than lenses without encountering the sag problem.

NEWTONIAN REFLECTOR

Isaac Newton designed a *reflecting telescope* that was free of chromatic aberration. His design is still used in many reflecting telescopes today. The *newtonian reflector* employs a concave objective mirror mounted at one end of a long tube. The other end of the tube is open to admit incoming light. A small, flat mirror is mounted at a 45° angle near the open end of the tube to reflect the focused light through an opening in the side of the tube containing the eyepiece (Fig. 19-11a).

The flat mirror obstructs some of the incoming light, slightly reducing the effective surface area of the objective mirror. As a typical example, suppose that a newtonian reflector has an objective mirror 20 cm in

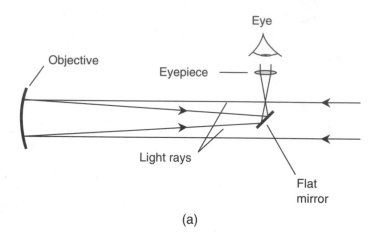

(a)

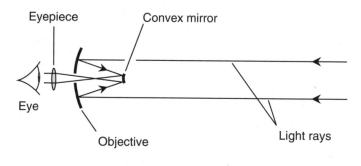

(b)

Fig. 19-11. The newtonian reflector (*a*) has an eyepiece set into the side of
the tube. In the cassegrain reflector (*b*), the eyepiece is in the center of the
objective mirror.

diameter. The total surface area of this mirror is approximately 314 centimeters squared (cm^2). If the eyepiece mirror is a 3-cm square, its total
area is 9 cm^2, which is about 3 percent of the total surface area of the
objective.

Newtonian reflectors have limitations. Some people find it unnatural
to "look sideways" at objects. If the telescope has a long tube, it is necessary to use a ladder to view objects at high elevations. These annoy-
ances can be overcome by using a different way to get the light to the
eyepiece.

CASSEGRAIN REFLECTOR

Figure 19-11*b* shows the design of the *cassegrain reflector.* The eyepiece mirror is mounted closer to the objective than it is in the newtonian design. It is not angled, but it is convex. The convexity of this mirror increases the effective focal length of the objective mirror. Light reflects from the convex mirror and passes through a small hole in the center of the objective containing the eyepiece.

The cassegrain reflector can be made with a physically short tube and an objective mirror with more curvature than that of a newtonian telescope having the same diameter. As a result, the cassegrain telescope is less massive and less bulky. Cassegrain reflectors with heavy-duty mountings are physically stable, and they can be used at low magnification to obtain a wide view of a large portion of the sky.

Telescope Specifications

Several parameters are significant when determining the effectiveness of a telescope for various applications. Here are the most important ones.

MAGNIFICATION

The *magnification,* also called *power* and symbolized $\times$, is the extent to which a telescope makes objects look closer. (Actually, telescopes increase the observed sizes of distant objects, but they do not look closer in terms of perspective.) The magnification is a measure of the factor by which the apparent angular diameter of an object is increased. A 20$\times$ telescope makes the Moon, whose disk subtends about 0.5° of arc as observed with the unaided eye, appear 10° of arc in diameter. A 180$\times$ telescope makes a crater on the Moon with an angular diameter of only 1 minute of arc ($^1\!/_{60}$ of a degree) appear 3° across.

Magnification is calculated in terms of the focal lengths of the objective and the eyepiece. If f_o is the effective focal length of the of the objective and f_e is the focal length of the eyepiece (in the same units as f_o), then the magnification factor m is given by this formula:

$$m = f_o/f_e$$

For a given eyepiece, as the effective focal length of the objective increases, the magnification of the telescope also increases. For a given objective, as the effective focal length of the eyepiece increases, the magnification of the telescope decreases.

RESOLVING POWER

The *resolution,* also called *resolving power,* is the ability of a telescope to separate two objects that are not in exactly the same place in the sky. It is measured in an angular sense, usually in seconds of arc (units of $\frac{1}{3600}$ of a degree). The smaller the number, the better is the resolving power.

The best way to measure a telescope's resolving power is to scan the sky for known pairs of stars that are appear close to each other in the angular sense. Astronomical data charts can determine which pairs of stars to use for this purpose. Another method is to examine the Moon and use a detailed map of the lunar surface to ascertain how much detail the telescope can render.

Resolving power increases with magnification, but only up to a certain point. The greatest image resolution a telescope can provide is directly proportional to the diameter of the objective lens or mirror, up to a certain maximum dictated by atmospheric turbulence. In addition, the resolving power depends on the acuity of the observer's eyesight (if direct viewing is contemplated) or the coarseness of the grain of the photographic or detecting surface (if an analog or digital camera is used).

LIGHT-GATHERING AREA

The light-gathering area of a telescope is a quantitative measure of its ability to collect light for viewing. It can be defined in centimeters squared (cm^2) or meters squared (m^2), that is, in terms of the effective surface area of the objective lens or mirror as measured in a plane perpendicular to its axis. Sometimes it is expressed in inches squared (in^2).

For a refracting telescope, given an objective radius of r, the light-gathering area A can be calculated according to this formula:

$$A = \pi r^2$$

where π is approximately equal to 3.14159. If r is expressed in centimeters, then A is in centimeters squared; if r is in meters, then A is in meters squared.

For a reflecting telescope, given an objective radius of r, the light-gathering area A can be calculated according to this formula:

$$A = \pi r^2 - B$$

where B is the area obstructed by the secondary mirror assembly. If r is expressed in centimeters and B is expressed in centimeters squared, then A is in centimeters squared; if r is in meters and B is in meters squared, then A is in meters squared.

ABSOLUTE FIELD OF VIEW

When you look through the eyepiece of a telescope, you see a circular patch of sky. Actually, you can see anything within a cone-shaped region whose apex is at the telescope (Fig. 19-12). The *absolute field of view* is the angular diameter q of this cone; q can be specified in degrees, minutes, and/or seconds of arc. Sometimes the angular radius is specified instead of the angular diameter.

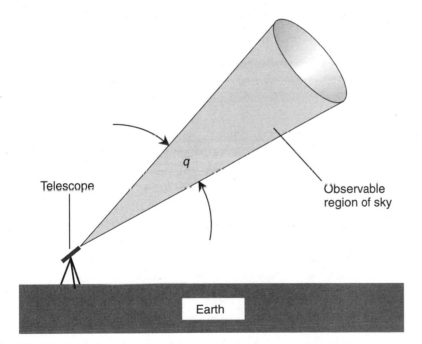

Fig. 19-12. A telescope's absolute field of view q is measured in angular degrees, minutes, and/or seconds of arc.

The absolute field of view depends on several factors. The magnification of the telescope is important. When all other factors are held constant, the absolute field of view is inversely proportional to the magnification. If you double the magnification, you cut the field in half; if you reduce the magnification to one-quarter of its previous value, you increase the field by a factor of 4.

The viewing angle—that is, the apparent field of view—provided by the eyepiece is important. Some types of eyepieces have a wide field, such as 60° or even 90°. Others have narrower apparent fields, in some cases less than 30°.

Another factor that affects the absolute field of view is the ratio of the objective diameter to its focal length. In general, the larger this ratio, the wider is the maximum absolute field of view that can be obtained with the telescope. Long, narrow telescopes have the smallest maximum absolute fields of view; short, fat ones have the widest maximum fields.

PROBLEM 19-5
How much more light can a refracting telescope with a 15.0-cm-diameter objective gather compared with a refracting telescope having an objective whose diameter is 6.00 cm? Express the answer as a percentage.

SOLUTION 19-5
Light-gathering area is proportional to the square of the objective's radius. Therefore, the ratio of the larger telescope's light-gathering area to the smaller telescope's light-gathering area is proportional to the square of the ratio of their objectives' diameters. Let's call the ratio k. Then in this case

$$k = 15.0/6.00 = 2.50$$

$$k^2 = 2.50^2 = 6.25$$

The larger telescope gathers 6.25 times, or 625 percent, as much light as the smaller one.

PROBLEM 19-6
Suppose that a telescope has a magnification factor of 100× with an eyepiece of 20.0 mm focal length. What is the focal length of the objective?

SOLUTION 19-6
Use the formula given in the section entitled, "Magnification." The value of f_o in this case is the unknown; f_e = 20.0 mm, and m = 100. Therefore

$$m = f_o/f_e$$

$$100 = f_o/20.0$$

$$f_o = 100 \times 20.0 = 2,000 \text{ mm}$$

Technically, we are justified in expressing the answer to only three significant digits. We can legitimately say that $f_o = 2.00$ m.

PROBLEM 19-7
Suppose that the absolute field of view provided by the telescope in Problem 19-6 is 20 arc minutes. If the 20-mm eyepiece is replaced with a 10-mm eyepiece that provides the same viewing angle as the 20-mm eyepiece, what happens to the absolute field of view provided by the telescope?

SOLUTION 19-7
The 10-mm eyepiece provides twice the magnification of the 20-mm eyepiece. Therefore, the absolute field of view of the telescope using the 10-mm eyepiece is half as wide or 10 arc minutes.

The Compound Microscope

Optical microscopes are designed to greatly magnify the images of objects too small to resolve with the unaided eye. Microscopes, in contrast to telescopes, work at close range. The design is in some ways similar to that of the telescope, but in other ways it differs. The simplest microscopes consist of single convex lenses. These can provide magnification factors of up to $10\times$ or $20\times$. In the laboratory, an instrument called the *compound microscope* is preferred because it allows for much greater magnification.

BASIC PRINCIPLE

A compound microscope employs two lenses. The objective has a short focal length, in some cases 1 mm or less, and is placed near the specimen or sample to be observed. This produces an image at some distance above the objective, where the light rays come to a focus. The distance (let's call it s) between the objective and this image is always greater than the focal length of the objective.

The eyepiece has a longer focal length than the objective. It magnifies the *real image* produced by the objective. In a typical microscope, illumination can be provided by shining a light upward through the sample if the sample is translucent. Some microscopes allow for light to be shone downward on opaque specimens. Figure 19-13 is a simplified diagram of a compound microscope showing how the light rays are focused and how the specimen can be illuminated.

Laboratory-grade compound microscopes have two or more objectives, which can be selected by rotating a wheel to which each objective is attached. This provides several different levels of magnification for a given eyepiece. In general, as the focal length of the objective becomes shorter, the magnification of the microscope increases. Some compound microscopes can magnify images up to about 2,500 times. A hobby-grade compound microscope can provide decent image quality at magnifications of up to about 1,000 times.

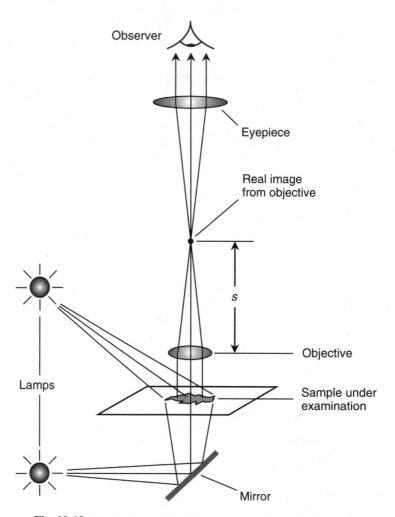

Fig. 19-13. Illumination and focusing in a compound optical microscope.

FOCUSING

A compound microscope is focused by moving the entire assembly, including both the eyepiece and the objective, up and down. This must be done with a precision mechanism because the *depth of field* (the difference between the shortest and the greatest distance from the objective at which an object is in good focus) is exceedingly small. In general, the shorter the focal length of the objective, the smaller is the depth of field, and the more critical is the focusing. High-magnification objectives have depths of field on the order of 2 μm (2×10^{-6} m) or even less.

If the eyepiece is moved up and down in the microscope tube assembly while the objective remains in a fixed position, the magnification varies. However, microscopes usually are designed to provide the best image quality for a specific eyepiece-to-objective separation, such as 16 cm (approximately 6.3 in).

If a bright enough lamp is used to illuminate the specimen under examination, and especially if the specimen is transparent or translucent so that it can be lit from behind, the eyepiece can be removed from the microscope and a decent image can be projected onto a screen on the ceiling of the room. A diagonal mirror can reflect this image to a screen mounted on a wall. This technique works best for objectives having long focal lengths, and hence low magnification factors.

MICROSCOPE MAGNIFICATION

Refer to Fig. 19-14. Suppose that f_o is the focal length (in meters) of the objective lens and f_e is the focal length (in meters) of the eyepiece. Assume that the objective and the eyepiece are placed along a common axis and that the distance between their centers is adjusted for proper focus. Let s represent the distance (in meters) from the objective to the real image it forms of the object under examination. The *microscopic magnification* (a dimensionless quantity denoted m in this context) is given by

$$m = [(s - f_o)/f_o] \, [(f_e + 0.25)/f_e]$$

The quantity 0.25 represents the average *near point* of the human eye, which is the closest distance over which the eye can focus on an object: approximately 0.25 m.

A less formal method of calculating the magnification of a microscope is to multiply the magnification of the objective by the magnification of the eyepiece. These numbers are provided with objectives and eyepieces and

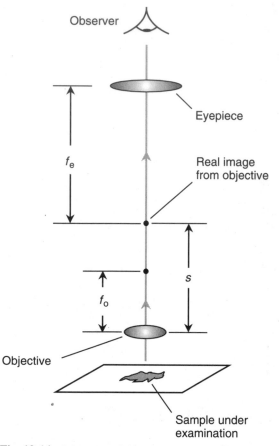

Fig. 19-14. Calculation of the magnification factor in a compound microscope. See text for details.

are based on the use of an air medium between the objective and the specimen, as well as on the standard distance between the objective and the eyepiece. If m_e is the power of the eyepiece and m_o is the power of the objective, then the power m of the microscope as a whole is

$$m = m_e m_o$$

NUMERICAL APERTURE AND RESOLUTION

In an optical microscope, the *numerical aperture* of the objective is an important specification in determining the resolution, or the amount of detail the microscope can render. This is defined as shown in Fig. 19-15.

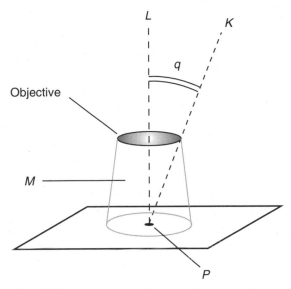

Fig. 19-15. Determination of the numerical aperture for a microscope objective. See text for details.

Let L be a line passing through a point P in the specimen to be examined, as well as through the center of the objective. Let K be a line passing through P and intersecting the outer edge of the objective lens opening. (It is assumed that this outer edge is circular.) Let q be the measure of the angle between lines L and K. Let M be the medium between the objective and the sample under examination. This medium M is usually air, but not always. Let r_m be the refractive index of M. Then the numerical aperture of the objective A_o is given by

$$A_o = r_m \sin q$$

In general, the greater the value of A_o, the better is the resolution. There are three ways to increase the A_o of a microscope objective of a given focal length:

- The diameter of the objective can be increased.
- The value of r_m can be increased.
- The wavelength of the illuminating light can be decreased.

Large-diameter objectives having short focal lengths, thereby providing high magnification, are difficult to construct. Thus, when scientists want to examine an object in high detail, they can use blue light, which has a relatively short wavelength. Alternatively, or in addition, the medium M between the objective and the specimen can be changed to something with a high index of refraction, such as clear oil. This shortens the wavelength of the illuminating beam that strikes the objective because it slows down the speed of light in M. (Remember the relation between the speed of an electromagnetic disturbance, the wavelength, and the frequency!) A side effect of this tactic is a reduction in the effective magnification of the objective lens, but this can be compensated for by using an objective with a smaller radius of surface curvature or by increasing the distance between the objective and the eyepiece.

The use of monochromatic light rather than white light offers another advantage. Chromatic aberration affects the light passing through a microscope in the same way that it affects the light passing through a telescope. If the light has only one wavelength, chromatic aberration does not occur. In addition, the use of various colors of monochromatic light (red, orange, yellow, green, or blue) can accentuate structural or anatomic features in a specimen that do not always show up well in white light.

PROBLEM 19-8
A compound microscope objective is specified as 10×, whereas the eyepiece is rated at 5×. What is the power m of this instrument?

SOLUTION 19-8
Multiply the magnification factor of the objective by that of the eyepiece:

$$m = (5 \times 10) \times = 50\times$$

Quiz

Refer to the text in this chapter if necessary. A good score is eight correct. Answers are in the back of the book.

1. A simple convex lens has a focal length that varies slightly depending on the wavelength of the light passing through it. When such a lens is used as the objective of a telescope, this effect results in
   (a) dispersion.
   (b) spherical aberration.
   (c) chromatic aberration.
   (d) nothing! The premise is wrong. A convex lens has the same focal length for all wavelengths of light passing through it.

2. Suppose that a microscope has an objective whose focal length is 1.00 mm and an eyepiece whose focal length is 25.0 mm. What is the magnification?
    (a) 25×
    (b) 625×
    (c) 0.0400×. This device doesn't magnify. It makes the specimen look smaller.
    (d) We need more information to calculate the magnification.

3. Suppose that a pane of crown glass, with a refractive index of 1.52, is immersed in water, which has a refractive index of 1.33. A ray of light traveling in the water strikes the glass at 45° relative to the normal and travels through the pane. What angle, relative to the normal, will the ray of light subtend when it leaves the pane and reenters the water?
    (a) 38°
    (b) 54°
    (c) 45°
    (d) No angle at all! The premise is wrong. The light will never enter the glass. It will be reflected when it strikes the glass surface.

4. Suppose that the numerical aperture of a microscope objective in air is 0.85. The medium between the lens and the specimen is replaced by water, which has a refractive index of 1.33. The numerical aperture of the objective
    (a) does not change.
    (b) increases to 1.13.
    (c) decreases to 0.639.
    (d) cannot be calculated from this information.

5. According to the law of reflection,
    (a) a ray of light traveling from a medium having a low refractive index to a medium having a higher refractive index is reflected at the boundary.
    (b) a ray of light traveling from a medium having a high refractive index to a medium having a lower refractive index is reflected at the boundary.
    (c) a ray of light always reflects from a shiny surface in a direction exactly opposite the direction from which it arrives.
    (d) none of the above.

6. A Cassegrain-type reflecting telescope has an objective mirror with a diameter of 300 mm and an eyepiece with a focal length of 30 mm. The magnification is
    (a) 100×.
    (b) 10×.
    (c) 9,000×.
    (d) impossible to calculate from this information.

7. A diverging lens
    (a) can collimate converging rays of light.
    (b) can focus the Sun's rays to a brilliant point.
    (c) is also known as a convex lens.
    (d) is ideal for use as the objective in a refracting telescope.

8. Suppose that the speed of red visible light in a certain transparent medium is 270,000 km/s. What, approximately, is the index of refraction for this substance with respect to red light?
   (a) 0.900
   (b) 1.11
   (c) 0.810
   (d) It cannot be calculated from this information.

9. As the magnification of a telescope is increased,
   (a) the image resolution decreases in direct proportion.
   (b) physical stability becomes more and more important.
   (c) the light-gathering area increases in direct proportion.
   (d) dimmer and dimmer objects can be seen.

10. What is the critical angle of light rays inside a gem whose refractive index is 2.4? Assume that the gem is surrounded by air.
    (a) 25°
    (b) 65°
    (c) 67°
    (d) 90°

CHAPTER 20

Relativity Theory

There are two aspects to Albert Einstein's relativity theory: the *special theory* and the *general theory*. The special theory involves relative motion, and the general theory involves acceleration and gravitation. However, before we get into relativity, let's find out what follows from the hypothesis that the speed of light is absolute, constant, and finite and that it is the highest speed anything can attain.

Simultaneity

When he became interested in light, space, and time, Einstein pondered the results of experiments intended to find out how the Earth moves relative to the supposed medium that carries electromagnetic (EM) waves such as visible light. Einstein came to believe that such a medium doesn't exist and that EM waves can travel through a perfect vacuum.

THE LUMINIFEROUS ETHER

In the 1800s, physicists determined that light has wavelike properties and in some ways resembles sound. But light travels much faster than sound. However, light can travel through a vacuum, whereas sound cannot. Sound waves require a material medium such as air, water, or metal to propagate. Most scientists thought light also must require some sort of medium, but what? What could exist everywhere, even in a jar from which all the air was pumped out? This mysterious medium was called *luminiferous ether,* or simply *ether.* It turned out to be nothing but a figment of the imagination.

If the ether exists, some scientists wondered, how could it pass right through everything, even the entire Earth, and get inside an evacuated chamber? How could the ether be detected? One idea was to see if the ether "blows" against the Earth as our planet orbits around the Sun, and as the Solar System orbits around the center of the Milky Way galaxy, and as our galaxy drifts through the cosmos. If there is an "ether wind," then the speed of light ought to be different in different directions. This, it was reasoned, should occur for the same reason a passenger on a fast-moving truck measures the speed of sound waves coming from the front as faster than the speed of sound waves coming from behind.

In 1887, an experiment was done by two physicists named Albert Michelson and Edward Morley in an attempt to find out how fast the "ether wind" is blowing and from what direction. The *Michelson-Morley experiment,* as it became known, showed that the speed of light is the same in all directions. This cast doubt on the ether theory. If the ether exists, then according to the results obtained by Michelson and Morley, it must be moving right along with the Earth. This seemed to be too great a coincidence. Attempts were made to explain away this result by suggesting that the Earth drags the ether along with itself. Einstein could not accept that. He took the results of the Michelson-Morley experiment at face value. Einstein believed that the Michelson-Morley experiment would have the same outcome for observers on the Moon, on any other planet, on a space ship, or anywhere in the universe.

THE SPEED OF LIGHT IS CONSTANT

Einstein rejected the notion of luminiferous ether. Instead, he proposed an axiom: In a vacuum, the speed at which light, or any other EM field, travels is an absolute constant. This is the case regardless of the motion of the observer with respect to the source. (In media other than a vacuum, such as glass, this axiom does not apply.) Armed with this axiom, Einstein set out to deduce what logically follows.

Einstein did all his work by using a combination of mathematics and daydreaming that he called "mind journeys." He wasn't an experimentalist but a theorist. There is a saying in physics: "One experimentalist can keep a dozen theorists busy." Einstein turned that inside out. His theories have kept thousands of experimentalists occupied.

NO ABSOLUTE TIME

One of the first results of Einstein's speed-of-light axiom is the fact that there can be no such thing as an absolute time standard. It is impossible to synchronize the clocks of two observers so that they will see both clocks as being in exact agreement unless both observers occupy the exact same point in space.

In recent decades we have built atomic clocks, and we claim they are accurate to within billionths of a second (where a billionth is 0.000000001 or 10^{-9}). But this has meaning only when we are right next to such a clock. If we move a little distance away from the clock, the light (or any other signal that we know of) takes some time to get to us, and this throws the clock reading off.

The speed of EM-field propagation, the fastest speed known, is approximately 3.00×10^8 m/s (1.86×10^5 mi/s). A beam of light therefore travels about 300 m (984 ft) in 1.00×10^{-6} s (1.00 μs). If you move a little more than the length of a football field away from a superaccurate billionth-of-a-second atomic clock, the clock will appear to be in error by 1.00 μs. If you go to the other side of the world, where the radio signal from that clock must travel 20,000 km (12,500 mi) to reach you, the time reading will be off by 0.067 s. If you go to the Moon, which is about 4.0×10^5 km (2.5×10^5 mi) distant, the clock will be off by approximately 1.3 s.

If scientists ever discover an energy field that can travel through space instantaneously regardless of the distance, then the conundrum of absolute time will be resolved. In practical scenarios, however, the speed of light is the fastest possible speed. (Some recent experiments suggest that certain effects can propagate faster than the speed of light over short distances, but no one has demonstrated this on a large scale yet, much less used such effects to transmit any information such as data from an atomic clock.) We can say that the speed of light is the speed of time. Distance and time are inextricably related.

POINT OF VIEW

Suppose that there are eight clocks in space arranged at the vertices of a gigantic cube. Each edge of the cube measures 1 light-minute, or approximately 1.8×10^7 km (1.1×10^7 mi), long, as shown in Fig. 20-1. We are given a challenge: Synchronize the clocks so that they agree within the limit of visibility, say, to within 1 second of each other. Do you suppose that this will be easy?

Because the clocks are so far apart, the only way we can ascertain what they say is to equip them with radio transmitters that send time signals. Alternatively, if we have a powerful enough telescope, we can observe them and read them directly by sight. In either case, the information that tells us what the clocks say travels to us at the speed of light. We get in our space ship and maneuver ourselves so that we are in the exact center of the cube, equidistant from all eight clocks. Then we proceed to synchronize them using remote-control, wireless two-way data communications equipment. Thank heaven for computers! The task is accomplished in a just a few minutes. It can't be done instantaneously, of course, because our command signals take the better part of a minute to reach the clocks from our central location, and then the signals coming back from the clocks take just as long to get to us so that we can see what they say. Soon, however,

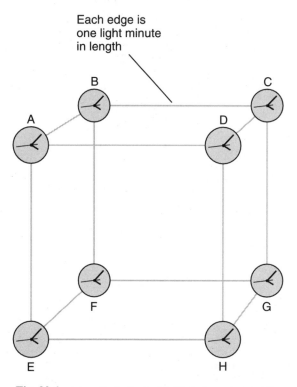

Each edge is
one light minute
in length

Fig. 20-1. A hypothetical set of eight clocks arranged at the vertices of a cube that measures 1 light-minute on each edge. How will we synchronize these clocks?

everything is in agreement. Clocks *A* through *H* all tell the same time to within a fraction of a second.

Satisfied with our work, we cruise out of the cube. We take a look back at the clocks. What do we see? The clocks have already managed to get out of sync. We take our ship back to the center of the cube to correct the problem. When we get there, however, there is no problem to correct! The clocks are all in agreement again.

You can guess what is happening here. The clock readings depend on how far their signals must travel to reach us. For an observer at the center of the cube, the signals from all eight clocks, *A* through *H,* arrive from exactly the same distance. However, this is not true for any other point in space. Therefore, the clocks can be synchronized only for that one favored point; if we go somewhere else, we will have to synchronize them all over again. This can be done, but then the clocks will be synchronized only when observed from the new favored vantage point. There is a unique *sync point*—the spot in space from which all eight clocks read the same—for each coordination of the clocks.

No sync point is more valid than any other from a scientific standpoint. If the cube happens to be stationary relative to some favored reference point such as the Earth, we can synchronize the clocks, for convenience, from that reference point. However, if the cube is moving relative to our frame of reference, we will never be able to keep the clocks synchronized. Time depends on where we are and on whether or not we are moving relative to whatever device we use to indicate the time. Time is not absolute, but relative, and there is no getting around it.

PROBLEM 20-1
Suppose that there is an atomic clock on the Moon (clock *M*), and that its time signals are broadcast by a powerful radio transmitter. This clock is set to precisely agree with another atomic clock in your home town on Earth (clock *E*), and it is also equipped with a radio transmitter. If you travel to the Moon, what will be the relative readings of the two clocks, as determined by listening to the radio signals?

SOLUTION 20-1
Radio signals travel through space at about 3.00×10^5 km/s. The Moon is about 4.0×10^5 km, or 1.3 light-seconds, away from Earth. The reading of clock *M* will be shifted approximately 1.3 s ahead in time (that is, earlier) because the time lag for its signals to reach you will be eliminated. The reading of clock *E* will be shifted about 1.3 s behind in time (that is, later) because a time lag will be introduced where previously there was none. When you get to the Moon, clock *M* will be approximately 2.6 s ahead of clock *E*.

Time Dilation

The relative location of an observer in space affects the relative readings of clocks located at different points. Similarly, relative motion in space affects the apparent rate at which time "flows." Isaac Newton hypothesized that time flows in an absolute way and that it constitutes a fundamental constant in the universe. Einstein showed that this is not the case; it is the speed of light, not time, that is constant. In order to understand why *relativistic time dilation* occurs based on Einstein's hypothesis, let's conduct a "mind experiment."

A LASER CLOCK

Suppose that we have a space ship equipped with a laser/sensor on one wall and a mirror on the opposite wall (Fig. 20-2). Imagine that the laser/sensor and the mirror are positioned so that the light ray from the laser must travel perpendicular to the axis of the ship, perpendicular to its walls, and (once we get it moving) perpendicular to its direction of motion. The laser and mirror are adjusted so that they are separated by 3.00 m. Because the speed of light in air is approximately 3.00×10^8 m/s, it takes 1.00×10^{-8} s, or 10.0 nanoseconds (10.0 ns), for the light ray to get across the ship from the laser to the mirror and another 10.0 ns for the ray to return to the sensor. The ray therefore requires 20.0 ns to make one round trip from the laser/sensor to the mirror and back again.

Our laser emits pulses of extremely brief duration, far shorter than the time required for the beam to get across the ship. We might even suppose that the beam emits just a few photons in each burst! We measure the time increment using an extremely sophisticated oscilloscope so that we can observe the pulses going out and coming back and measure the time lag between them. This is a special clock; its timekeeping ability is based on the speed of light, which Einstein proposed is constant no matter from what point of view it is observed. There is no better way to keep time.

CLOCK STATIONARY

Suppose that we start up the ship's engines and get moving. We accelerate with the eventual goal of reaching nearly the speed of light. Suppose that we manage to accelerate to a sizable fraction of the speed of light, and then we shut off the engines so that we are coasting through space. You ask, "Relative to what

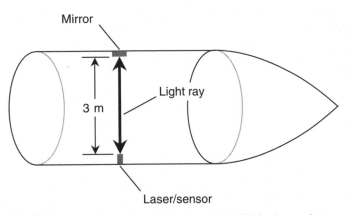

Fig. 20-2. A space ship equipped with a laser clock. This is what an observer
in the ship always sees.

are we moving?" This, as we shall see, is an important question! For now, suppose that we measure speed with respect to the Earth.

We measure the time it takes for the laser to go across the ship and back
again. We are riding along with the laser, the mirror, and all the luxuries of
a small spacecraft. We find that the time lag is still exactly the same as it
was when the ship was not moving relative to Earth; the oscilloscope still
shows a delay of 20.0 ns. This follows directly from Einstein's axiom. The
speed of light has not changed because it cannot. The distance between the
laser and the mirror has not changed either. Therefore, the round trip takes
the same length of time as it did before we got the ship moving.

If we accelerate so that the ship is going 60 percent, then 70 percent, and
ultimately 99 percent of the speed of light, the time lag will always be 20.0 ns
as measured from a *reference frame,* or point of view, inside the ship.

At this point, let's add another axiom to Einstein's: In free space, light
beams always follow the shortest possible distance between two points.
Normally, this is a straight line. You ask, "How can the shortest path
between two points in space be anything other than a straight line?" This is
another good question. We'll deal with it later in this chapter. For now, note
that light beams appear to follow straight lines through free space as long
as the observer is not accelerating relative to the light source.

CLOCK IN MOTION

Imagine now that we are outside the ship and are back on Earth. We are
equipped with a special telescope that allows us to see inside the ship as it

whizzes by at a significant fraction of the speed of light. We can see the laser, the mirror, and even the laser beam itself because the occupants of the space vessel have temporarily filled it with smoke to make the viewing easy for us. (They have pressure suits on so that they can breathe.)

What we see is depicted in Fig. 20-3. The laser beam still travels in straight lines, and it still travels at 3.00×10^8 m/s relative to us. This is true because of Einstein's axiom concerning the speed of light and the fact that light rays always appear to travel in straight lines as long as we are not accelerating. However, the rays have to travel farther than 3.00 m to get across the ship. The ship is going so fast that by the time the ray of light has reached the mirror from the laser, the ship has moved a significant distance forward. The same thing happens as the ray returns to the sensor from the mirror. As a result of this, it will seem to us, as we watch the ship from Earth, to take more than 20.0 ns for the laser beam to go across the ship and back.

As the ship goes by, time appears to slow down inside it, as seen from a "stationary" point of view. Inside the ship, however, time moves at normal speed. The faster the ship goes, the greater is this discrepancy. As the speed of the ship approaches the speed of light, the *time dilation factor* can become large indeed; in theory, there is no limit to how great it can become. You can visualize this by imagining Fig. 20-3 stretched out horizontally so that the light rays have to travel almost parallel to the direction of motion, as seen from the "stationary" reference frame.

FORMULA FOR TIME DILATION

There exists a mathematical relationship between the speed of the space ship in the foregoing "mind experiment" and the extent to which time is

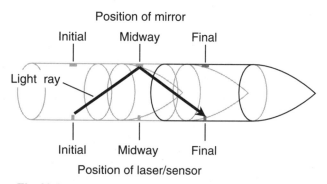

Fig. 20-3. This is what an external observer sees as the laser clock–equipped space ship whizzes by at a sizable fraction of the speed of light.

dilated. Let t_{ship} be the number of seconds that appear to elapse on the moving ship as precisely 1 s elapses as measured by a clock next to us as we sit in our Earth-based observatory. Let u be the speed of the ship as a fraction of the speed of light. Then

$$t_{ship} = (1 - u^2)^{1/2}$$

The time dilation factor (call it k) is the reciprocal of this; that is,

$$k = 1/[(1 - u^2)^{1/2}]$$
$$= (1 - u^2)^{-1/2}$$

In these formulas, the number 1 represents a mathematically exact value and is good to any number of significant digits.

Let's see how great the time dilation factor is if the space ship is traveling at 1.50×10^8 m/s. In this case, $u = 0.500$. If 1.00 s passes on Earth, then according to an earthbound observer,

$$t_{ship} = (1.00 - 0.500^2)^{1/2}$$
$$= (1.00 - 0.250)^{1/2}$$
$$= 0.750^{1/2}$$
$$= 0.866 \text{ s}$$

That is, 0.866 s will seem to pass on the ship as 1.00 s passes as we measure it while watching the ship from Earth. This means that the time dilation factor is 1.00/0.866, or approximately 1.15. Of course, on the ship, time will seem to "flow" normally.

Just for fun, let's see what happens if the ship is going 2.97×10^8 m/s. In this case, $s = 0.990$. If 1.00 s passes on Earth, then we, as earthbound observers, will see this:

$$t_{ship} = (1.00 - 0.990^2)^{1/2}$$
$$= (1.00 - 0.98)^{1/2}$$
$$= 0.0200^{1/2}$$
$$= 0.141 \text{ s}$$

That is, 0.141 s will seem to pass on the ship as 1.00 s passes on Earth. The time dilation factor k in this case is 1.00/0.141, or approximately 7.09. Time "flows" more than seven times more slowly on a ship moving at 99 percent of the speed of light than it "flows" on Earth—from the reference frame of someone on Earth.

As you can imagine, this has implications for time travel. According to the special theory of relativity, if you could get into a space ship and travel fast enough and far enough, you could propel yourself into the future. You might travel to a distant star, return to Earth in what seemed to you to be only a few months, and find yourself in the year 5000 A.D. When science-fiction writers realized this in the early 1900s just after Einstein published his work, they had a bonanza with it.

PROBLEM 20-2
Why don't we notice relativistic time dilation on short trips by car, train, or airplane? When we land, the clocks are still synchronized (except for time-zone differences in some cases). Why isn't the traveling clock behind the earth-based ones?

SOLUTION 20-2
Theoretically, they are. However, the difference is too small to be noticed. The time dilation factor is exceedingly small unless a vessel travels at a significant fraction of the speed of light. Its effect at normal speeds can't be measured unless atomic clocks, accurate to a minuscule fraction of 1 s, are used to measure the time in both reference frames.

PROBLEM 20-3
What speed is necessary to produce a time dilation factor of $k = 2.00$?

SOLUTION 20-3
Use the formula for time dilation, and let u be the unknown. Then u can be found, step by step, this way:

$$k = (1 - u^2)^{-1/2}$$

$$2.00 = (1 - u^2)^{-1/2}$$

$$0.500 = (1 - u^2)^{1/2}$$

$$0.250 = 1 - u^2$$

$$-0.750 = -u^2$$

$$u^2 = 0.750$$

$$u = (0.750)^{1/2} = 0.866$$

That is, the speed is 86.6 percent of the speed of light, or 2.60×10^8 m/s.

Spatial Distortion

Relativistic speeds—that is, speeds high enough to cause significant time dilation—cause objects to appear foreshortened in the direction of their motion. As with time dilation, *relativistic spatial distortion* occurs only from the point of view of an observer watching an object speed by at a sizable fraction of the speed of light.

POINT OF VIEW: LENGTH

If we travel inside a space ship, regardless of its speed, everything appears normal as long as our ship is not accelerating. We can cruise along at 99.9 percent of the speed of light relative to the Earth, but if we are inside a space ship, the ship is always stationary relative to us. Time, space, and mass appear normal from the point of view of passengers on a relativistic space journey. However, as we watch the space ship sail by from the vantage point of the Earth, its length decreases as its speed increases. Its diameter is not affected. The extent to which this happens is the same as the extent to which time slows down.

Let L be the apparent length of the moving ship as a fraction of its length when it is standing still relative to an observer. Let u be the speed of the ship as a fraction of the speed of light. Then

$$L = (1 - u^2)^{1/2}$$

This effect is shown in Fig. 20-4 for various relative forward speeds. The foreshortening takes place entirely in the direction of motion. This produces apparent physical distortion of the ship and everything inside, including the passengers. It's sort of like those mirrors in fun houses that are concave in only one dimension and reflect your image "all scrunched up." As the speed of the ship approaches the speed of light, its observed length approaches zero.

SUPPOSITIONS AND CAUTIONS

This is a curious phenomenon. You might wonder, based on this result, about the shapes of photons, the particles of which visible light and all other EM radiation are comprised. Photons travel at the speed of light. Does this mean that they are infinitely thin, flat disks or squares or triangles

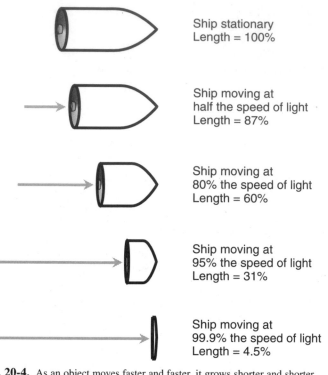

Ship stationary
Length = 100%

Ship moving at
half the speed of light
Length = 87%

Ship moving at
80% the speed of light
Length = 60%

Ship moving at
95% the speed of light
Length = 31%

Ship moving at
99.9% the speed of light
Length = 4.5%

Fig. 20-4. As an object moves faster and faster, it grows shorter and shorter
along the axis of its motion.

hurtling sidelong through space? No one has ever seen a photon, so no one knows how they are shaped. It is interesting to suppose that they are two-dimensional things and as such have zero volume. However, if they have zero volume, how can we say they exist?

Scientists know a lot about what happens to objects as they approach the speed of light, but we must not extrapolate and try to say what would happen if the speed of light could be attained by a material thing. We will see shortly that no physical object (such as a space ship) can reach the speed of light, so the notion of a real object being squeezed down to zero thickness is nothing more than an academic fantasy. As for photons, comparing them with material particles such as bullets or baseballs is an unjustified intuitive leap. We cannot bring a photon to rest, nor can we shoot a bullet or throw a baseball at the speed of light. As they might say in certain parts of the country, "Baseballs and photons ain't the same animals."

PROBLEM 20-4

Suppose that a space ship measures 19.5 m long at rest. How long will it look if it zooms by at a speed of 2.40×10^8 m/s?

SOLUTION 20-4

First, convert the speed to a fraction of the speed of light and call this fraction u:

$$u = (2.40 \times 10^8)/(3.00 \times 10^8)$$

$$= 0.800$$

Then use the formula for spatial distortion to find L, the fraction of its at-rest length:

$$L = (1 - u^2)^{1/2}$$

$$= (1 - 0.800^2)^{1/2}$$

$$= (1 - 0.640)^{1/2}$$

$$= 0.360^{1/2}$$

$$= 0.600$$

Finally, multiply the at-rest length of the vessel, 19.5 m, by 0.600 to get 11.7 m. This is how long the vessel will appear to be as it whizzes by at 2.40×10^8 m/s.

Mass Distortion

Another interesting effect of relativistic speeds is an increase in the masses of objects as they move faster and faster. This increase occurs to the same extent as the decrease in length and the slowing down of time.

POINT OF VIEW: MASS

If we travel inside a space ship, regardless of its speed, the masses of all the objects in the ship with us appear normal as long as our ship is not accelerating. However, from the vantage point of Earth, the mass of the ship and the masses of all the atoms inside it increase as its speed increases.

Let m be the mass of the moving ship as a multiple of its mass when it is stationary relative to an observer. Let u be the speed of the ship as a fraction of the speed of light. Then

$$m = 1/(1 - u^2)^{1/2}$$
$$= (1 - u^2)^{-1/2}$$

This is the same as the factor k that we defined a little while ago. It is always greater than or equal to 1.

Look again at Fig. 20-4. As the space ship moves faster, it "scrunches up." Imagine now that it also becomes more massive. The combination of smaller size and greater mass is a "double whammo" in regard to the density of the ship.

Suppose that the *rest mass* (the mass when stationary) of our ship is 10 metric tons. When it speeds by at half the speed of light, its mass increases to a little more than 11 metric tons. At 80 percent of the speed of light, its mass is roughly 17 metric tons. At 95 percent of the speed of light, the ship masses about 32 metric tons. At 99.9 percent of the speed of light, the ship's mass is more than 220 metric tons. And so it can go indefinitely. As the speed of the ship approaches the speed of light, its mass grows larger and larger without limit.

SPEED IS SELF-LIMITING

It is tempting to suppose that the mass of an object, if it could be accelerated all the way up to the speed of light, would become infinite. After all, as u approaches 1 (or 100 percent), the value of m in the preceding formula increases without limit. However, it's one thing to talk about what happens as a measured phenomenon or property approaches some limit; it is another matter entirely to talk about what happens when that limit is reached, assuming that it can be reached.

No one has ever seen a photon at rest. No one has ever seen a space ship moving at the speed of light. No finite amount of energy can accelerate any real object to the speed of light, and it is because of relativistic mass increase that this is so. Even if it were possible, the mass-increase factor, as determined by the preceding formula, would be meaningless. We would have to divide by zero to calculate it, and division by zero is not defined in mathematics.

The more massive a speeding space ship becomes, the more powerful is the rocket thrust necessary to get it moving faster. As a space ship approaches the speed of light, its mass becomes gigantic. This makes it harder and harder to give it any more speed. Using integral calculus, astronomers and physicists have proven that no finite amount of energy can propel a space ship to the speed of light.

HIGH-SPEED PARTICLES

You've heard expressions such as *electron rest mass,* which refers to the mass of an electron when it is not moving relative to an observer. If an electron is observed whizzing by at relativistic speed, it has a mass greater than its rest mass and thus will have momentum and kinetic energy greater than is implied by the formulas used in classical physics. This fact, unlike spatial distortion, is more than mere fodder for "mind experiments." When electrons move at high enough speed, they attain properties of much more massive particles and acquire some of the properties of x-rays or gamma rays such as are emitted by radioactive substances. There is a name for high-speed electrons that act this way: *beta particles.*

Physicists take advantage of the relativistic effects on the masses of protons, helium nuclei, neutrons, and other subatomic particles. When these particles are subjected to powerful electrical and magnetic fields in a device called a *particle accelerator,* they get moving so fast that their mass increases because of relativistic effects. When the particles strike atoms of matter, the nuclei of those target atoms are fractured. When this happens, energy can be released in the form of infrared (IR), visible light, ultraviolet (UV), x-rays, and gamma rays, as well as a potpourri of exotic particles.

If astronauts ever travel long distances through space in ships moving at speeds near the speed of light, relativistic mass increase will be a practical concern. While their own bodies won't seem to be more massive from their own point of view and the things inside the ship will appear normal to them, the particles whizzing by outside will become more massive in a real and dangerous way. It is scary enough to think about what will happen when a 1-kg meteoroid strikes a space ship traveling at 99.9 percent of the speed of light. However, that 1-kg stone will mass more than 22 kg when u = 0.999. In addition, every atom outside the ship will strike the vessel's "prow" at relativistic speed, producing deadly radiation of the same sort that occurs in high-energy particle accelerators.

EXPERIMENTAL CONFIRMATION

Relativistic time dilation and mass increase have both been measured under controlled conditions, and the results concur with Einstein's formulas stated earlier. Therefore, these effects are more than mere tricks of the imagination.

To measure time dilation, a superaccurate atomic clock was placed on board an aircraft, and the aircraft was sent up in flight to cruise around for awhile at several hundred kilometers per hour. Another atomic clock was kept at the place where the aircraft took off and landed. Although the aircraft's speed was only a tiny fraction of the speed of light and the resulting time dilation therefore was exceedingly small, it was large enough to measure. When the aircraft arrived back at the terminal, the clocks, which had been synchronized (when placed right next to each other, of course!) before the trip began, were compared. The clock that had been on the aircraft registered a time slightly earlier than the clock that had been resting comfortably on Earth.

To measure mass increase, particle accelerators are used. It is possible to determine the mass of a moving particle based on its known rest mass and the kinetic energy it possesses as it moves. When the mathematics is done, Einstein's formula is always shown to be correct.

PROBLEM 20-5
Suppose that a small meteoroid, whose mass is 300 milligrams (300 mg), strikes the shell of a space vessel moving at 99.9 percent of the speed of light. What is the apparent mass of the meteoroid?

SOLUTION 20-5
Use the preceding formula for relativistic mass increase, considering $u = 0.999$. Then the mass is multiplied by a factor m as follows:

$$m = (1 - 0.999^2)^{-1/2}$$

$$= (1 - 0.998)^{-1/2}$$

$$= 0.002^{-1/2}$$

$$= 1/(0.002)^{1/2}$$

$$= 1/0.0447$$

$$= 22.4$$

The mass of the meteoroid when it strikes the vessel is 300×22.4 mg, or 6.72 grams (g).

General Relativity

There is no absolute standard for location in the universe, nor is there an absolute standard for velocity. Another way of saying this is that any reference frame is just as valid as any other as long as acceleration does not take place. The notions of "the center of the universe" and "at rest" are relative. If we measure position or velocity, we must do so with respect to something, such as the Earth or the Sun or a space ship coasting through the void.

ACCELERATION IS DIFFERENT!

Einstein noticed something special about accelerating reference frames compared with those that are not accelerating. This difference is apparent if we consider the situation of an observer who is enclosed in a chamber that is completely sealed and opaque.

Suppose that you are in a space ship in which the windows are covered up and the radar and navigational equipment have been placed on standby. There is no way for you to examine the surrounding environment and determine where you are, how fast you are moving, or in what direction you are moving. However, you can tell whether or not the ship is accelerating. This is so because acceleration always produces a force on objects inside the ship.

When the ship's engines are fired and the vessel gains speed in the forward direction, all the objects in the ship (including your body) perceive a force directed backward. If the ship's retro rockets are fired so that the ship slows down (decelerates), everything in the ship perceives a force directed forward. If rockets on the side of the ship are fired so that the ship changes direction without changing its speed, this too is a form of acceleration and will cause everything inside the ship to perceive a sideways force. Some examples are illustrated in Fig. 20-5.

The greater the acceleration, or change in velocity, to which the space ship is subjected, the greater is the force on every object inside it. If m is the mass of an object in the ship (in kilograms) and a is the acceleration of the ship (in meters per second per second), then the force F (in newtons) is their product:

$$F = ma$$

This is one of the most well-known formulas in physics. You should recall it from Chapter 7.

This *acceleration force* occurs even when the ship's windows are covered up, the radar is switched off, and the navigational equipment is placed on standby. There is no way the force can be blocked out. In this way,

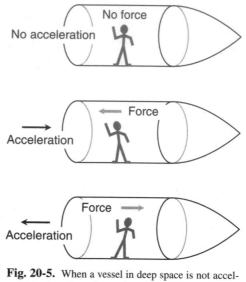

Fig. 20-5. When a vessel in deep space is not accelerating, there is no force on the objects inside. When the ship accelerates, there is always a force on the objects inside.

Einstein reasoned, it is possible for interstellar travelers to determine whether or not their ship is accelerating. Not only this, but they can calculate the magnitude of the acceleration as well as its direction. When it comes to acceleration, there are, in a certain sense, absolute reference frames in the cosmos.

THE EQUIVALENCE PRINCIPLE

Imagine that our space ship, instead of accelerating in deep space, is set down on the surface of a planet. It might be tail-downward, in which case the force of gravity pulls on the objects inside as if the ship were accelerating in a forward direction. It might be nose-downward so that gravity pulls on the objects inside as if the ship were decelerating. It could be oriented some other way, and the force of gravity would pull on the objects inside as if the ship were changing course in a lateral direction. Acceleration can consist of a change in speed, a change in direction, or both.

If the windows are kept covered, the radar is shut off, and the navigational aids are placed on standby, how can passengers in such a vessel tell whether the force is caused by gravity or by acceleration? Einstein's answer: They cannot tell the difference.

From this notion came the *equivalence principle*. The so-called acceleration force is exactly the same as gravitation. Einstein reasoned that the two forces act in an identical way on everything, from people's perceptions to atoms and from light rays to the fabric of space-time. This is the basis of the theory of general relativity.

SPATIAL CURVATURE

Imagine that you are in a space ship traveling through deep space. The ship's rockets are fired, and the vessel accelerates at an extreme rate. Suppose that the laser apparatus described earlier in this chapter is in the ship, but instead of a mirror on the wall opposite the laser, there is a screen. Before the acceleration begins, you align the laser so that it shines at the center of the screen (Fig. 20-6). What will happen when the rockets are fired and the ship accelerates?

In a real-life scenario, the spot from the laser will not move on the screen enough for you to notice. This is so because any reasonable (that is, non-life-threatening) rate of acceleration will not cause sufficient force to affect the beam. However, let's suspend our disbelief and imagine that we can accelerate the vessel at any rate, no matter how great, without being squashed against the ship's rear wall. If we accelerate fast enough, the ship pulls away from the laser beam as the beam travels across the ship. We, looking at the situation from inside the ship, see the light beam follow a curved path (Fig. 20-7). A stationary observer on the outside sees the light

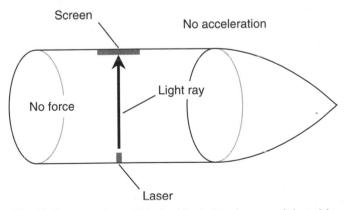

Fig. 20-6. As seen from within the ship, the laser beam travels in straight line across the vessel when it is not accelerating.

beam follow a straight path, but the vessel pulls out ahead of the beam (Fig. 20-8).

Regardless of the reference frame, the ray of light always follows the shortest possible path between the laser and the screen. When viewed from any nonaccelerating reference frame, light rays appear straight. When observed from accelerating reference frames, however, light rays appear curved. The shortest distance between the two points at opposite ends of the laser beam in Fig. 20-7 is, in fact, curved. The apparently straight path is in reality longer than the curved one, as seen from inside the accelerating vessel! It is this phenomenon that has led some people to say that "space is curved" in a powerful acceleration field. Because of the principle of equivalence, powerful gravitation causes the same *spatial curvature.*

For spatial curvature to be as noticeable as it appears in Figs. 20-7 and 20-8, the vessel must accelerate at an extremely large pace. The standard unit of acceleration is the meter per second per second, or *meter per second squared* (m/s^2). Astronauts and aerospace engineers also express acceleration in units called *gravities* (symbolized g), where one gravity ($1 \, g$) is the acceleration that produces the same force as the gravitational field of the Earth at the surface, approximately 9.8 m/s^2. (Don't confuse the abbreviation for gravity or gravities with the abbreviation for grams! Pay attention to the context if you see a unit symbolized g.) The drawings of Figs. 20-7 and 20-8 show the situation for an acceleration of many thousands of gravities. If you weigh 150 pounds on Earth, you will weigh many tons in a ship accelerating at a rate, or in a gravitational field of such intensity, so as to cause that much spatial curvature.

Is this a mere academic exercise? Are there actually gravitational fields powerful enough to "bend light rays" significantly? Yes. They exist near the event horizons of black holes.

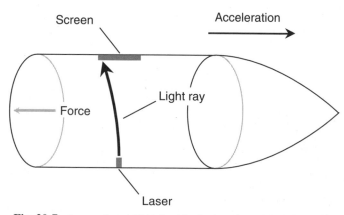

Fig. 20-7. As seen from within the ship, the laser beam travels in curved path across the vessel when it is accelerating at a high rate.

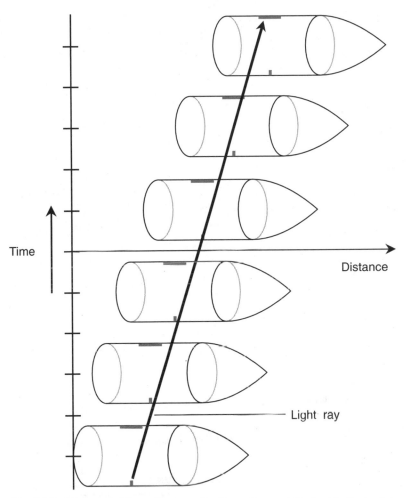

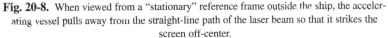

Fig. 20-8. When viewed from a "stationary" reference frame outside the ship, the acceler-ating vessel pulls away from the straight-line path of the laser beam so that it strikes the screen off-center.

TIME DILATION CAUSED BY ACCELERATION OR GRAVITATION

The spatial curvature caused by intense acceleration or gravitation produces an effective slowing down of time. Remember the fundamental axiom of spe-cial relativity: The speed of light is constant, no matter what the point of view. The laser beam traveling across the space ship, as shown in some of the illustrations in this chapter, always moves at the same speed. This is one thing about which all observers in all reference frames must agree.

The path of the light ray as it travels from the laser to the screen is longer in the situation shown by Fig. 20-7 than in the situation shown by Fig. 20-6. This is partly because the ray takes a diagonal path rather than traveling straight across. In addition, however, the path is curved. This increases the time interval even more. From the vantage point of a passenger in the space ship, the curved path shown in Fig. 20-7 represents the shortest possible path the light ray can take across the vessel between the point at which it leaves the laser and the point at which it strikes the screen.

The laser device itself can be turned slightly, pointing a little bit toward the front of the ship; this will cause the beam to arrive at the center of the screen (Fig. 20-9) instead of off-center. However, the path of the beam is still curved and is still longer than its path when the ship is not accelerating (see Fig. 20-6). The laser represents the most accurate possible timepiece because it is based on the speed of light, which is an absolute constant. Thus time dilation is produced by acceleration not only as seen by observers looking at the ship from the outside but also for passengers within the vessel itself. In this respect, acceleration and gravitation are even more powerful "time dilators" than relative motion.

Suspending our disbelief again, and assuming that we could experience such intense acceleration force (or gravitation) without being physically crushed, we will actually perceive time as slowing down inside the vessel under conditions such as those that produce spatial curvature such as shown in Figs. 20-7, 20-9 or 20-10. Clocks will really seem to run more slowly, even from reference frames inside the ship. In addition, everything inside the ship will appear bent out of shape.

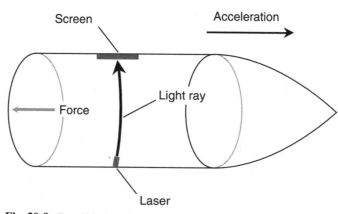

Fig. 20-9. Even if the laser is turned so that the light ray hits the center of the screen, the path of the ray is curved when the ship accelerates at a high rate.

OBSERVATIONAL CONFIRMATION

When Einstein developed his general theory of relativity, some of the paradoxes inherent in special relativity were resolved. (These paradoxes have
been avoided here because discussing them would only confuse you.) In
particular, light rays from distant stars were observed as they passed close
to the Sun to see whether or not the Sun's gravitational field, which is quite
strong near its surface, would bend the light rays. This bending would be
observed as a change in the apparent position of a distant star in the sky as
the Sun passes close to it (Fig. 20-10).

The problem with this type of observation was, as you might guess, the
fact that the Sun is far brighter than any other star in the sky, and the Sun's
light normally washes out the faint illumination from distant stars.
However, during a total solar eclipse, the Sun's disk is occulted by the
Moon. In addition, the angular diameter of the Moon in the sky is almost
exactly the same as that of the Sun, so light from distant stars passing close
to the Sun can be seen by earthbound observers during a total eclipse.
When this experiment was carried out, the apparent position of a distant
star was indeed offset by the presence of the Sun, and this effect took place
to the same extent as Einstein's general relativity formulas said it should.

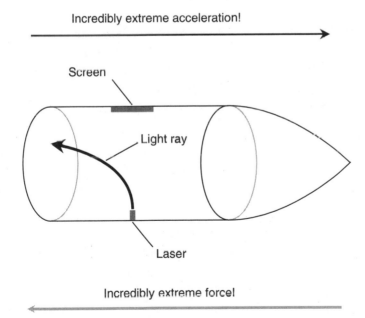

Fig. 20-10. If the acceleration is great enough, the spatial curvature
becomes extreme.

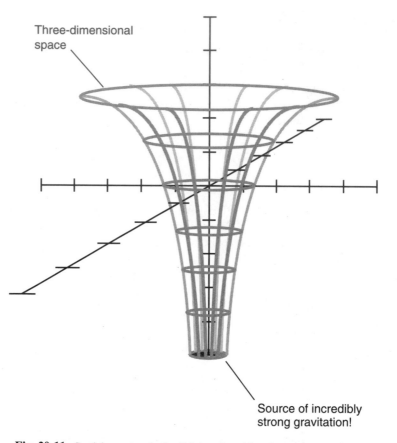

Three-dimensional space

Source of incredibly strong gravitation!

Fig. 20-11. Spatial curvature in the vicinity of an object that produces an intense gravitational field.

More recently, the light from a certain quasar has been observed as it passes close to a suspected black hole. On its way to us, the light from the quasar follows multiple curved paths around the dark, massive object. This produces several images of the quasar, arranged more or less in the form of a "plus sign" or "cross" with the dark object at the center.

The curvature of space in the presence of a strong gravitational field has been likened to a funnel shape (Fig. 20-11), except that the surface of the funnel is three-dimensional rather than two-dimensional. The shortest distance in three-dimensional space between any two points near the gravitational source is always a curve with respect to four-dimensional space. This is impossible for most (if not all) people to envision directly without "cheating" by taking away one dimension. The mathematics is straightforward enough, though, and observations have shown that it correctly explains the phenomenon.

AND SO...

We have conducted "mind experiments" in this chapter, many of which require us to suspend reality. In real life, scenarios such as these would kill anyone attempting to make the observations. So why is relativity theory important? If space is bent and time is slowed by incredibly powerful gravitational fields, so what?

General relativity plays an important role in the development of theories concerning the geometry and evolution of the universe. On a vast scale, gravitation acquires a different aspect than on a local scale. A small black hole, such as that surrounding a collapsed star, is dense and produces gravitation strong enough to destroy any material thing crossing the event horizon. However, if a black hole contains enough mass, its density is not necessarily great. Black holes with quadrillions of solar masses can exist, at least in theory, without life-threatening forces at any point near their event horizons. If such a black hole is ever found, and if we develop space ships capable of intergalactic flight, we will be able to cross its event horizon unscathed, leave this universe, and enter another—forever.

Quiz

Refer to the text in this chapter if necessary. A good score is eight correct. Answers are in the back of the book.

1. A common unit of acceleration is the
   (a) meter per second.
   (b) kilometer per second.
   (c) kilometer per hour.
   (d) gravity.

2. Suppose that you have a spherical ball with mass of a hundred grams (100 g) at rest. If you throw the ball at three-quarters of the speed of light, what will its mass become, as measured from a stationary point of view?
   (a) 100 g
   (b) 133 g
   (c) 151 g
   (d) It cannot be calculated from this information.

3. Suppose that the ball in quiz question 2 has an apparent diameter, as measured laterally (sideways to the direction of its motion), of a hundred millimeters (100 mm) when it is speeding along at three-quarters of the speed of light. What will its diameter be when it comes to rest?
   (a) 100 mm
   (b) 133 mm
   (c) 151 mm
   (d) It cannot be calculated from this information.

4. If a space ship is slowing down, that is, losing speed in the forward direction, the perceived force inside the ship is directed
   (a) toward the rear.
   (b) toward the front.
   (c) toward the side.
   (d) nowhere; there is no acceleration force.

5. The Michelson-Morley experiment
   (a) showed that the speed of light depends on the direction in which it is measured.
   (b) showed that the speed of light depends on the velocity of the observer.
   (c) showed that the speed of light does not depend on the direction in which it is measured.
   (d) proved that the ether passes right through the Earth.

6. If you are in a spacecraft accelerating at 9.8 m/s^2 through interplanetary space, you will feel the same force you would feel if you were sitting still on the surface of the Earth. This is an expression of
   (a) a complete falsehood! Traveling through space is nothing at all like being on Earth.
   (b) the fact that the speed of light is absolute, finite, and constant and is the fastest known speed.
   (c) Einstein's equivalence principle.
   (d) the results of the Michelson-Morley experiment.

7. Suppose that you see a space ship whiz by at the speed of light. What is the time dilation factor k that you observe when you measure the speed of a clock inside that ship and compare it with the speed of a clock that is stationary relative to you?
   (a) 1
   (b) 0
   (c) Infinity
   (d) It is not defined

8. Some light beams will follow curved paths
   (a) under no circumstances.
   (b) when measured inside a space ship that is coasting at high speed.
   (c) when measured in the presence of an extreme gravitational field.
   (d) when measured from a reference frame that is not accelerating.

9. Suppose that you get on a space ship and travel toward the star Sirius at 150,000 km/s, which is approximately half the speed of light. If you measure the speed of the light arriving from Sirius, what figure will you obtain?
   (a) 150,000 km/s
   (b) 300,000 km/s
   (c) 450,000 km/s
   (d) It cannot be calculated from this information.

10. Clocks in different locations are impossible to synchronize from every possible reference frame because
   (a) the speed of light is absolute, finite, and constant and is the fastest known speed.
   (b) the speed of light depends on the location of the reference frame from which it is measured.
   (c) the speed of light depends on velocity of the reference frame from which it is measured.
   (d) there is no such thing as a perfect clock.

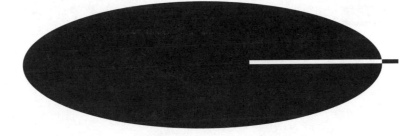

Test: Part Three

Do not refer to the text when taking this test. A good score is at least 37 correct. Answers are in the back of the book. It is best to have a friend check your score the first time so that you won't memorize the answers if you want to take the test again.

1. A wave has a frequency of 10^5 Hz. What is its period in microseconds?
   (a) 10^{-5}
   (b) 10^{-2}
   (c) 10
   (d) 1,000
   (e) It depends on the speed of propagation.

2. A cassegrain reflector
   (a) has a concave objective mirror and a flat secondary mirror.
   (b) has a concave objective mirror and a convex secondary mirror.
   (c) has a longer tube than an equivalent Newtonian reflector.
   (d) suffers from lens sag if the diameter of the objective is too large.
   (e) has the eyepiece mounted in the side of the tube.

3. Localized doses of radiation are sometimes used to
   (a) control insects.
   (b) improve female fertility.
   (c) treat cataracts.
   (d) treat cancerous growths.
   (e) alleviate nausea.

4. Which of the following statements is *false?*
   (a) A convex lens can bring parallel light rays to a focus.
   (b) A concave mirror can collimate light rays from a point source.
   (c) A convex mirror can be used as the main objective in a reflecting telescope.
   (d) A concave lens makes close-up objects appear smaller.
   (e) A convex lens can be used as the main objective in a refracting telescope.

5. Two waves mix together. Their frequencies are 5.00 MHz and 500 kHz, respectively. Which of the following is a beat frequency resulting from the heterodyning of these two signals?
   (a) 2.50 MHz
   (b) 2.50 GHz
   (c) 10.0 MHz
   (d) 4.50 MHz
   (e) None of the above

6. Regardless of whether a reference frame is accelerating or not, light rays always
   (a) travel in straight lines.
   (b) follow the shortest possible path between two points in space.
   (c) travel in curved paths.
   (d) are repelled by gravitational fields.
   (e) travel fastest in the direction of motion.

7. Suppose that a sample of a substance contains a huge number of radioactive atoms. You measure the length of time required for 10,000 of them to degenerate and then average all the results. The average decay time thus derived is called the
   (a) half-life.
   (b) mean life.
   (c) decay life.
   (d) ionizing life.
   (e) degeneration life.

8. When light from the Sun shines on a spherical reflective object such as a steel ball bearing, the reflected rays
   (a) are all parallel.
   (b) diverge.
   (c) converge.
   (d) focus to a point hot enough to start fires.
   (e) behave unpredictably.

9. Imagine that a space ship whizzes by so fast that clocks on board seem, as seen from our point of view on Earth, to be running at one-third their normal speed. Suppose that you mass 60 kg on Earth and have a friend riding on the ship who also masses 60 kg on Earth. If your friend measures her mass while traveling on the ship, what will she observe it to be?
   (a) 20 kg
   (b) 60 kg
   (c) 180 kg
   (d) 540 kg
   (e) It cannot be determined without more information.

10. Electromagnetic (EM) waves at radio frequencies
    (a) are visible to the unaided eye.
    (b) are a form of ionizing radiation that can cause genetic mutations.
    (c) can propagate long distances through the air.
    (d) are completely blocked by the atmosphere.
    (e) produce curvature of space-time.

11. Which of the following statements is *false?*
    (a) The visible-light range is a small part of the EM spectrum.
    (b) Gamma rays have high penetrating power.
    (c) Extremely-low-frequency (ELF) radiation is a form of radioactivity.
    (d) Microwaves are longer than infrared (IR) waves.
    (e) X-rays are shorter than visible-light waves.

12. In a compound microscope,
    (a) the objective is a convex lens and the eyepiece is a concave lens.
    (b) both the objective and the eyepiece are concave lenses.
    (c) the objective is a concave mirror and the eyepiece is a convex lens.
    (d) the eyepiece has a longer focal length than the objective.
    (e) the eyepiece has a shorter focal length than the objective.

13. Carbon dating is commonly used to estimate
    (a) the size of the universe.
    (b) the temperature on the surface of the Sun.
    (c) the rate at which the ozone layer is deteriorating.
    (d) the altitude of the ionosphere.
    (e) none of the above.

14. Fill in the blank in the following sentence: "When a space ship moves at a speed approaching the speed of light relative to an observer, that observer will see a clock on the ship appear to _____."
    (a) run too fast
    (b) stop
    (c) run too slowly
    (d) run infinitely fast
    (e) run at normal speed

15. The so-called double-slit experiment demonstrates that
    (a) sound waves can turn corners.
    (b) electrons and positrons have opposite electric charge.
    (c) visible light has wavelike properties.
    (d) matter consists mostly of empty space.
    (e) electromagnetic fields can travel through a vacuum.

16. Suppose that a space ship whizzes by so fast that clocks on board seem, as seen from our point of view, to be running at half speed. If the rest mass of the ship is 50 metric tons, what will be the mass of the ship from our point of view as it whizzes by?
    (a) 25 metric tons
    (b) 50 metric tons
    (c) 100 metric tons
    (d) 400 metric tons
    (e) It cannot be determined without more information.

17. An EM field has a wavelength of 120 m. This is
    (a) an ELF field.
    (b) a radio wave.
    (c) microwave energy.
    (d) infrared energy.
    (e) x-ray energy.

18. Time travel into the future might be possible by taking advantage of
    (a) relativistic time dilation.
    (b) relativistic mass distortion.
    (c) relativistic spatial distortion.
    (d) the gravitational pull of the Earth.
    (e) nothing! Time travel into the future is theoretically impossible.

19. Which of the following statements is *false?*
    (a) A glass prism bends green light more than it bends orange light.
    (b) The focal length of a simple glass convex lens is shorter for green light than for orange light.
    (c) Dispersion occurs when white light passes through a simple glass lens.
    (d) The index of refraction of glass depends on the color of the light shining through it.
    (e) All of the above statements are true.

20. All waves, no matter what the medium, have three distinct properties that depend on each other. These three properties are
    (a) frequency, wavelength, and propagation speed.
    (b) frequency, amplitude, and propagation speed.
    (c) amplitude, waveform, and period.
    (d) wavelength, amplitude, and period.
    (e) waveform, propagation speed, and amplitude.

21. The maximum diameter of a refracting telescope is limited, in practice, by
    (a) lens sag.
    (b) spherical aberration.
    (c) paraboloidal aberration.
    (d) focal length.
    (e) dispersion.

22. Infrared radiation at low or moderate intensity
    (a) can be felt as heat or warmth on the skin.
    (b) is reflected by the Earth's ionosphere.
    (c) appears red or orange in color.
    (d) has extreme penetrating power.
    (e) can cause slow radioactive decay.

23. Suppose that two superaccurate atomic clocks, called clock *A* and clock *B*, are synchronized on Earth so that they agree exactly. Now imagine that clock *B* is placed aboard a space vessel and sent to Mars and back. The clock readings are compared after the ship returns. What do we find?
    (a) Clocks *A* and *B* still agree precisely.
    (b) Clock *A* is behind clock *B*.
    (c) Clock *A* is ahead of clock *B*.
    (d) Any of the above, depending on the extent to which the ship accelerated during its journey.
    (e) None of the above.

24. Fill in the blank in the following sentence: "As the focal length of the objective lens in a compound microscope is reduced, and if all other factors remain constant, _____."
    (a) the magnification decreases
    (b) the field of view increases
    (c) the resolution decreases
    (d) the magnification increases
    (e) nothing changes

25. Suppose that you tune your car radio to a station in the AM broadcast band and you can hear that station even when driving through a valley or ravine. This effect is caused by
    (a) wave diffraction.
    (b) refraction.
    (c) total internal reflection.
    (d) a manifestation of Snell's law.
    (e) Earth's magnetic field.

26. Einstein's principle of equivalence states that
    (a) gravitational force is just like acceleration force.
    (b) force equals mass times acceleration.
    (c) the speed of light is constant, no matter what.
    (d) the speed of light is the highest possible speed.
    (e) the shortest distance between two points is a straight line.

27. Which of the following types (a, b, c, or d) of radiation is not ionizing?
    (a) Extremely low frequency
    (b) X rays
    (c) Gamma rays

   (d) Primary cosmic particles
   (e) All of the above types of radiation are ionizing.

28. A corpuscle of visible light is called
   (a) a positron.
   (b) an electron.
   (c) a neutron.
   (d) a photon.
   (e) an illumitron.

29. Suppose that a certain material transmits light at a speed of 150,000 km/s.
    What is its index of refraction, accurate to three significant figures?
   (a) 0.500
   (b) 0.805
   (c) 1.00
   (d) 1.24
   (e) 2.00

30. The Earth's ionosphere
   (a) blocks all radio waves coming from space.
   (b) shields us from the Sun's x-rays.
   (c) can give rise to damaging ultraviolet (UV) radiation.
   (d) exists in layers at various altitudes above the surface.
   (e) disappears during the daytime.

31. Suppose that there are two sound waves traveling in the air, wave A and wave B.
    The frequency of wave A is 500 Hz. The frequency of wave B is 2,500 Hz.
    What can be said about these waves?
   (a) Wave B travels five times as fast as wave A.
   (b) Wave A travels five times as fast as wave B.
   (c) The amplitude of wave B is five times the amplitude of wave A.
   (d) The amplitude of wave A is five times the amplitude of wave B.
   (e) The two waves are harmonically related.

32. Spatial distortion can be caused by all of the following *except*
   (a) acceleration.
   (b) gravitation.
   (c) high relative speed.
   (d) black holes.
   (e) the solar wind.

33. Fill in the blank in the following statement to make it true: "The light from a point
    source can be _____ by a convex lens."
   (a) minimized
   (b) diffracted
   (c) collimated
   (d) reflected
   (e) blocked

34. High levels of UV radiation, either short term or long term, are known or believed to cause all the following *except*
    (a) long-term suppression of the immune system in humans.
    (b) fluorescence of certain materials.
    (c) skin cancer.
    (d) cataracts in the eyes.
    (e) ozone depletion.

35. When Michelson and Morley measured the speed of light in various directions, they discovered that
    (a) the speed of light is slowest in the direction in which the Earth travels through space.
    (b) the speed of light is fastest in the direction in which the Earth travels through space.
    (c) the Earth drags the luminiferous ether along with itself.
    (d) the speed of light is the same in all directions.
    (e) the speed of light cannot be accurately determined.

36. Which of the following statements is *true* for an EM wave in free space? Assume that propagation speed is expressed in meters per second, period is expressed in seconds, frequency is expressed in hertz, and wavelength is expressed in meters.
    (a) Propagation speed is equal to frequency times wavelength.
    (b) Propagation speed is equal to frequency divided by wavelength.
    (c) Propagation speed is equal to frequency times period.
    (d) Period is equal to propagation speed divided by wavelength.
    (e) Frequency is equal to propagation speed times wavelength.

37. Suppose that the path of the light from a distant quasar passes by an extremely massive, dense, dark object that is closer to us. Multiple images of the quasar appear around the dark object. This is the result of
    (a) time dilation.
    (b) red shift.
    (c) spatial curvature.
    (d) spherical aberration.
    (e) chromatic aberration.

38. The term *auroral propagation* refers to
    (a) reflection of radio waves by the aurorae.
    (b) the tendency for aurorae to occur after a solar flare.
    (c) a tendency for the aurorae to occur near the geomagnetic poles.
    (d) the strange movements commonly observed in the aurorae.
    (e) effects of the aurorae on living tissue.

39. Imagine a solid sphere of glass that is perfectly transparent and perfectly uniform with a spherical hollow space in the exact center. Imagine a light bulb, call it lamp *A*, with a point-source filament located at the center of the spherical hollow space and therefore also at the center of the whole sphere of glass.

Imagine a second light bulb in the open air, also with a point-source filament; call it lamp *B*. How do the rays of light from the two lamps compare in their behavior?

(a) The rays from both lamps radiate outward in straight lines and in exactly the same way.

(b) The rays from lamp *A* are reflected totally inside the cavity within the sphere of glass, but the rays from lamp *B* radiate outward in straight lines.

(c) The rays from lamp *A* converge to a point somewhere outside the sphere of glass, but the rays from lamp *B* radiate outward in straight lines.

(d) The rays from lamp *A* diverge more when they emerge from the sphere of glass compared with the rays from lamp *B* that do not have to pass through the glass.

(e) It is impossible to say without more information.

40. A cordless telephone is advertised as operating at a frequency of 900 MHz. This is the same as a frequency of
(a) 9.00×10^5 Hz.
(b) 0.900 GHz.
(c) 9.00×10^{-4} GHz.
(d) 0.900 THz.
(e) 9.00×10^8 kHz.

41. A common source of ELF radiation is
(a) uranium.
(b) a light bulb.
(c) a wire carrying direct current.
(d) a solar cell.
(e) an electrical utility line.

42. In a compound microscope, the adverse effects of chromatic aberration can be practically eliminated by
(a) using a mirror for the objective rather than a lens.
(b) illuminating the specimen from behind.
(c) illuminating the specimen from in front.
(d) increasing the distance between the objective and the eyepiece.
(e) using monochromatic light to illuminate the specimen.

43. The occurrence of time dilation, as predicted by Einstein's special theory of relativity, was demonstrated by
(a) placing atomic clocks in gravitational fields of various intensities.
(b) measuring the red shift of light as it passes near the Sun on its way to us from distant stars.
(c) observing the curvature of space in a fast-moving vessel.
(d) comparing the reading of an atomic clock on an aircraft with the reading of a similar clock on Earth.

(e) none of the above; it cannot be demonstrated at speeds attainable using present technology.

44. A wave that concentrates all its energy at a single frequency has a shape that can be described as
    (a) square.
    (b) rectangular.
    (c) triangular.
    (d) sawtooth.
    (e) sinusoidal.

45. Relativistic spatial distortion, resulting from high relative velocity, occurs
    (a) only at speeds faster than the speed of light.
    (b) only when objects accelerate.
    (c) only along the axis of relative motion.
    (d) only for extremely dense or massive objects.
    (e) only within black holes.

46. The depth of field in a compound microscope at high magnification
    (a) is essentially infinite.
    (b) is large, on the order of several kilometers.
    (c) is small, on the order of a few micrometers.
    (d) depends on the illumination level.
    (e) depends on the type of eyepiece used.

47. Suppose that a light beam consists of photons, all of which contain the same amount of energy. If the frequency of the light waves is multiplied by 5 (so that it becomes UV), what will happen to the energy contained in each photon?
    (a) It will not change.
    (b) It will become 5 times as great.
    (c) It will become 25 times as great.
    (d) It will become $^1/_5$ as great.
    (e) It will become $^1/_{25}$ as great.

48. A rising Moon sometimes appears reddish because
    (a) red light is scattered by the atmosphere more than other light.
    (b) red light is scattered by the atmosphere less than other light.
    (c) dust in the air always has a reddish hue.
    (d) atoms in the air are actually red.
    (e) an optical illusion occurs.

49. A telescope can be made using
    (a) a concave objective lens and a convex eyepiece lens.
    (b) a convex objective lens and a concave eyepiece lens.
    (c) a convex objective mirror and a concave eyepiece lens.
    (d) a concave objective lens and a concave eyepiece lens.
    (e) any of the above.

50. Suppose that there are two sound waves, wave A and wave B, traveling through the air. They have identical frequency, say, 800 Hz. Wave A is a square wave, and wave B is a sawtooth. What can be said about these waves?

(a) Waves A and B both concentrate all their energy at 800 Hz.

(b) Wave A has a different timbre than wave B.

(c) Waves A and B travel through the air at different speeds.

(d) Waves A and B have different amplitudes.

(e) Waves A and B have different wavelengths.

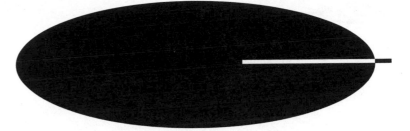

Final Exam

This exam contains questions from material in Parts One, Two, and Three but not from Part Zero. Do not refer to the text when taking this exam. A good score is at least 75 correct. Answers are in the back of the book. It is best to have a friend check your score the first time so that you won't memorize the answers if you want to take the test again.

1. A pellet gun shoots a projectile whose mass is 0.125 g at a speed of 100 m/s. Neglecting the effect of gravity, what is the magnitude of the momentum vector of the object as it leaves the gun?
    (a) 0.00125 kg · m/s
    (b) 0.0125 kg · m/s
    (c) 0.125 kg · m/s
    (d) 1.25 kg · m/s
    (e) 12.5 kg · m/s

2. Suppose that you are on board a spacecraft in a weightless environment. You attach a ball to a string and whirl the ball around your body at a constant angular speed. The acceleration vector of the ball
    (a) points in the same direction as the ball is moving at any given instant in time.
    (b) always points inward toward you.
    (c) always points outward away from you.
    (d) is zero because the velocity is not changing.
    (e) points in a direction perpendicular to the plane in which the ball revolves.

3. It takes a certain amount of energy to change a sample of solid matter to its liquid state, assuming that the matter is of the sort that can exist in either of these two states. This quantity varies for different substances and is called the
    (a) liquefaction energy.
    (b) melting energy.
    (c) heat of fusion.
    (d) thawing energy.
    (e) critical point.

4. Unless acted on by an outside force, an object moving with uniform velocity
   (a) gradually comes to rest.
   (b) abruptly comes to rest.
   (c) falls to the Earth.
   (d) continues to move at that velocity.
   (e) has no momentum.

5. Two-thirds of an ac cycle is the equivalent of
   (a) 60 degrees of phase.
   (b) 120 degrees of phase.
   (c) 180 degrees of phase.
   (d) 240 degrees of phase.
   (e) 270 degrees of phase.

6. Three inductors, each with a value of 30 μH, are connected in parallel. There is no mutual inductance among them. The net inductance of the parallel-connected set is
   (a) 30 μH.
   (b) 90 μH.
   (c) 10 μH.
   (d) dependent on the frequency of the ac passing through it.
   (e) impossible to determine.

7. Suppose that a jar is filled with liquid, and then another liquid that does not react chemically with the first liquid is introduced into the jar. Gradually, the two liquids blend together. This process is
   (a) dispersion.
   (b) diffusion.
   (c) a change of state.
   (d) fusion.
   (e) molecular averaging.

8. In an electrical circuit, the complex number $5 - j7$ represents
   (a) 5 ohms of reactance and 7 farads of capacitance.
   (b) 5 ohms of resistance and 7 ohms of capacitive reactance.
   (c) 5 ohms of resistance and 7 ohms of inductive reactance.
   (d) 5 ohms of reactance and 7 henrys of inductance.
   (e) none of the above.

9. When a semiconductor diode is reverse-biased with a steady dc voltage less than the avalanche voltage,
   (a) it conducts well.
   (b) it conducts some of the time.
   (c) it conducts poorly or not at all.
   (d) it has low resistance.
   (e) it has high inductance.

10. Protons and neutrons
    (a) have nearly the same mass, but protons have electric charge while neutrons do not.
    (b) have vastly different masses and equal but opposite electric charges.
    (c) have vastly different masses but identical electric charges.
    (d) have zero mass, no electric charge, and travel at the speed of light.
    (e) annihilate if they collide.

11. Suppose that a certain computer has a clock speed specified in gigahertz. If you wanted to talk about the clock speed in terahertz instead, you would use a number
    (a) a thousand times larger.
    (b) a million times larger.
    (c) a thousand times smaller.
    (d) a million times smaller.
    (e) the same size.

12. Even if the voltage across a sample is large, the current through it is small if
    (a) the ohmic value is low.
    (b) the conductance is high.
    (c) electrons pass easily from atom to atom.
    (d) the material is a conductor such as copper or silver.
    (e) the resistance is high.

13. One microwatt is the equivalent of
    (a) 10^{-6} joule-seconds.
    (b) 10^{-6} joules per second.
    (c) 10^{-6} ampere-seconds.
    (d) 10^{-6} amperes per second.
    (e) 10^{-6} ergs per second.

14. If you place a kettle of water on a hot stove, heat is transferred from the burner to the kettle and from the kettle to the water. This is an example of
    (a) convection.
    (b) conduction.
    (c) evaporation.
    (d) condensation.
    (e) radiation.

15. A specific impedance can be completely defined in terms of
    (a) a scalar quantity.
    (b) a vector quantity.
    (c) a magnetic field.
    (d) an electrical field.
    (e) a combination of voltage and current.

16. Demodulation is the process of
    (a) recovering the data from a signal carrier.
    (b) impressing data onto a signal carrier.
    (c) conversion of ac to dc.
    (d) conversion of dc to ac.
    (e) elimination of unwanted fluctuations in a signal.

17. When the relative hardness of two substances is determined according to the Mohs scale, the following rule or rules apply:
    (a) A substance always scratches something softer than itself; a substance never scratches anything harder than itself.
    (b) A substance cannot scratch anything softer than itself.
    (c) Substances scratch each other only when one substance is much harder than the other.
    (d) Harder substances are shattered when struck by softer substances.
    (e) None of the above are true.

18. Extremely-low-frequency (ELF) radiation is not
    (a) produced by computer monitors.
    (b) a form of electromagnetic field.
    (c) a form of ionizing radiation.
    (d) a subject of any interest to scientists or engineers.
    (e) produced by fluctuating electrical or magnetic fields.

19. The current path in a field-effect transistor is called the
    (a) gate.
    (b) channel.
    (c) base.
    (d) substrate.
    (e) collector.

20. The foot-pound-second (fps) system of units
    (a) is preferred by scientists in Europe.
    (b) is preferred by scientists in the United States.
    (c) is used by some lay people.
    (d) is also known as the International System.
    (e) is based on quantities related by powers of 10.

21. Under what circumstances is the shortest distance between two points a curve rather than a straight line?
    (a) Under no circumstances
    (b) When space is "flat"
    (c) When two observers are moving relative to each other at constant speed
    (d) In the presence of a powerful gravitational field
    (e) Whenever clocks are not synchronized

22. Which of the statements (a, b, c, d, or e) below is *false?*
    (a) The current into any point in a dc electric circuit is the same as the current going out of that point.

(b) In a simple dc circuit, the current is proportional to the voltage divided by the resistance.

(c) If the resistance of a component in a dc circuit remains constant while the voltage across it decreases, the current across it decreases.

(d) If the voltage across a resistor in a dc circuit doubles while the resistance remains constant, the power dissipated in the resistor doubles.

(e) All the statements are true.

23. The emitter in an *npn* bipolar transistor
    (a) must always be connected to an electrical ground.
    (b) always provides the output signal.
    (c) is a thin *p*-type layer sandwiched in between two *n*-type layers.
    (d) consists of *n*-type semiconductor material.
    (e) usually acts as the control electrode.

24. When a vector quantity **v** (such as velocity) is divided by a scalar quantity k, the result is
    (a) a vector whose direction is the same as that of **v** and whose magnitude is $1/k$ that of **v**.
    (b) a scalar equal to v/k.
    (c) a vector whose direction is opposite that of **v** and whose magnitude is $1/k$ that of **v**.
    (d) a scalar equal to $-v/k$.
    (e) meaningless; a vector cannot be divided by a scalar.

25. The rate at which time "flows" depends on
    (a) absolute position in space.
    (b) the intensity of the gravitational field.
    (c) the speed of light.
    (d) the intensity of the Earth's magnetic field.
    (e) all of the above.

26. Different elements can join together, sharing electrons. When this happens, the result is
    (a) an ion.
    (b) a compound.
    (c) an isotope.
    (d) an electrical conductor.
    (e) antimatter.

27. Suppose that a gas is placed in a sealed rigid chamber under high pressure. Suddenly, a significant amount of the gas is allowed to escape. What happens?
    (a) Inside the chamber, the gas gets cooler.
    (b) Inside the chamber, the gas gets warmer.
    (c) Inside the chamber, the volume of gas decreases.
    (d) Inside the chamber, the gas molecules speed up.
    (e) The gas condenses as it leaves the chamber.

28. Full-duplex two-way communication by radio, in which either station operator can interrupt the other in an instant, will be impossible between stations on Earth and any interstellar space vessel
    (a) because of relativistic time dilation.
    (b) because of the difference in relative velocity of the two stations.
    (c) because the speed of EM wave propagation is only 3×10^8 m/s.
    (d) because of red shift.
    (e) No! The premise is wrong; such communication is possible.

29. Which of the following statements is *false?*
    (a) A mass can be weightless.
    (b) Force is proportional to mass times acceleration.
    (c) Every action is attended by an equal and opposite reaction.
    (d) Mass is a vector quantity.
    (e) Velocity has both magnitude and direction.

30. A sample of liquid has a volume of 1.200 liters. Its mass is 2.400 kilograms. What is its density?
    (a) 2.000 kg/m^3
    (b) 20.00 kg/m^3
    (c) 200.0 kg/m^3
    (d) 2,000 kg/m^3
    (e) It cannot be determined without more information.

31. The geomagnetic field is produced by
    (a) the solar wind.
    (b) the aurorae.
    (c) the ionosphere.
    (d) molten iron circulating inside the Earth.
    (e) radioactivity.

32. Which of the following statements is *false?*
    (a) Speed is a scalar quantity, but velocity is a vector quantity.
    (b) Velocity consists of a speed component and a direction component.
    (c) An object can have constant speed but changing velocity.
    (d) Speed is the magnitude of a velocity vector.
    (e) None of the above; they are all true.

33. Fahrenheit and Celsius readings are the same
    (a) at 0° in either scale.
    (b) at 4° in either scale.
    (c) at 100° in either scale.
    (d) at −40° in either scale.
    (e) at no point in either scale.

34. Suppose that you connect a lightbulb across a common 6.0-V lantern battery. The bulb lights up (and does not burn out) and draws 1.5 W of power. What is the current through the bulb?
    (a) 9.0 A
    (b) 4.0 A
    (c) 0.38 A
    (d) 0.25 A
    (e) It cannot be calculated from this information.

35. Inductance is the opposition of an electrical component to ac by temporarily storing some of the electrical energy in the form of
    (a) heat.
    (b) a magnetic field.
    (c) an electrical field.
    (d) visible light.
    (e) momentum.

36. Avogadro's number, equal to approximately 6.02×10^{23}, is a unit of material quantity also known as a
    (a) mole.
    (b) candela.
    (c) metric ton.
    (d) gross.
    (e) myriad.

37. According to Einstein's special theory of relativity, the luminiferous ether
    (a) passes through matter as easily as it does through space.
    (b) affects the speed of light, depending on the location of the observer.
    (c) affects the speed of light, depending on the motion of the observer.
    (d) is the absolute standard for motion in the universe.
    (e) does not necessarily exist at all.

38. A lodestone orients itself in a particular direction because of
    (a) interaction between it and the Earth's magnetic field.
    (b) gravitational effects.
    (c) tidal effects.
    (d) the Earth's rotation.
    (e) the fact that it is irregularly shaped.

39. Capacitances in series add like
    (a) inductances in series.
    (b) resistances in parallel.
    (c) voltages in parallel.
    (d) magnetic fields in series.
    (e) none of the above.

40. A volume of 1 milliliter (1 ml) is the same as a volume of
    (a) 1 mm^3.
    (b) 1 cm^3.
    (c) 1 mm^2.
    (d) 1 cm^2.
    (e) none of the above.

41. The objective in a reflecting telescope is
    (a) a convex mirror.
    (b) a concave mirror.
    (c) a convex lens.
    (d) a concave lens.
    (e) a planoconcave lens.

42. A compound microscope has
    (a) a convex objective and a convex eyepiece.
    (b) a concave objective and a concave eyepiece.
    (c) a concave objective and a convex eyepiece.
    (d) a convex objective and a concave eyepiece.
    (e) a single compound lens.

43. The sum of all the voltages, as you go around a dc circuit from a fixed point and return there from the opposite direction, and taking polarity into account,
    (a) depends on the current.
    (b) depends on the number of components.
    (c) is zero.
    (d) is positive.
    (e) is negative.

44. A certain appliance consumes energy at a rate of 1,200 joules per minute. This is the same as saying that the appliance consumes
    (a) 20 ergs.
    (b) 20 newtons.
    (c) 20 meters per second squared.
    (d) 20 kilograms per second squared.
    (e) 20 watts.

45. A converging lens
    (a) brings parallel light rays to a focus.
    (b) spreads parallel light rays out.
    (c) makes converging light rays parallel.
    (d) can do any of a, b, and c.
    (e) can do none of a, b, or c.

46. Which of the following is *not* a variable in an electromagnetic wave?
    (a) Frequency
    (b) Amplitude
    (c) Period

    (d) Resolution

    (e) Wavelength

47. The term *phase opposition* for two sine waves having the same frequency means that they differ in phase by

    (a) zero.

    (b) $\pi/2$ radians.

    (c) π radians.

    (d) $3\pi/2$ radians.

    (e) 2π radians.

48. In every respect, acceleration force manifests itself in precisely the same way as

    (a) the force caused by gravitation.

    (b) the force caused by time dilation.

    (c) the force caused by high speeds.

    (d) the force caused by constant relative motion.

    (e) none of the above.

49. Diffraction makes it possible for

    (a) ocean waves to interact so that they amplify each other's effects.

    (b) sound waves to travel faster than they normally would.

    (c) harmonic radio waves to be generated.

    (d) monochromatic light to be turned into white light.

    (e) sound waves to propagate around corners.

50. An asset of integrated-circuit (IC) technology is the fact that it

    (a) allows for miniaturization.

    (b) allows for modular construction of circuits, devices, and systems.

    (c) consumes less power than equivalent circuits made of discrete components.

    (d) optimizes task speed by minimizing the distances among individual active components.

    (e) does all of the above.

51. Mechanical work can be defined in terms of

    (a) the product of force and mass.

    (b) the product of force and speed.

    (c) the product of force and acceleration.

    (d) the product of force and displacement.

    (e) the product of force and energy.

52. In the presence of unusual solar activity, the "northern lights" often reflect radio waves at some frequencies. This is called

    (a) a geomagnetic storm.

    (b) auroral propagation.

    (c) total internal reflection.

    (d) sporadic-E propagation.

    (e) space-wave propagation.

53. A constant force of 3.00 N is applied to a mass of 6.00 kg in deep space, far away from the gravitational influence of any star or planet. What is the magnitude of the acceleration?
    (a) It cannot be determined from this information.
    (b) 0.500 m/s^2
    (c) 0.667 m/s^2
    (d) 1.50 m/s^2
    (e) 2.00 m/s^2

54. Square, ramp, sawtooth, and triangular waves
    (a) are infinitely high in frequency.
    (b) are composites of sinusoids in specific proportions.
    (c) have no defined wavelengths.
    (d) travel faster than sine waves in the same medium.
    (e) all contain energy at a single frequency.

55. The magnetic flux density in a specific spot near a current-carrying wire is
    (a) inversely proportional to the current in the wire.
    (b) directly proportional to the current in the wire.
    (c) inversely proportional to the square of the current in the wire.
    (d) directly proportional to the square of the current in the wire.
    (e) constant regardless of the current in the wire.

56. The unit of ionizing radiation representing one nuclear transition per second is the
    (a) hertz.
    (b) meter per second.
    (c) becquerel.
    (d) ampere.
    (e) joule.

57. Two atoms are scrutinized. Their nuclei have the same number of protons, but one nucleus has two neutrons more than the other. These atoms represent
    (a) the same element and the same isotope.
    (b) different elements but the same isotope.
    (c) the same element but different isotopes.
    (d) different elements and different isotopes.
    (e) an impossible situation; this scenario cannot occur.

58. Fill in the blank in the following sentence to make it *true:* "For any natural source of ionizing radiation, such as radium or uranium, there is a _____ that is a function of the radiation intensity versus time."
    (a) wavelength
    (b) frequency
    (c) period
    (d) decay curve
    (e) bandwidth

59. Suppose that an ac voltage has a dc voltage superimposed and that the dc voltage exceeds the peak amplitude of the ac voltage. The polarity of the resulting wave
    (a) does not change, and its amplitude does not vary.
    (b) does not change, although its amplitude varies.
    (c) changes at the same frequency as that of the ac wave.
    (d) changes at half the frequency as that of the ac wave.
    (e) cannot be described without more information.

60. The density of an elemental gas such as helium can be defined in terms of
    (a) the number of atoms per unit displacement.
    (b) the number of atoms per unit area.
    (c) the number of atoms per unit mass.
    (d) the number of atoms per unit volume.
    (e) the number of atoms per unit weight.

61. A constant such as e or π, presented as a plain number without any units associated, is called
    (a) a physical constant.
    (b) a relative constant.
    (c) an absolute constant.
    (d) a qualitative constant.
    (e) a dimensionless constant.

62. Suppose that a radio wave has a frequency of 60 MHz. What is the wavelength in a vacuum? Consider the propagation speed of an electromagnetic wave in free space to be 3.00×10^8 m/s.
    (a) 5.0 m
    (b) 20 m
    (c) 180 m
    (d) 1.8×10^4 m
    (e) It cannot be calculated from this information.

63. Suppose that a length of wire 10.00 m long is hung from a tree during a warm afternoon when the temperature is 35°C. During the predawn hours, the temperature has dropped to 20°C, and the wire measures 9.985 m in length. What is the coefficient of linear expansion of this wire?
    (a) It can't be determined from this information.
    (b) 10^{-5}/°C
    (c) 10^{-4}/°C
    (d) 0.001/°C
    (e) 0.01/°C.

64. Total internal reflection can occur for a light beam
    (a) striking a pane of glass from the outside.
    (b) passing through a pane of glass at a right angle.
    (c) striking the surface of a prism at a grazing angle from the inside.

    (d) traveling from one place to another through a vacuum.

    (e) under no circumstances.

65. The Earth's ionosphere can affect

    (a) the propagation of light waves.

    (b) the conductivity of copper wires.

    (c) the intensity of a solar flare.

    (d) the propagation of radio waves at certain frequencies.

    (e) the apparent magnitudes of distant stars.

66. Impulse is equivalent to

    (a) a change in kinetic energy.

    (b) a change in potential energy.

    (c) a change in momentum.

    (d) a change in mass.

    (e) a change in speed.

67. Fill in the blank in the following sentence to make it *true:* "If wave X begins a small fraction of a cycle earlier than wave Y, then wave X _____ wave Y in phase."

    (a) leads

    (b) lags

    (c) opposes

    (d) is in quadrature with

    (e) is coincident with

68. Which of the following components can be used to build a voltage amplifier?

    (a) an electrochemical cell

    (b) an electric generator

    (c) a photovoltaic cell

    (d) a field-effect transistor

    (e) a diode

69. When water is electrolyzed,

    (a) the protons are stripped from the atomic nuclei.

    (b) hydrogen and oxygen combine.

    (c) hydrogen and oxygen atoms are separated from each other.

    (d) new isotopes are formed.

    (e) nuclear fission occurs.

70. Given an object at a specific and constant temperature, the number representing the Celsius temperature of the object is

    (a) approximately 273 greater than the number representing the Kelvin temperature.

    (b) approximately 273 less than the number representing the Kelvin temperature.

    (c) approximately 460 greater than the number representing the Kelvin temperature.

(d) approximately 460 less than the number representing the Kelvin temperature.

(e) approximately $^5/_9$ of the number representing the Kelvin temperature.

71. Complex-number impedance is comprised of
  (a) resistance, capacitance, and inductance.
  (b) reactance, capacitance, and inductance.
  (c) capacitive and inductive reactance.
  (d) resistance, capacitive reactance, and inductive reactance.
  (e) reactance, capacitive resistance, and inductive resistance.

72. Which of the following is a scalar quantity?
  (a) Displacement
  (b) Velocity
  (c) Mass
  (d) Acceleration
  (e) All of the above

73. In a complete standing-wave cycle, there are
  (a) two loops and two nodes.
  (b) two loops and one node.
  (c) two nodes and one loop.
  (d) infinitely many loops and nodes.
  (e) no loops or nodes.

74. The term *black light* refers to
  (a) radio waves.
  (b) infrared radiation.
  (c) ultraviolet radiation.
  (d) visible light of extremely low intensity.
  (e) visible light with no definable wavelength.

75. An important characteristic of a solid is its density relative to that of pure liquid water at 4°C. This is known as
  (a) specific density.
  (b) specific mass.
  (c) specific weight
  (d) specific volume.
  (e) specific gravity.

76. The magnitude of the kinetic energy in a moving object is directly proportional to the square of its
  (a) mass.
  (b) acceleration.
  (c) weight.
  (d) displacement.
  (e) speed.

77. Retentivity is a measure of how well a substance can
    (a) reverse its magnetic polarity.
    (b) become temporarily magnetized.
    (c) become permanently magnetized.
    (d) be demagnetized.
    (e) concentrate magnetic lines of flux.

78. The prefix *pico-* in front of a unit refers to
    (a) 10^{-18} of that unit.
    (b) 10^{-15} of that unit.
    (c) 10^{-12} of that unit.
    (d) 10^{12} times that unit.
    (e) 10^{18} times that unit.

79. In an electromagnetic field, the direction of wave travel is
    (a) parallel to the magnetic lines of flux.
    (b) parallel to the electrical lines of flux.
    (c) parallel to the electrical and the magnetic lines of flux.
    (d) parallel to neither the electrical nor the magnetic lines of flux.
    (e) dependent on the wavelength.

80. Four resistors are connected in series. Two of them have resistances of 120 Ω, and the other two have resistances of 150 Ω. The total resistance is
    (a) 540 Ω
    (b) 270 Ω
    (c) 133 Ω
    (d) 33.3 Ω
    (e) It cannot be calculated from this information.

81. A telescope has an objective lens with a diameter of 250 mm. An eyepiece with a focal length of 10 mm is used. What is the magnification?
    (a) 25×
    (b) 2,500×
    (c) 250×
    (d) 10×
    (e) It cannot be calculated from this information.

82. In a rigid chamber filled with an elemental gas such as oxygen at a constant temperature,
    (a) the pressure is proportional to the number of atoms of gas in the chamber.
    (b) the pressure is inversely proportional to the number of atoms of gas in the chamber.
    (c) the pressure does not depend on the number of atoms of gas in the chamber.
    (d) the pressure is proportional to the intensity of the gravitational field in which the chamber exists.
    (e) none of the above are true.

83. According to the special theory of relativity, a space ship can exceed the speed of light
    (a) if it turns into antimatter.
    (b) if time is made to run backwards.
    (c) if antigravity propulsion systems are used.
    (d) if space is curved in the vicinity of the vessel.
    (e) under no circumstances.

84. Heterodynes occur at frequencies equal to
    (a) the frequencies of the input waves.
    (b) even multiples of the frequencies of the input waves.
    (c) odd multiples of the frequencies of the input waves.
    (d) whole-number multiples of the frequencies of the input waves.
    (e) the sum and difference of the frequencies of the input waves.

85. Mathematically, 1 ohm is the equivalent of
    (a) 1 volt per ampere.
    (b) 1 ampere per volt.
    (c) 1 watt-second.
    (d) 1 watt per second.
    (e) 1 joule per second.

86. To convert hertz to radians per second, you must
    (a) multiply by 2π.
    (b) multiply by π.
    (c) divide by 2π.
    (d) divide by π.
    (e) do nothing; the two units are equivalent.

87. Fill in the blank in the following sentence to make it *true:* "The general theory of relativity is concerned with acceleration and gravitation, and the special theory of relativity is concerned with _____."
    (a) black holes and white dwarfs
    (b) relative motion
    (c) space travel
    (d) absolute motion
    (e) the gravitational collapse of the universe

88. Fill in the blank to make the following sentence *true:* "Hydrogen is changed into _____ inside the Sun; this process is responsible for the Sun's energy output."
    (a) lithium
    (b) copper
    (c) antimatter
    (d) helium
    (e) ionized form

89. A typical solar battery consists of
    (a) zinc-carbon or alkaline cells.
    (b) photovoltaic cells.
    (c) bipolar transistors.
    (d) field-effect transistors.
    (e) any of the above.

90. A substance in which the electrons move easily from atom to atom is
    (a) an electrical solid.
    (b) an electrical liquid.
    (c) an electrical gas.
    (d) an electrical insulator.
    (e) an electrical conductor.

91. Suppose that you have a sealed, rigid container filled with air. The speed with
    which the molecules in the container move around depends directly on the
    (a) number of atoms inside the container.
    (b) mass of the air inside the container.
    (c) temperature of the air inside the container.
    (d) volume of the container.
    (e) length of time the container has been sealed.

92. A simple objective lens refracts orange light to a slightly different extent than
    it refracts blue light. This is observed in a telescope as
    (a) partial internal refraction.
    (b) selective refraction.
    (c) chromatic aberration.
    (d) coma.
    (e) astigmatism.

93. If 60-Hz ac is passed through a coil of wire, the resulting magnetic field
    (a) has constant polarity.
    (b) reverses polarity every $1/60$ second.
    (c) reverses polarity every $1/120$ second.
    (d) is zero.
    (e) permanently magnetizes the wire.

94. Consider a straight current-carrying wire passing through a flat sheet of
    paper at a right angle (that is, the wire is perpendicular to the paper). What
    are the general shapes of the magnetic lines of flux in the plane containing
    the paper?
    (a) Straight, parallel lines
    (b) Straight lines radiating out from the wire
    (c) Hyperbolas centered at the wire
    (d) Circles centered at the wire
    (e) It is impossible to know without more information

95. The radian is a unit of
    (a) radioactivity.
    (b) angular measure.
    (c) temperature.
    (d) electric current.
    (e) electric voltage.

96. Fill in the blank to make the following sentence *true:* "The inductive reactance of a coil is _____ the peak amplitude of the ac passing through it."
    (a) directly proportional to
    (b) directly proportional to the square of
    (c) inversely proportional to
    (d) inversely proportional to the square of
    (e) unrelated to

97. According to the law of conservation of momentum, when multiple objects collide in an ideal system,
    (a) each object has the same momentum after the collision as before.
    (b) each object transfers its momentum to any object with which it collides.
    (c) every object in the system has the same momentum.
    (d) the total system momentum does not change as the collision takes place.
    (e) all of the above are true.

98. All the energy in an ac signal is contained at a single wavelength in
    (a) a square wave.
    (b) a sawtooth wave.
    (c) a ramp wave.
    (d) a rectangular wave.
    (e) none of the above.

99. Specific heat is defined in terms of
    (a) degrees Celsius per kilogram.
    (b) calories per gram.
    (c) grams per degree Celsius.
    (d) calories per gram per degree Celsius.
    (e) degrees Celsius per gram per calorie.

100. Suppose that two atoms are analyzed. Atom X has 12 protons, 14 neutrons, and 12 electrons. Atom Y has 12 protons, 12 neutrons, and 10 electrons. Which of the following statements is *true?*
    (a) X and Y are the same isotope of the same element; Y is an ion and X is not.
    (b) X and Y are different isotopes of the same element, and both are ions.
    (c) X and Y are different elements; Y is an ion and X is not.
    (d) X and Y are different isotopes of the same element; Y is an ion and X is not.
    (e) X and Y are the same isotope of different elements; Y is an ion and X is not.

Answers to Quiz, Test, and Exam Questions

Chapter 1

1. c	2. b	3. a	4. d	5. b
6. c	7. b	8. a	9. d	10. a

Chapter 2

1. c	2. b	3. a	4. b	5. c
6. a	7. d	8. b	9. a	10. d

Chapter 3

1. b	2. b	3. c	4. c	5. a
6. d	7. b	8. c	9. a	10. d

Chapter 4

1. c	2. b	3. d	4. c	5. a
6. b	7. b	8. b	9. a	10. b

Chapter 5

1. a	2. a	3. b	4. d	5. a
6. d	7. c	8. a	9. d	10. c

Test: Part Zero

1. c	2. d	3. b	4. a	5. e
6. b	7. d	8. b	9. a	10. c
11. a	12. e	13. b	14. a	15. e
16. e	17. c	18. a	19. c	20. a

21. c	22. d	23. c	24. d	25. c
26. e	27. c	28. b	29. c	30. e
31. c	32. b	33. a	34. b	35. e
36. d	37. a	38. e	39. d	40. c
41. e	42. a	43. b	44. a	45. c
46. c	47. e	48. c	49. c	50. a

Chapter 6

| 1. b | 2. d | 3. a | 4. b | 5. c |
| 6. c | 7. c | 8. b | 9. d | 10. d |

Chapter 7

| 1. c | 2. d | 3. a | 4. c | 5. b |
| 6. b | 7. d | 8. c | 9. a | 10. b |

Chapter 8

| 1. c | 2. b | 3. d | 4. d | 5. c |
| 6. a | 7. a | 8. c | 9. b | 10. b |

Chapter 9

| 1. b | 2. d | 3. b | 4. c | 5. c |
| 6. a | 7. a | 8. b | 9. a | 10. d |

Chapter 10

| 1. b | 2. a | 3. d | 4. d | 5. a |
| 6. c | 7. a | 8. b | 9. b | 10. c |

Chapter 11

1. b	2. a	3. d	4. b	5. c
6. c	7. a	8. b	9. d	10. c

Test: Part One

1. a	2. c	3. d	4. a	5. b
6. b	7. a	8. b	9. c	10. e
11. e	12. b	13. a	14. b	15. d
16. d	17. d	18. c	19. e	20. a
21. e	22. e	23. b	24. d	25. c
26. a	27. e	28. e	29. a	30. c
31. d	32. a	33. b	34. a	35. b
36. c	37. b	38. e	39. a	40. a
41. d	42. a	43. d	44. d	45. a
46. c	47. e	48. d	49. d	50. e

Chapter 12

1. d	2. c	3. b	4. c	5. a
6. c	7. d	8. a	9. b	10. a

Chapter 13

1. b	2. c	3. a	4. c	5. a
6. a	7. d	8. b	9. d	10. d

Chapter 14

1. c	2. d	3. b	4. c	5. c
6. d	7. b	8. d	9. c	10. b

Chapter 15

1. d	2. c	3. a	4. a	5. b
6. a	7. c	8. b	9. a	10. a

Chapter 16

1. a	2. c	3. d	4. d	5. c
6. d	7. a	8. b	9. a	10. d

Test: Part Two

1. a	2. d	3. b	4. d	5. c
6. a	7. d	8. c	9. e	10. b
11. e	12. a	13. e	14. d	15. a
16. c	17. c	18. d	19. d	20. b
21. e	22. a	23. a	24. a	25. a
26. e	27. e	28. c	29. e	30. a
31. b	32. c	33. c	34. c	35. c
36. b	37. c	38. b	39. a	40. a
41. c	42. e	43. e	44. b	45. a
46. d	47. e	48. d	49. c	50. b

Chapter 17

1. d	2. c	3. d	4. c	5. a
6. a	7. b	8. a	9. d	10. c

Chapter 18

1. c	2. d	3. d	4. b	5. b
6. d	7. c	8. a	9. c	10. c

Chapter 19

1. c	2. d	3. c	4. b	5. d
6. d	7. a	8. b	9. b	10. a

Chapter 20

1. d	2. c	3. a	4. b	5. c
6. c	7. d	8. c	9. b	10. a

Test: Part Three

1. c	2. b	3. d	4. c	5. d
6. b	7. b	8. b	9. b	10. c
11. c	12. d	13. e	14. c	15. c
16. c	17. b	18. a	19. e	20. a
21. a	22. a	23. c	24. d	25. a
26. a	27. a	28. d	29. e	30. d
31. e	32. e	33. c	34. e	35. d
36. a	37. c	38. a	39. a	40. b
41. e	42. e	43. d	44. e	45. c
46. c	47. b	48. b	49. b	50. b

Final Exam

1. b	2. b	3. c	4. d	5. d
6. c	7. b	8. b	9. c	10. a
11. c	12. e	13. b	14. b	15. b
16. a	17. a	18. c	19. b	20. c
21. d	22. d	23. d	24. a	25. b
26. b	27. a	28. c	29. d	30. d
31. d	32. e	33. d	34. d	35. b
36. a	37. e	38. a	39. b	40. b

Answers

41. b	42. a	43. c	44. e	45. a
46. d	47. c	48. a	49. e	50. e
51. d	52. b	53. b	54. b	55. b
56. c	57. c	58. d	59. b	60. d
61. e	62. a	63. c	64. c	65. d
66. c	67. a	68. d	69. c	70. b
71. d	72. c	73. a	74. c	75. e
76. e	77. c	78. c	79. d	80. a
81. e	82. a	83. e	84. e	85. a
86. a	87. b	88. d	89. b	90. e
91. c	92. c	93. c	94. d	95. b
96. e	97. d	98. e	99. d	100. d

Suggested Additional References

Books

Gautreau, Ronald, and William Savin, *Schaum's Outline of Modern Physics.* McGraw-Hill, New York, 1999.

Halpern, Alvin, *3000 Solved Problems in Physics.* McGraw-Hill, New York, 1988.

Kuhn, Karl F., *Basic Physics: A Self-Teaching Guide,* 2d ed. Wiley, New York, 1996.

Seaborn, James B., *Understanding the Universe: An Introduction to Physics and Astrophysics.* Springer Verlag, New York, 1997.

Web Sites

Encyclopedia Britannica Online, www.britannica.com.

Eric's Treasure Troves of Science, www.treasure-troves.com.

INDEX

ABOUT THE AUTHOR

Stan Gibilisco is one of McGraw-Hill's most diverse and best-selling authors. Known for his clear, user-friendly, and entertaining writing style, Mr. Gibilisco's depth of knowledge and ease of presentation make him an excellent choice for a book such as *Physics Demystified.* His previous titles for McGraw-Hill include: *The TAB Encyclopedia of Electronics for Technicians and Hobbyists, Teach Yourself Electricity and Electronics,* and the *Illustrated Dictionary of Electronics. Booklist* named his book, *The McGraw-Hill Encyclopedia of Personal Computing* one of the Best References of 1996.